Lecture Notes in Statistics

Edited by J. Berger, S. Fienberg, J. Gani, and K. Krickeberg

45

J.K. Ghosh
Editor

Statistical Information and Likelihood

A Collection of Critical Essays by Dr. D. Basu

Springer-Verlag
New York Berlin Heidelberg London Paris Tokyo

J.K. Ghosh
Indian Statistical
Calcutta 700 035
India

Mathematics Sul

Library of Congres

Ghosh, J.K.
Statistical inforr
(Lecture notes i
Bibliography: p.
1. Estimation th
J.K. II. Title. III. S
statistics (Springer
QA276.8.B393 19

Camera-ready text prepared by the editor.
Printed and bound by Edwards Brothers, Inc., Ann Arbor, Michigan.
Printed in the United States of America.

9 8 7 6 5 4 3 2 1

ISBN 0-387-96751-6 Springer-Verlag New York Berlin Heidelberg
ISBN 3-540-96751-6 Springer-Verlag Berlin Heidelberg New York

DEDICATED
To the Fond and Colourful
Memories of

SIR RONALD AYLMER FISHER (1890–1962)
and
PROFESSOR JERZY NEYMAN (1894–1981)

FOREWORD

It is an honor to be asked to write a foreword to this book, for I believe that it and other books to follow will eventually lead to a dramatic change in the current statistics curriculum in our universities.

I spent the 1975-76 academic year at Florida State University in Tallahassee. My purpose was to complete a book on Statistical Reliability Theory with Frank Proschan. At the time, I was working on total time on test processes. At the same time, I started attending lectures by Dev Basu on statistical inference. It was Lehmann's hypothesis testing course and Lehmann's book was the text. However, I noticed something strange — Basu never opened the book. He was obviously not following it. Instead, he was giving a very elegant, measure theoretic treatment of the concepts of sufficiency, ancillarity, and invariance. He was interested in the concept of information — what it meant — how it fitted in with contemporary statistics. As he looked at the fundamental ideas, the logic behind their use seemed to evaporate. I was shocked. I didn't like priors. I didn't like Bayesian statistics. But after the smoke had cleared, that was all that was left.

Basu loves counterexamples. He is like an art critic in the field of statistical inference. He would find a counterexample to the Bayesian approach if he could. So far, he has failed in this respect.

In 1979, Basu wrote the following : "It is about 12 years now that I finally came to the sad conclusion that most of the statistical methods that I had learned from pioneers like Karl Pearson, Ronald Fisher and Jerzy Neyman and survey practitioners like Morris Hanson, P. C. Mahalanobis and Frank Yates are logically untenable." I believe he is right. Read Basu.

Richard Barlow

PREFACE

These essays on foundations of Statistical Inference have been brought together in a single volume to honour Dr Basu, as he is affectionately known to his colleagues in the Indian Statistical Institute. The essays are among the most significant contributions of our time to questions of foundation. The sequence in which they have been arranged, in consultation with Dr. Basu, makes it possible to read them (at least Parts I and II) as a single contemporary discourse on the likelihood principle, the paradoxes that attend its violation and the radical deviation from classical statistical practices that its adoption would entail. One may also read them, with the aid of the author's notes, as a record of a personal quest.

If one accepts the likelihood principle, one is led almost inevitably to a Bayesian position. Basu is a Bayesian too. However, he holds this view in a tentative rather than dogmatic way, even though his rejection of almost all classical statistics is complete. This tentative Bayesian view appears in several chapters but specially in chapters VII and XVII where it is evident in the position that he adopts regarding elimination of nuisance parameters or partial likelihood. It also appears in his reluctance, expressed in many private discussions, to put a prior on a large or infinite dimensional parameter space.

The relation of Part III to the earlier parts needs to be made clear. The first three chapters are a rounding off of some of the earlier discussions. Three, namely chapters XXII, XXIII and XXIV present a few well-known results of Basu which appear directly or indirectly in many of the examples in Part I. Chapters XX and XXI are on an entirely different aspect of foundations. They examine the impact of choice of loss functions on the notion of efficient estimates in small and large samples. Chapter XX contains a beautiful construction which shows the existence of a best unbiased estimate depends crucially on convexity of the loss function. Chapter XXI contains the germ of a seminal idea which took its final shape in the hands of Bahadur. Both are forerunners of concern about robustness under varying loss functions.

The following biographical details came out in course of an interview with him in October 1987.

Basu had his first course of Statistics from Professor M.C. Chakrabarty in the honours programme in Mathematics at Dacca University in what is now Bangladesh. This was around 1945. He got his master's degree in Mathematics also from Dacca and taught there from 1947 to 1948 when he moved to Calcutta. This was the time of independence, and partition of India when the whole subcontinent was in turmoil. Basu spent some time in an insurance company, trying to become an actuary. Realising that this was not his vocation, he returned to Dacca as a Senior

Lecturer in 1950. After a few weeks he came to Calcutta again and joined the Indian Statistical Institute as a research scholar under Professor C. R. Rao. After submitting his thesis in 1953 he went to Berkeley as a Fullbright scholar. He came back a complete Neyman-Pearsonian. His first doubts began to form when he learnt about ancillaries and conditional inference from Fisher during Fisher's visit to the Institute in 1955. From then onwards he began examining very critically both the Neyman-Pearsonian and the Fisherian framework, his examination eventually forcing him to a Bayesian point of view, via the likelihood route, which had been opened up but then abandoned by Barnard and Birnbaum. His final conversion, if one may call it so, took place in a rather amusing way in 1968. Though not a Bayesian yet, he had been invited by Professor H. K. Nandi to be the first speaker in a Bayesian symposium at the Indian Science Congress held in Benaras in January, 1968. Basu says his lecture convinced none but himself; in course of preparing for that lecture he became a Bayesian. The polemical papers on foundation began to appear after 1968.

The papers are being reproduced more or less as they were originally published, except for the fact that occasionally a long paper has been broken up into more than one chapter and the reference style has been made uniform. A few inconsistencies in spellings and references remain. References to earlier chapters are sometimes in the form of references to the corresponding papers or parts thereof. Moreover, all references to contemporary or past events remain as they originally appeared.

Bringing out this volume would not have been possible without the help and co-operation of many people. Professor Barlow was kind enough to write a foreword. Dr. Basu has provided a summing up and author's notes at my request. I would like to thank all my colleagues in the Institute who spent a lot of time looking through the manuscript but a special mention must be made of Tapas Samanta and Sumitra Purkayastha who put in an enormous amount of work. My thanks are due to Subhas Dutta for typing the manuscript and to Prandhan Nandy for taking care of all the correspondence related to the publication of the monograph and much else.

J. K. Ghosh

Indian Statistical Institute
Calcutta 700 035

November, 1987

ACKNOWLEDGEMENT

I am grateful to

Professor G. A. Barnard
Professor O. Barndorff-Nielsen
Professor D. R. Cox
Professor A. P. Dempster
Professor A. W. F. Edwards
Professor V. P. Godambe
Professor David V. Hinkley
Professor J.D. Kalbfleisch
Professor Oscar Kempthorne
Professor J. C. Koop
Professor David A. Lane
Professor S. L. Lauritzen
Professor J. N. K. Rao
Professor R. Royall
Professor D. B. Rubin

for their permission to include their discussion on Dr. Basu's papers and to

Academic Press
American Statistical Association
Concordia University
Holt, Rinehart and Winston
Institute of Mathematical Statistics
International Statistical Institute
John Wiley & Sons Limited
North Holland Publishing Company
Statistical Publishing Society
University of California Press
University of North Carolina
University of Sao Paulo

for giving permission to reproduce material published by them. The exact source is indicated in the Reference List at the end of the volume.

J. K. G.

CONTENTS

PART II

SURVEY SAMPLING AND RANDOMIZATION

A SUMMING UP

I wish I could rewrite the 19 articles (that make up this collection) carefully piecing them together in the form of a more readable monograph. But even if I had the time and the inclination, my failing eyesight and a certain lack of will did not let that happen. Now, looking through the manuscript, I am embarrased at the frequent repetitions of the same ideas and phrases in the introductions to the various chapters. However, the redeeming feature of the present format is that the Reader can open the book almost anywhere. The book is named after the main article that spans three chapters. Three of the 24 chapters present some dissenting viewpoints as discussions.

I used to be fascinated by Number Theory and wanted to be a mathematician like my father, N. M. Basu and my favourite Professor T. Vijayaraghavan. The unfortunate political turmoil of the mid-forties in the Indian Subcontinent made that dream unrealizable. In September, 1950 I somehow got into the Indian Statistical Institute, Calcutta and embarked on a voyage of discovery in Mathematical Statistics. Professor Abraham Wald's short (and tragic) visit to India in December, 1950 gave a real fillip to my interest in the subject. In 1953-54 I visited Berkeley for a year as a post doctoral research fellow and was duly influenced by Professor Jerzy Neyman and his brilliant young colleagues of that time. In 1954 I returned to India quite convinced that Statistics was after all a part of Mathematics.

It took me the greater part of the next two decades to realize that statistics deals with the mental process of induction and is therefore essentially anti-mathematics. How can there be a deductive theory of induction? The Institute of Mathematical Statistics ought to change its name!

I met Professor R. A. Fisher in the winter of 1954-55 and had had several brief encounters with him over the next five years. The astonishing 50 years (1912-62) of Fisher's creative writings in statistics have been neatly divided into two nearly equal halves by my friend Oscar Kempthorne as the Prefiducial Period (1912-35) and the Postfiducial Period (1936-62) which contained such ideas as Choice of the Reference Set and Recovery of Ancillary Information. Recently, the Berkeley School has rediscovered the Conditionality Argument of Fisher. This is evidenced by the late Professor Kiefer's writings on Conditional Confidence Statements and Professor Lehmann's 1981 paper (An Interpretation of Completeness and Basu's Theorem) and the inclusion of a chapter on Conditional Tests in the second edition of his Testing Statistical Hypotheses. With his reference set argument Sir Ronald was trying to find a via media between the two poles of Statistics — Berkeley and Bayes. My efforts to understand this Fisher compromise eventually led me to the Likelihood

Principle.

I wholeheartedly agree with De Finetti that Probability Does Not Exist. It is a normative model for a coherent mode of thought. But there need not be only one model. Indeed there are several. Quantum Probability is an abarrent example.

Since probability does not exist outside the mind, the same must be true of the likelihood. The likelihood function is after all a bunch of conditional probabilities. And we have to admit that the notion of utility (or loss) dwells only in the mind.

What have I got to offer in this book? I am afraid, nothing but a whole set of negative propositions like the above. However, in all humility let me draw the attention of the would be readers of the book to the ancient Vedic saying : "neti, neti, . . . , iti", which means : The Ultimate Reality is beyond human comprehension. The only available route towards understanding The Reality is the way of "not this, not this, . . . , this."

D. Basu

PART I
INFORMATION AND LIKELIHOOD

CHAPTER I

RECOVERY OF ANCILLARY INFORMATION

0. Notes

Editor's Note : This article first appeared in Contributions to Statistics, a contributory volume published by Pergamon Press in 1964 and presented to Professor P.C. Mahalanobis on his 70th birthday. It was then reproduced in Sankhya (1964).

Author's Note : It is over thirtyfive years now that I came across the notion of ancillary information. This is how it happened. In those days the question that used to be asked most frequently in Indian university examinations in Statistics was the following: "Prove that the Maximum Likelihood Estimator is sufficient, consistent and efficient". The question used to puzzle me no end until I came up with some counterexamples to satisfy myself that none of the above assertions about the ML method is true in general. One of my examples was as follows.

Example : With n observations on a random variable X that is uniformly distributed over the closed interval $[\theta, 2\theta]$, the likelihood function is $1/\theta^n$ over the interval $[M/2, m]$ and is zero outside the interval, where m and M are respectively the smallest and the largest sample observation. How good is the ML estimator $\hat{\theta} = M/2$ in this case? The minimum (positive) likelihood estimator m is not too unreasonable an estimator of θ. The pair $\hat{\theta}$ and m constitute the munimum sufficient statistic and they are asymptotically independent (conditionally on θ) as $n \to \infty$. To marginalize the data to the ML estimator $\hat{\theta}$ will clearly entail a substantial loss of information in this case.

In this example, θ is a scale parameter and both $\hat{\theta}$ and m are equivariant estimators of θ. The mean squared error of m is exactly four times larger than that of $\hat{\theta}$. So I tried the weighted average $T = (4\hat{\theta} + m)/5 = (2M + m)/5$ as an estimator of θ and found that the ratio $E(T - \theta)^2/E(\hat{\theta} - \theta)^2$ is always less than unity and that the limiting value of the ratio is 12/25 as n tends to infinity. In this case the ML estimator can hardly be called efficient in an asymptotic sense.

At that time I used to share an office with Professor C. R. Rao who was then an ardent supporter of everything Fisherian including the ML method. He soon pointed out to me that I had been wrong in calling the ML estimator $\hat{\theta}$ inefficient and that my mistake was to compute the mean squared error of $\hat{\theta}$ unconditionally. The ratio $Y = M/m$ is an ancillary statistic since θ is a scale parameter and Y is scale invariant. Furthermore, Y is the ancillary complement to $\hat{\theta}$ in the sense that jointly they are sufficient. So following Sir Ronald we ought to look at the average performance characteristics of $\hat{\theta}$ conditionally with Y held fixed at its observed level.

It took me quite a while to understand the ancillarity argument. Only then I recognized where I went wrong. I ought to have altered my example slightly so that the conditionality argument was no longer available! Suppose X was uniformly distributed over the interval $[\theta, \theta^2]$, where $\theta > 1$ is the parameter. The likelihood would then be spread over the interval $[\sqrt{M}, m]$ and the ML estimate would have been $\sqrt{M}$. Clearly there would be a substantial loss of information if we marginalized the data to the insufficient estimate $\sqrt{M}$. But where is the ancillary complement to the ML estimate in this case?

In 1963, when I wrote this article, the light of likelihood had not yet dawned on me. That is why, like a sinking man, I was trying then to catch at the straw of "performable experiments"!

RECOVERY OF ANCILLARY INFORMATION

1. Introduction

The main upsurge of the late Professor R.A. Fisher's theory of Statistical Inference took place within a brief span of about 10 years (1920-30). It was during this period that Fisher came out with the notions of likelihood, fiducial probability, information and intrinsic accuracy, sufficiency, ancillary statistics and recovery of information - concepts around which the superstructure of the theory was built.

Many eminent statisticians and mathematicians have made detailed studies of some particular aspect of Fisher's theory and some of these studies gave rise to important streams of fundamental research in statistical theory. Basu (1959), made a very much localised study of the notion of ancillary statistics from a purely mathematical point of view. This note is a follow up study from the statistical angle. Here we discuss the very controversial subject matter of 'recovery of ancillary information' through proper choice of 'reference sets'. For the purpose of pinpointing our attention to the basic issues raised, we restrict ourselves to the one parameter setup only.

In the one parameter setup, Fisher defines an ancillary statistic as one whose probability (sampling) distribution is free of the parameter and which, in conjunction with the maximum likelihood estimator of θ (the parameter), is sufficient. The use of ancillary statistics has been recommended in two different inference situations namely the point estimation problems and the testing of hypotheses problems.

In point estimation problems the use of a suitably chosen ancillary statistic is recommended in situations where the maximum likelihood estimator T of θ is not a sufficient statistic. The use of T as an estimator of θ will then entail a certain loss of information which, according to Fisher, may be meaningfully (at least in the large sample case) measured, in some situations, as follows. The information contained in the whole sample X is defined as

$$I(\theta) = E\left[-\frac{\partial^2}{\partial\theta^2} \log f(X|\theta) \middle| \theta \right]$$

where $f(x|\theta)$ is the frequency or density function for X. Similarly the information contained in a statistic T (which may be vector-valued) is measured by the function

$$\mathcal{J}(\theta) = E\left[-\frac{\partial^2}{\partial\theta^2} \log g(T \mid \theta) \mid \theta \right]$$

where $g(t \mid \theta)$ is the frequency or density function for the estimator T. The difference

$$\lambda(\theta) = I(\theta) - \mathcal{J}(\theta)$$

may then be taken as a measure of the information lost. Under certain regularity conditions the following results hold true.

(1) $\lambda(\theta) \geqslant 0$ for all values of θ ;

(2) $\lambda(\theta) \equiv 0$ if and only if T is a sufficient statistic;

(3) if T together with the statistic Y is sufficient for θ then the information contained in the pair (T, Y) is $I(\theta)$;

(4) if Y is ancillary and the pair (T, Y) is sufficient then

$$I(\theta) = E\left[\mathcal{J}(\theta \mid Y) \mid \theta \right]$$

where $\mathcal{J}(\theta|y)$ is the conditional amount of information contained in T under the condition that Y takes the value y, i.e.

$$\mathcal{J}(\theta|y) = E\left[-\frac{\partial^2}{\partial\theta^2} \log f(T|y,\theta) \mid Y = y, \theta \right]$$

where $f(t|y, \theta)$ is the conditional frequency (density) function for T under the condition $Y = y$. The relation (4) follows directly from the observation that the joint frequency (density) function $h(t,y|\theta)$ of (T, Y) may be factorized as

$$h(t,y|\theta) = g(y) f(t|y,\theta)$$

where $g(y)$ is the θ-free (Y being ancillary) marginal frequency (density) function for Y.

Now, consider a situation where the maximum likelihood estimator T is not sufficient but where we are able to find another statistic Y whose marginal distribution is θ -free and which complements T in the sense that the pair (T, Y) is jointly sufficient. The statistic Y by itself contains no information about θ. But in a sense it 'summarises in itself' the quantum of information $\lambda(\theta)$ that is lost in the use of T as an estimator of θ . The problem is how to recover this apparent loss of information. According to Fisher it is a mistake to calculate the information content of T with reference to the whole sample space, i.e. with reference to the marginal distribution of T. The appropriate measure of the information content of T is $\mathcal{J}(\theta|Y)$ not $\mathcal{J}(\theta)$. Having observed T and Y we should consider the conditional distribution of T with the observed value of Y as the condition. We take as our 'reference set' not the whole sample space but the subset of those sample points that could give rise to the observed value of the ancillary statistic Y.

The following two quotations from Fisher brings out the analogy between the sample size and the ancillary statistic.

"Having obtained a criterion for judging the merits of an estimate in the real case of finite samples, the important fact emerges that, though sometimes the best estimate we can make exhausts the information in the sample, and is equivalent for all future purposes to the original data, yet sometimes it fails to do so, but leaves a measurable amount of the information unutilized. How can we supplement our estimate so as to utilize these too? It is shown that some, or sometimes all of the lost information may be recovered by calculating what I call ancillary statistics, which themselves tell us nothing about the value of the parameter, but, instead, tell us how good an estimate we have made of it. Their function is, in fact, analogous to the part which the size of our sample is always expected to play, in telling us what reliance to place on the result." (Fisher, 1935c).

"When sufficient estimation is possible, there is no problem, but the exhaustive treatment of the cases in which no sufficient estimate exists is now seen to be an urgent requirement. This at present is in the interesting stage of being possible sometimes, though, so far as we know, not always. I have spoken of the sufficient estimates as containing in themselves the whole of the information provided by the data. This is not strictly accurate. There is always one piece of additional, or ancillary, information which we require, in conjunction with even a sufficient estimate, before this can be utilized. That piece of information is the size of the sample or, in general, the extent of the observational record. We always need to know this in order to know how reliable our estimate is. Instead of taking the size of the sample for granted, and saying that the peculiarity of the cases where sufficient estimation is possible lies in the fact that the estimate then contains all the further informations required, we might equally well have inverted our statement, and, taking the estimate of maximum likelihood for granted, have said that the peculiarity of these cases was that, in addition, nothing more than the size of the sample was needed for its complete interpretation. This reversed aspect of the problem is the more fruitful of the two, once we have satisfied ourselves that, when information is lost, this loss is minimised by using the estimate of maximum likelihood. The cases in which sufficient estimation is impossible are those in which, in utilizing this estimate, other ancillary information is required from the sample beyond the mere number of observations which compose it. The function which this ancillary information is required to perform is to distinguish among samples of the same size those from which more or less accurate estimates can be made; or, in general, to distinguish among samples having different likelihood functions, even though they may be maximised at the same value. Ancillary information never modifies the value of our estimate; it determines its precision." (Fisher, 1936).

Often the 'extent of observational record' is planned in advance and is taken for granted in the subsequent analysis of the data. If we take n independent observations on a normal variable with unknown mean θ and known standard deviation, we see no need to bother about any characteristic of the sample other than the sample mean $\bar{x}$; but yet the fact remains that without some knowledge about n the maximum likelihood estimator $\bar{x}$ of θ will be hardly of any use to any statistician. Along with the information that the sample is drawn from a normal population and the observed value of $\bar{x}$, we need to know the value of the sample size n. The 'reliability' of the estimator $\bar{x}$ is interpreted in terms of its average performance in repeated sampling with the fixed sample size n.

What happens if 'chance' plays (or is allowed to) a part in the determination of n? Suppose we toss a true coin and, depending on whether the outcome is a 'head' or a 'tail', we draw a sample of size 10 or 100. It is easily verified that the sample mean $\bar{x}$ is still the maximum likelihood estimator of θ but that it no longer is a sufficient statistic. It is the pair ($\bar{x}$, n), where n is the (variable) sample size, that is sufficient for θ. Here n is an ancillary statistic taking the two values 10 and 100 with equal probabilities. Now, having drawn a sample of size n (which is either 10 or 100) and having estimated θ by the sample mean $\bar{x}$, how does the statistician report the 'reliability', the precision, the information content of the estimate? Of course a sample of size 10 will lead to a less reliable estimate than a sample of size 100. Having drawn a sample of size 10 should the statistician turn a blind eye to the actual smallness of the sample size and try to figure out the long run performance of his estimation procedure in a hypothetical series of experimentations in which 50% of the cases he draws sample of size 10 and the other 50% of the cases the sample size is 100? What should be the reference set for judging the performance characteristic of the estimator— the 10 dimensional Euclidean space R_{10} or the union of R_{10} and R_{100} ? The author agrees with Fisher that, having drawn a sample with the ancillary statistic $n = 10$, the statistician should judge (if at all he must) the performance of the maximum likelihood estimator $\bar{x}$ in the conditional sample space (restricted reference set) R_{10}. However, the author feels that Fisher, in his writings on ancillary statistics and choice of reference sets, has pushed the above analogy with the sample size a little too far, thereby giving rise to some logical difficulties the real nature of which will be discussed later.

In problems of testing Fisher uses ancillary statistics for the determination of the 'true' level of significance. Having selected the test criterion — a measure of the extent to which the observed sample departs from the expected one under the null hypothesis — the level of significance is the probability (under the null hypothesis) of getting a sample with a larger criterion score than the one actually obtained. In the presence of a suitable ancillary statistic Y, Fisher recommends that the level of significance of a test should be computed by referring to the conditional sample space determined by the set of all possible samples for which the value of Y is the one presently observed.

The following example worked out in Fisher (1956, pp. 163-69) is reproduced here with quotations with the idea of bringing out the essential features of the method of 'Recovery of Information' (as envisaged by Fisher in the context of point estimation) through proper choice of 'reference sets'.

Example: Let us suppose that we have N pairs of independent observations on the pair (X, Y) of positive random variables with joint probability density function

$$p(x,y|\theta) = e^{-(\theta x + \frac{y}{\theta})}, \quad x > 0, y > 0, \theta > 0 .$$

Let (X_i, Y_i), i = 1,2, ..., N, be the N paris of observations and let

$$T = \sqrt{\Sigma Y_i/\Sigma X_i} \quad \text{and} \quad U = \sqrt{(\Sigma X_i)(\Sigma Y_i)}.$$

It is easy to check that

(i) T is the maximum likelihood estimator of θ,

(ii) T is not sufficient for θ,

(iii) the pair (T,U) is jointly sufficient for θ,

(iv) U is an ancillary statistic, i.e. the marginal distribution of U is θ-free.

"Since the likelihood cannot be expressed in terms of only θ and T, there will be no sufficient estimate, and some information will be lost if the sample is replaced by the estimate T only."

The amount of information supplied by the whole sample (of N pairs of observations) is

$$I(\theta) = 2N/\theta^2$$

while the amount of information contained in the statistic T is

$$J(\theta) = \frac{2N}{\theta^2}\,\frac{2N}{2N+1}\,.$$

"The loss of information is less than half the value of a single pair of observations, and never exceeds one third of the total available. Nevertheless its recovery does exemplify very well the mathematical processes required to complete the logical inference."

Here, U is the ancillary statistic and so we have to consider the conditional distribution of T given U. From this conditional distribution the conditional information content of T works out as

$$J(\theta|U) = \frac{2N}{\theta^2}\,\frac{K_1(2U)}{K_o(2U)}$$

where K_o and K_1 are Bessel functions.

Let us note that the information content $J(\theta|U)$ "depends upon the value of U actually available, but has an average value, when variations of U are taken into account, of

$$2N/\theta^2\,,$$

the total amount expected on the average from N observations, none is now lost. The information is recovered and inference completed by replacing the distribution of T for given size of sample N, by the distribution of T given U, which indeed happens not to involve N at all. In fact, U has completely replaced N as a means of specifying the precision to be ascribed to the estimate. In both cases the estimate T is the same, the calculation of U enables us to see exactly how precise it is, not on the average, but for the particular value of U supplied by the sample."

2.. The Sample Size Analogy

This section is devoted to a discussion on the Fisher analogy between the sample size and an ancillary statistic. The reliability of an estimate is assessed in terms of the average performance of the experimental procedure in a hypothetical series of experimentations with the sample size n fixed at the level actually obtained in the sample at hand. Fisher always interpreted the reliability (information content, variance etc.) of an estimate in terms of the average performance of some well defined experimental (estimation) procedure in some hypothetical sequence of experimentations. When Fisher talks of the reliability of an estimate the adjective 'reliability' is used only as a transferred epithet and is actually meant to be attached to the estimation procedure that has given rise to the estimate.

In our example of the previous section, the statistician (for some unknown reasons) decided to choose between a sample of size 10 or 100 on the basis of the flip of a coin. Here a random choice is being made between two sampling experiments $\mathcal{E}_{10}$ and $\mathcal{E}_{100}$ with the associated sample spaces R_{10} and R_{100} and the corresponding probability distributions. More generally, suppose $\mathcal{E}_v$ is the sampling experiment corresponding to a sample of size v and suppose the observed sample size n is determined by a θ-free (parameter-free) chance mechanism. Once we recognize that the estimate T of θ is generated by the random choice $\mathcal{E}_n$ from the family $\{\mathcal{E}_v\}$ of available experimental procedures we ought to transfer the reliability index of the chosen experiment $\mathcal{E}_n$ to the estimate T. When the statistician is forced to make a selection from a family of available experimental procedures he should report the reliability of the procedure actually selected by him. The following example is somewhat more realistic than the above example.

Example : Suppose, from a finite population of N units, we draw a sample of size s with replacements. Let $X_1, X_2, \ldots, X_N$ be the population values and $x_1, x_2, \ldots, x_s$ the s sample values (arranged in order of their appearances). The problem is to estimate the population mean $\overline{X} = (X_1 + \ldots + X_N)/N$. The sample mean $\overline{x} = (x_1 + \ldots + x_s)/s$ has a standard deviation of $\sigma/\sqrt{s}$ where σ^2 is the population variance. But will it be correct to recognize s as the true sample size? Since the sample was drawn with replacements, it is plausible that some of the population units came repeatedly in the sample. In realistic sample survey situations, we can usually distinguish between the population units. Consider the extreme situation where all the s sample units happen to be the same (population unit). Confronted with a situation like this the statistician would surely hesitate to report the reliability of his estimate as $\sigma/\sqrt{s}$. In this extreme case the honest statistician will have to admit that he had drawn an unlucky sample whose effective size is only 1 (and not s) and report the reliability of his estimate as σ (and not $\sigma/\sqrt{s}$). More generally, consider the situation where n is the number of distinct units in the sample and $x_1^*, x_2^*, \ldots, x_n^*$ are the corresponding sample values (arranged say in an increasing order of their population unit-indices). It is easy to see that the probability distribution of n involves only N and s, and since they are known constants, the statistic n is ancillary. If we define $\overline{x}^*$ as

$$\overline{x}^* = (x_1^* + \ldots + x_n^*)/n$$

then [see Basu (1958) for a detailed discussion on this] $\overline{x}^*$ is better than $\overline{x}$ as an unbiased estimator of $\overline{X}$. All right, we agree to estimate $\overline{X}$ by $\overline{x}^*$ and forget all about $\overline{x}$. But our troubles are not yet over. What is the standard deviation of $\overline{x}^*$? For a fixed n, the conditional distribution of $(x_1^*, \ldots, x_n^*)$ is the same as that of a simple random sample of size n from the population of N values. We see then that the ancillary statistic n is really a sample size. When we draw a sample of size s with replacements and agree to take note of only the statistic $(x_1^*, x_2^*, \ldots, x_n^*)$, we are then in effect drawing a simple random sample of variable size n. If we denote by $\mathcal{E}_i$ (i = 1,2, ..., s) the experimental procedure of drawing a simple random sample of size i from the population of N units and by p_i the parameter free probability that n = i, then the above experimental procedure is the same as that of selecting one of the experiments $\mathcal{E}_1, \mathcal{E}_2, \ldots, \mathcal{E}_s$ with probabilities $p_1, p_2, \ldots, p_s$ (and then carrying out the experiment $\mathcal{E}_n$ so selected). From what we have said earlier, it then follows that we should assess the reliability of

$\bar{x}^*$ in terms of that of the experiment $\mathscr{E}_n$ actually selected, i.e. $V(\bar{x}^*)$ should be reported as

$$\frac{N-n}{N-1}\,\frac{\sigma^2}{n} \qquad \dots (2.1)$$

and not as

$$E\left(\frac{N-n}{N-1}\,\frac{\sigma^2}{n}\right) \qquad \dots (2.2)$$

Let us repeat once again that when reporting the reliability of an estimate we are actually saying something about the long term average performance of some well defined estimation procedure. Both (2.1) and (2.2) are reliabiluty indices—(2.1) for the experimental procedure $\mathscr{E}_n$ with a fixed n and (2.2) for the experimental procedure where n is allowed to vary (in the parameter-free manner described earlier) from trial to trial. We may briefly summarise the basic Fisherian point of view (in the present context) as follows :

(a) Suppose a whole family $\{\mathscr{E}_y\}$, where y runs over an index set $\mathscr{Y}$, of statistical experiments is available any one of which may be meaningfully performed for the purpose of making a scientific inference about some physical quantity θ.

(b) And suppose that the experiment that the statistician actually performs is recognized to be equivalent to the two stage experiment of first selecting at random a point Y in $\mathscr{Y}$ and then performing the experiment $\mathscr{E}_Y$.

(c) Suppose, further, that the probability distribution of Y in $\mathscr{Y}$ is θ - free.

Under the above conditions, the true 'reference set' for the statistician is the experiment $\mathscr{E}_Y$ (with its associated sample space and probability distributions) and not the two stage experiment described in (b) above. The reliability (information content, standard deviation, significance level etc.) of the inference made about θ should be assessed in terms of the average performance characteristic of the inference procedure in a long hypothetical sequence of independent repetitions under identical conditions of the experiment $\mathscr{E}_Y$ (where Y is supposed to be held fixed at the particular point that obtains in the present instance).

Under the above circumstances Fisher would like the statistician to say something to the following effect : "It was rather silly of me to let chance have a hand in the determination of the experiment $\mathcal{E}_Y$ for me. But I now recognize that I have performed the experiment $\mathcal{E}_Y$ and see no reason whatsoever to fuss about the other experiments in the family $\{\mathcal{E}_y\}$ that might have been handed down to me by chance. The inference that I make about θ is the most appropriate one for the experiment $\mathcal{E}_Y$, and the assessment of the reliability of my inference is made with reference to the experiment $\mathcal{E}_Y$ alone."

Now, let us consider a general inference situation and see whether the above arguments hold in the presence of an ancillary statistic. Let $\mathcal{E}$ be an arbitrary statistical experiment performed with a view to elicit some information about a physical quantity θ. From the mathematical standpoint we are then concerned with the trio $(\mathcal{X}, \mathcal{A}, \mathcal{P})$ where $\mathcal{X} = \{x\}$ is the sample space and $\mathcal{A} = \{A\}$ is the σ-field of events on which the family $\mathcal{P} = \{P_\theta\}$ of probability measures is defined. For the sake of simplicity, we ignore the possibility of nuisance parameters and we assume that different possible values of θ are associated with different probability distributions. Now, let Y be an ancillary statistic taking values in the space $\mathcal{Y}$. Corresponding to each point y in $\mathcal{Y}$ we then have (under some regularity conditions) a trio $(\mathcal{X}_y, \mathcal{A}_y, \mathcal{P}_y)$, where $\mathcal{X}_y$ is the subset of points of $\mathcal{X}$ for which $Y = y$ and $\mathcal{P}_y = \{P_\theta^y\}$ is the family of conditional probability distributions on a σ-field $\mathcal{A}_y$ of subsets of $\mathcal{X}_y$.

We have only to imagine a conceptual experiment $\mathcal{E}_y$ that gives rise to the trio $(\mathcal{X}_y, \mathcal{A}_y, \mathcal{P}_y)$ and the analogy that we have been trying to drive home is complete. The statistical experiment $\mathcal{E}$ is then equivalent to the two stage experiment of first observing the random variable Y (whose distribution is θ-free) and then performing the conceptual experiment $\mathcal{E}_Y$ leading ultimately to a point X in $\mathcal{X}$.

Why the insistence on Y being an ancillary statistic? The sample X that we arrive at through the experiment $\mathcal{E}$ (or the equivalent two stage breakdown Y - $\mathcal{E}_Y$), is the only source of our information about θ. Now, according to Fisher, the likelihood function $L(\theta)$ for the sample X is the sole basis for making any judgement about θ. Nothing else need be taken cognizance of. Let us observe that the likelihood function $L(\theta)$ is the same (excepting for a θ-free multiplicative constant) whether we consider X to be generated by the experiment $\mathcal{E}$ or the conceptual experiment $\mathcal{E}_Y$. This, according to the author, is the main explanation as to why in the above Y-decomposition of the probability structure

$(\mathcal{X}, \mathcal{A}, \mathcal{P})$ into the family $\{\mathcal{X}_y, \mathcal{A}_y, \mathcal{P}_y\}$ the statistic Y has to have a θ-free distribution. If Y be ancillary then the choice of the 'reference set' $(\mathcal{X}_Y, \mathcal{A}_Y, \mathcal{P}_Y)$ does not affect the likelihood scale.

3. A Logical Difficulty

The decomposition of the probability structure $(\mathcal{X}, \mathcal{A}, \mathcal{P})$ into the family of probability structures $\{\mathcal{X}_y, \mathcal{A}_y, \mathcal{P}_y\}$ depends on the ancillary statistic Y. Which ancillary statistic Y to work with? It was noted that each of two statistics Y_1 and Y_2 may individually be ancillary but jointly not so. Thus, in case of a controversy as to which one of the two ancillaries Y_1 and Y_2 should determine the 'reference set' one cannot solve the dilemma by referring to the conditional probability strcture conditioned by the observed values of both Y_1 and Y_2. Consider the following example :

Example : Let (X, Y) have a bivariate normal distribution with zero means, unit standard deviations, and unknown correlation coefficient θ. If (X_i, Y_i), $i = 1,2 \ldots, n$, be n pairs of independent observations on (X, Y) then we see at once that the set of n observations $(X_1, X_2, \ldots, X_n)$ on X is an ancillary statistic. Similarly $(Y_1, Y_2, \ldots, Y_n)$ also is ancillary. But the two ancillary statistics together is the whole data and is obviously sufficient. In regression studies the statistician often ignores the sampling variations in the observed X-values and treats them as preselected exprimental constants. If, in the above situation he is justified in similarly treating the observed Y-values also, then how do we define the sampling error for the estimate of θ?

Fisher recommended the choice of the 'reference set' with the help of an ancillary Y that complements the maximum likelihood estimator T in the sense that T is not sufficient but the pair (T, Y) is. This, however, is not a sufficient specification for Y. The two statistics Y_1 and Y_2 may each be ancillary complements to the maximum likelihood estimator T and lead to different 'reference sets' and different reliability indices for the estimator T. The following very simple example clearly brings out the above possibility.

Example : Suppose we have a biased die about which we have enough information to assume the following probability distribution :

scores :	1	2	3	4	5	6
probability distrubutions	$\frac{1-\theta}{12}$	$\frac{2-\theta}{12}$	$\frac{3-\theta}{12}$	$\frac{1+\theta}{12}$	$\frac{2+\theta}{12}$	$\frac{3+\theta}{12}$

where the parameter θ can take any value in the closed interval [0, 1]. Let $\mathcal{E}$ stand for the experiment of rolling the die only once leading to the observed score X. It is easily seen that the maximum likelihood estimator T is defined as follows:

X:	1	2	3	4	5	6
T(X) :	0	0	0	1	1	1

and leads to the partition (1, 2, 3), (4, 5, 6) of the sample space. Here X is the minimal sufficient statistic and T is not sufficient. In this example there are six non-equivalent[1] ancillary complements to T. They may be listed as follows:

X	1	2	3	4	5	6
$Y_1(X)$	0	1	2	0	1	2
$Y_2(X)$	0	1	2	0	2	1
$Y_3(X)$	0	1	2	1	0	2
$Y_4(X)$	0	1	2	2	0	1
$Y_5(X)$	0	1	2	1	2	0
$Y_6(X)$	0	1	2	2	1	0

Each of the six statistics $Y_1, Y_2, \ldots, Y_6$ is a maximal[2] ancillary.

The statistic T induces the partition (1,2,3) and (4,5,6) whereas Y_1 induces the partition (1,4), (2, 5) and (3, 6). Since each Y_1 - partition intersects each T-partition in a one point set, it follows that the pair (T, Y_1) is equivalent to X which is the minimal sufficient statistic. Thus, Y_1 is an ancillary complement to T. In

1 Two statistics are said to be equivalent if they lead to the same partition of the sample space.

2 See Basu (1959) for the definition of a maximal ancillary.

like manner, we prove that each of the other Y_i's is an ancillary complement to T. The joint probability distribution for T and Y_1 is described in the following table:

T \ Y_1	0	1	2	Total
0	$\frac{1-\theta}{12}$	$\frac{2-\theta}{12}$	$\frac{3-\theta}{12}$	$\frac{2-\theta}{4}$
1	$\frac{1+\theta}{12}$	$\frac{2+\theta}{12}$	$\frac{3+\theta}{12}$	$\frac{2+\theta}{4}$
total	$\frac{1}{6}$	$\frac{1}{3}$	$\frac{1}{2}$	1

The Y_1-decomposition of the experiment of rolling the biased die once is as follows:

"Choose one of the three pairs (1,4), (2,5) and (3,6) with probabilities 1/6, 1/3 and 1/2 respectively. Then select one from the chosen pair with probabilities

$\frac{1-\theta}{2}$ and $\frac{1+\theta}{2}$, if the chosen set is (1,4),

or, $\frac{2-\theta}{4}$ and $\frac{2+\theta}{4}$, if the chosen set is (2,5),

or, $\frac{3-\theta}{6}$ and $\frac{3+\theta}{6}$, if the chosen set is (3,6)".

The physical experiment of rolling the biased die once may then be imagined to be equivalent to the two stage (conceptual) experiment of first choosing (in a θ -free manner) one of three biased coins and then tossing the selected coin once.

Let us observe that we have six different decompositions of $\mathcal{E}$ corresponding to the six different ancillary complements to T. How do we recover the information lost in T ?

Let us suppose that the observed value of X is 5. The corresponding values of T, Y_1, Y_2 and Y_3 are respectively 1,1,2 and 0. The conditional distributions of the maximum likelihood estimator T under the three conditions $Y_1 = 0$, $Y_2 = 2$ and $Y_3 = 0$ are as follows :

range of T :		0	1
conditional probability distribution of T under the condition	$Y_1 = 1$	$\frac{(2-\theta)}{4}$	$\frac{(2+\theta)}{4}$
	$Y_2 = 2$	$\frac{(3-\theta)}{5}$	$\frac{(2+\theta)}{5}$
	$Y_3 = 0$	$\frac{(1-\theta)}{3}$	$\frac{(2+\theta)}{3}$

Thus, in this situation we find that different choices of ancillary statistics lead to different 'reference sets' and different reliability indices for the estimator T. There exists no unique way of recovering the ancillary information.

4. Conceptual Statistical Experiments

The author believes that the real trouble lies in our failure to recognize the difference between a real (performable) and a conceptual (non-performable) statistical experiment.[1] Every real experiment gives rise to a probability structure $(\mathcal{X}, \mathcal{A}, \mathcal{P})$ but the converse is not true. In Section 2 we saw how the probability structure $(\mathcal{X}, \mathcal{A}, \mathcal{P})$, generated by the experiment $\mathcal{E}$, may be decomposed into the family $\{(\mathcal{X}_y, \mathcal{A}_y, \mathcal{P}_y) ; y \epsilon \mathcal{Y}\}$ of probability structures and there we conceived of an experiment $\mathcal{E}_y$ corresponding to $(\mathcal{X}_y, \mathcal{A}_y, \mathcal{P}_y)$. In general, the experiments $\mathcal{E}_y$ are non-performable. If the statistician selects (on the flip of a coin) between a sample of size 10 and one of 100, he is making a random choice between two performable statistical experiments. But in the second example of the previous section the statistician can only conceive (in six different ways) of a breakdown of the experiment of once rolling the biased die into a two stage experiment of first making a (θ-free) random selection between three biased coins and then tossing the selected coin once. In this example the statistician has a die to experiment with; but where are the coins?[2]

---

1 The author does not think it necessary to enter into a discussion on the performability of a statistical experiment.

2 Of course, the experiment of rolling the die repeatedly until, say, either 2 or 5 appears (and then observing only the final score) is essentially equivalent to tossing once a biased coin with probabilities $(2-\theta)/4$ and $(2+\theta)/4$. But who is interested in such a wasteful experiment? The author would classify such experiments under the conceptual (non-performable) category.

When the sample size n is determined in a θ-free manner the statistician may be justified in regarding n as a prefixed experimental constant and contemplating on the long term performance characteristic of the experimental procedure with the sample size fixed at the level actually obtained. More generally, when the statistician is presented with a θ-free choice between a family $\{\mathcal{E}_y\}$ of performable experimental procedures then it would be correct to treat the Y actually obtained as a predetermined experimental constant. His sole concern should be the experiment $\mathcal{E}_Y$ he is actually presented with and none of the other members of the family $\{\mathcal{E}_y\}$. Difficulty arises in the attempted generalization to non-performable meaningless experiments. The kind of situations where an experiment $\mathcal{E}$ may be decomposed into a family $\{\mathcal{E}_y\}$ of real experiments are very rare indeed. The author is not aware of any example where such a decomposition (into real experiments) may be effected in more than one way. The elementary examples considered in Section 3 establish the possibility of a multiplicity of ancillary decompositions into conceptual experiments. The ancillary argument of Fisher cannot be extended to such cases. The sample size analogy for the ancillary statistic appears to be a false one. We end this discourse with an example where there exists an essentially unique maximum ancillary decomposition of the experiment but yet the ancillary argument leads us to a rather curious and totally unacceptable 'reference set'.

Example : Let X be an observable random variable with uniform probability distribution over the interval $[\theta, \theta + 1)$ where $0 \leqslant \theta < \infty$. For the sake of simplicity we consider the case of a single observation on X. The sample space $\mathcal{X}$ is the half line $[0, \infty)$. The likelihood function, for the observation X, is

$$L(\theta) = \begin{cases} 1 & \text{if } X - 1 < \theta \leqslant X \\ 0 & \text{otherwise} \end{cases}$$

Thus, the integer part $[X]$ of X has as good a claim to be considered a maximum likelihood estimator for θ as any other point in the interval $(X - 1, X]$. It is easy to check that $[X]$ is not a sufficient statistic. Does there exist an ancillary complement to $[X]$?

Consider the fractional part $\varphi(X) = X - [X]$ of X. It is not difficult to see that $\varphi(X)$ is an ancillary statistic with uniform probability distribution over the unit interval $[0, 1)$. Indeed, it is possible to show that $\varphi(X)$ is an essentially maximum ancillary in the sense that every ancillary statistic is essentially a function of $\varphi(X)$. As $X = [X] + \varphi(X)$, it follows at once that the pair $([X], \varphi(X))$ is equivalent to X and hence $\varphi(X)$ is the ancillary complement to $[X]$.

For a given $\theta = [\theta] + \varphi(\theta)$, the observation $X = [X] + \varphi(X)$ lies in the interval $[\theta, \theta + 1)$, i.e., with probability one,

$$\theta = [\theta\ ] + \varphi(\theta) \leqslant [X] + \varphi(X) < [\theta + 1] + \varphi(\theta) = \theta + 1.$$

From the above, it follows that

$$[X] = \begin{cases} [\theta] & \text{if } \varphi(X) \geqslant \varphi(\theta) \\ [\theta + 1] & \text{if } \varphi(X) < \varphi(\theta) \end{cases} \qquad \ldots \qquad (4.1)$$

Since $\varphi(X)$ has a uniform distribution over $[0, 1)$, it follows that the marginal distribution of $[X]$ is concentrated at the two points $[\theta]$ and $[\theta + 1] = [\theta] + 1$ with probabilities $1 - \varphi(\theta)$ and $\varphi(\theta)$ respectively. Hence

$$\begin{aligned} E([X]|\theta) &= [\theta](1 - \varphi(\theta)) + ([\theta] + 1)\varphi(\theta) \\ &= [\theta] + \varphi(\theta) \\ &= \theta \end{aligned}$$

i.e $[X]$ is an unbiased estimator of θ . And

$$V([X]|\theta\ ) = \varphi(\theta)\ (1 - \varphi(\theta)).$$

Now, since $\varphi(X)$ is the ancillary complement to $[X]$, let us see what 'reference set' it leads us to. Given $\varphi(X)$, the sample $X = [X] + \varphi(X)$ can vary over the restricted set

$$\varphi(X),\ 1 + \varphi(X),\ 2 + \varphi(X),\ \ldots\ldots\ldots$$

From the relation (4.1) it is now clear that, for any fixed θ , the conditional distribution of $X = [X] + \varphi(X)$, given $\phi(X)$, is degenerate at the point

$$[\theta] + \varphi(X), \quad \text{if } \varphi(\theta) \leqslant \varphi(X),$$

or, at the point

$$[\theta + 1] + \varphi(X), \text{ if } \varphi(\theta) > \varphi(X).$$

Thus, the 'reference set', corresponding to an observed value of the ancillary statistic $\varphi(X)$, is a one point, degenerate probability structure. The conditional distribution of the maximum likelihood estimator $[X]$, given $\varphi(X)$ and θ , is degenerate at the point $[\theta]$ or $[\theta + 1]$ depending on whether $\varphi(\theta) \leqslant \varphi(X)$ or $\varphi(\theta) > \varphi(X)$. Acceptance of this 'reference set' will alter the status of $[X]$ from a statistical variable to an unknown constant.

Writing $Y = \varphi(X)$, the two stage Y-decomposition of the experiment $\mathcal{E}$ of making a single observation on X will then be as follows :

(i) Select a number Y at random (with uniform probability distribution) in the unit interval [0, 1)

(ii) Determine the value of $[\theta]$ and whether

$$\text{(a)} \quad \varphi(\theta) \leqslant Y \quad \text{or}$$

$$\text{(b)} \quad \varphi(\theta) > Y$$

and then write
$$X = \begin{cases} [\theta] + Y & \text{in case of (a)} \\ [\theta] + 1 + Y & \text{in case of (b)} \end{cases}$$

The second stage of the Y-decomposition is clearly nonperformable.

CHAPTER II

STATISTICAL INFORMATION AND LIKELIHOOD

Part I : Principles

0. Notes

Editor's Note : The material covered in Chapters II, III, and IV were first exposited by Basu in a series of seminars at the University of Sheffield, England in the Fall of 1972 and then presented as a three part essay at a symposium on Statistical Inference held at Aarhus University, Denmark in the Summer of 1973. The paper was first published with discussions in the proceedings of the Aarhus symposium and then republished in Sankhya A, 1975.

Author's Note : I came to know of the Likelihood Principle from George Barnard over 30 years ago when he spent several weeks with us in the Indian Statistical Institute, Calcutta. At that time the Principle appeared to me to be a most curious one and in conflict with all that I knew of Statistics then. I vividly recall a quiet afternoon at Amrapali, the Mahalanobis residence in Calcutta, when I had had a long discussion with George about the Principle.

My main contention then was that we cannot make a likelihood based inference in standard nonparametric setups simply because we have no likelihood function in those cases. I also pointed out to George that in the Survey Sampling setup the likelihood function looks so stale, flat and unprofitable. My admiration for George was greatly enhanced when he looked really puzzled with my examples and promised to think about them.

Years later, the table was really turned. My growing discomfiture with the Neyman-Pearson-Wald (NPW) kind of Statistics was finally resolved when I recognized (in 1967) the universality of the likelihood principle. But when I was ready to accept the likelihood antithesis to the NPW thesis, I found George running scared of the Principle that he had so boldly enunciated in the fifties.!

Like Barnard, Birnbaum also eventually shied away from the likelihood principle. While Barnard was not prepared to give up his pet pivotal quantities, Birnbaum found the inherent contradictions between the likelihood principle and the so called Confidence Principle (of the NPW thesis) quite unacceptable.

In Part I of this essay I tried to pave the deserted trail blazed by Allan Birnbaum.

STATISTICAL INFORMATION AND LIKELIHOOD

PART I : PRINCIPLES

1. Statistical Information

The key word in Statistics is information. After all, this is what the subject is all about. A problem in statistics begins with a state of nature, a parameter of interest ω , about which we do not have enough information. In order to generate further information about ω , we plan and then perform a statistical experiment $\mathcal{E}$. This generates the sample x. By the term 'statistical data' we mean such a pair ($\mathcal{E}$, x) where $\mathcal{E}$ is a well defined statistical experiment and x the sample generated by a performance of the experiment. The problem of data analysis is to extract 'the whole of the relevant information' — an expression made famous by R.A. Fisher — contained in the data ($\mathcal{E}$, x) about the parameter ω . But, what is information? No other concept in statistics is more elusive in its meaning and less amenable to a generally agreed definition.

To begin with, let us agree to the use of the notation

$$\text{Inf}\ (\mathcal{E}, x)$$

only as a pseudo-mathematical short hand for the ungainly expression : 'the whole of the relevant information about ω contained in the data ($\mathcal{E}$, x)'. At this point an objection may well be raised to the following effect : The concept of information in the data ($\mathcal{E}$, x) makes sense only in the context of (i) the 'prior-information' q (about ω and other related entities) that we must have had to begin with and (ii) the particular 'inferential problem' Π (about ω) that made us look for further information.

While agreeing with the criticism that it is more realistic to look upon 'information in the data' as a function with four arguments Π , q, $\mathcal{E}$ and x, let us hasten to point out that at the moment we are concerned with variations in $\mathcal{E}$ and x only and so we are holding fixed the other two elements of Π and q. That Inf($\mathcal{E}$, x) may depend very critically on x, is well illustrated by the following simple example.

Example 1 : Suppose an urn contains 100 tickets that are numbered consecutively as $\omega + 1$, $\omega + 2$, ...$\omega + 100$ where ω is an unknown number. Let $\mathcal{E}_n$ stand for the statistical experiment of drawing a simple random sample of n tickets from the urn and then recording the sample as a set of n numbers $x_1 < x_2 < \dots < x_n$. If at the planning stage of the experiment, we are asked to choose between the two experiments $\mathcal{E}_2$ and $\mathcal{E}_{25}$ then, other things being equal, we shall no doubt prefer $\mathcal{E}_{25}$ to $\mathcal{E}_2$. Consider now the hypothetical situation where $\mathcal{E}_2$ has been performed resulting in the sample x = (17,115). How good is Inf($\mathcal{E}_2$,x)? A quick analysis of the data will reveal that ω has to be an integer and must satisfy both the inequalities.

$$\omega + 1 \leqslant 17 \leqslant \omega + 100 \text{ and } \omega + 1 \leqslant 115 \leqslant \omega + 100.$$

In other words, Inf($\mathcal{E}_2$, x) tells us categorically that ω is 15 or 16. Now, contrast the above with another hypothetical situation where $\mathcal{E}_{25}$ has been performed and has yielded the sample x' = (17, 20, ..., 52), where 17 and 52 are respectively the smallest and the largest number drawn. With Inf($\mathcal{E}_{25}$, x') we can now only assert that ω is an integer that lies somewhere in the interval [- 48, 16]. While it is clear that, in some average sense, the experiment $\mathcal{E}_{25}$ is 'more informative' than $\mathcal{E}_2$, it is equally incontrovertible that the particular sample (17, 115) from experiment $\mathcal{E}_2$ will tell us a great deal more about the parameter than will the sample (17, 20, ..., 52) from $\mathcal{E}_{25}$. To be more specific, with

$$\text{Inf}\ \{\mathcal{E}_n, (x_1, x_2, \dots, x_n)\}$$

we know without any shadow of doubt that the true value of ω must belong to the set

$$A = \{x_1 - 1, x_1 - 2, \dots, x_1 - m\}$$

where $m = 100 - (x_n - x_1)$. In the present case the likelihood function (for the parameter ω) is 'flat' over the set A and is zero outside (a situation that is typical of all survey sampling setups) and this means that the sample $(x_1, x_2, \dots, x_n)$ from experiment $\mathcal{E}_n$ 'supports' each of the points in the set A with equal intensity. Therefore, it seems reasonable to say that we may identify the information supplied by the data $\{\mathcal{E}_n, (x_1, x_2, \dots, x_n)\}$ with the set A and quantify the magnitude of the information by the statistic $m = 100 - (x_n - x_1)$ - the smaller the number m is, the more precise is our specification of the unknown ω. Once the experiment $\mathcal{E}_n$ is performed and the sample $(x_1, x_2, \dots, x_n)$ recorded, the magnitude of the information obtained depends on the integer m (which varies from sample to sample) rather than on the constant n.

Among contemporary statisticians there seems to be a complete lack of consensus about the meaning of the term 'statistical information' and the manner in which such an important notion may be meaningfully formalized. As a first step towards finding the greatest common factor among the various opinions held on the subject, let us make a beginning with the following loosely phrased operational definition of equivalence of two bits of statistical information.

Definition : By the equality or equivalence of Inf($\mathcal{E}_1, x_1$) and Inf($\mathcal{E}_2, x_2$) we mean the following :

(a) the experiment $\mathcal{E}_1$ and $\mathcal{E}_2$ are 'related' to the same parameter of interest ω, and

(b) 'everything else being equal', the outcome x_1 from $\mathcal{E}_1$ 'warrants the same inference' about ω as does the outcome x_2 from $\mathcal{E}_2$.

We plan to make an evaluation of several guidelines that have been suggested from time to time for deciding when two different bits of information ought to be regarded as equivalent. But before we proceed with that project, let us agree on a few definitions.

2. Basic Definitions and Relations

In contrast to the situation regarding the notion of statistical information, there exists a general consensus of opinion regarding a mathematical framework for the notion of a statistical experiment. We formalize a statistical experiment $\mathcal{E}$ as a triple ($\mathcal{X}, \Omega$, p) where

(i) $\mathcal{X}$, the sample space, is the set of all the possible samples (outcomes) x that a particular performance of $\mathcal{E}$ may give rise to,

(ii) Ω, the parameter space, is the set of all the possible values of an entity ω that we call the universal parameter or the state of nature, and

(iii) $p = p(x \mid \omega)$, the probability function, is a map $p : \mathcal{X} \times \Omega \to [0, 1]$ that satisfies the identity

$$\sum_{x \in \mathcal{X}} p(x|\omega) \equiv 1 \qquad \text{for all} \quad \omega \in \Omega$$

To avoid being distracted by measurability conditions, we stipulate from the beginning that both $\mathcal{X}$ and Ω are finite* sets. There is no loss of generality in the further assumption that

*I hold firmly to the view that this contingent and cognitive universe of ours is in reality only finite and, therefore, discrete. In this essay we steer clear of the logical quicksands of 'infinity' and the 'infinitesimal'. Infinite and continuous models will be used in the sequel, but they are to be looked upon as mere approximations to the finite realities.

$$\sum_{\omega\in\Omega} p(x|\omega) > 0 \qquad \text{for all} \quad x\in\mathcal{X}.$$

It will frequently happen that we are not really interested in ω itself, but rather in some characteristic $\theta = \theta(\omega)$ of the universal parameter. In such cases we call θ the parameter of interest and denote its range of values by Θ. If there exists a set Φ of points ϕ such that we can write

$$\Omega = \Theta \times \Phi \qquad \text{and} \qquad \omega = (\theta,\phi),$$

we then call $\phi = \phi(\omega)$ the nuisance parameter.

With reference to an experiment $\mathcal{E} = (\mathcal{X}, \Omega, p)$, we define a statistic T as a map $T : \mathcal{X} \to \mathcal{J}$ of $\mathcal{X}$ into a space $\mathcal{J}$ of points t. Every point $t\in\mathcal{J}$ defines a subset $\mathcal{X}_t = \{x|T(x) = t\}$ of $\mathcal{X}$ and the family $\{\mathcal{X}_t|t\in\mathcal{J}\}$ of all these subsets defines a partition of $\mathcal{X}$. Conversely, every partition of $\mathcal{X}$ is induced by some suitably defined statistic. It is convenient to visualize a statistic T as a partition of the sample space $\mathcal{X}$.

Given $\mathcal{E}$ and a statistic T, we define the marginal experiment $\mathcal{E}_T$ as

$$\mathcal{E}_T = (\mathcal{J}, \Omega, p_T)$$

where the map $p_T : \mathcal{J}\times \Omega \to [0, 1]$ is given by

$$p_T(t|\omega) = \sum_{x\in\mathcal{X}_t} p(x|\omega).$$

Operationally, we may define $\mathcal{E}_T$ as 'perform $\mathcal{E}$ and then observe only $T = T(x)$.'

Still taking T as above, we may define, for each $t\in\mathcal{J}$, a (conceptual) experiment

$$\mathcal{E}_t^T = (\mathcal{X}_t, \Omega, p_t^T)$$

where the map $p_t^T : \mathcal{X}_t \times \Omega \to [0, 1]$ is given by the formula

$$p_t^T(x|\omega) = p(x|\omega) / \sum_{x'\in\mathcal{X}_t} p(x'|\omega)$$

for all $x\in\mathcal{X}_t$ and $\omega\in\Omega$. [The usual care needs to be taken about a possible zero denominator here.] We call $\mathcal{E}_t^T$ the conditional experiment given that $T(x) = t$. The experiment $\mathcal{E}_t^T$ may be loosely characterized as : 'Reconstruct the sample x from the information that $T(x) = t$'. [In Section 4 we examine the question whether such a reconstruction is operationally meaningful.] With each statistic T we may then associate a conceptual decomposition of the experiment $\mathcal{E}$ into a two stage experiment: 'First perform $\mathcal{E}_T$ and then perform $\mathcal{E}_t^T$ where t is the outcome of $\mathcal{E}_T$.'

We now briefly list a set of wellknown definitions and theorems.

Definition 1 (A partial order) : The statistic $T : \mathcal{X} \to \mathcal{J}$ is wider or larger than the statistic $T' : \mathcal{X} \to \mathcal{J}'$, if for each $t' \varepsilon \mathcal{J}'$ there exists a $t \varepsilon \mathcal{J}$ such that $\mathcal{X}_t \subset \mathcal{X}_{t'}$, that is, if the partition of $\mathcal{X}$ induced by T is a sub-partition of the one induced by T'.

Definition 2 (Noninformative experiments): An experiment $\mathcal{E} = (\mathcal{X}, \Omega, p)$ is statistically trivial or noninformative (about the universal parameter ω) if, for each $x \varepsilon \mathcal{X}$, the function $\omega \to p(x|\omega)$ is a constant.

Definition 3 (Ancillary statistic) : $T : \mathcal{X} \to \mathcal{J}$ is an ancillary statistic (w.r.t. ω) if the marginal experiment $\mathcal{E}_T$ is noninformative (about ω).

Definition 4 (Sufficient statistic) : T is a sufficient statistic (for ω) if, for all $t \varepsilon \mathcal{J}$, the conditional experiment $\mathcal{E}_t^T$ is noninformative (about ω).

Definition 5 (Likelihood function) : When an experiment $\mathcal{E} = (\mathcal{X}, \Omega, p)$ is performed resulting in the outcome $x \varepsilon \mathcal{X}$, the function $\omega \to p(x|\omega)$ is called the likelihood function generated by the data $(\mathcal{E}, x)$ and is variously denoted in the sequel as L, $L(\omega)$, $L(\omega|x)$ or $L(\omega|\mathcal{E}, x)$.

Definition 6 (Equivalent likelihoods) : Two likelihood functions L_1 and L_2 defined on the same parameter space Ω [but possibly corresponding to two different pairs $(\mathcal{E}_1, x_1)$ and $(\mathcal{E}_2, x_2)$ respectively] are said to be equivalent if there exists a constant $c > 0$ such that $L_1(\omega) = cL_2(\omega)$ for all $\omega \varepsilon \Omega$. [The constant c may, of course, depend on $\mathcal{E}_1$, $\mathcal{E}_2$, x_1 and x_2]. We write $L_1 \sim L_2$ to indicate the equivalence of the likelihood functions.

Definition 7 (Standardized likelihood) : Each likelihood function L on Ω gives rise to an equivalent standardized likelihood function $\bar{L}$ on Ω defined as

$$\bar{L}(\omega) = L(\omega) / \sum_{\omega' \varepsilon \Omega} L(\omega').$$

Note that our assumptions about Ω and p preclude the possibilities of the denominator being zero or infinite.

Theorem 1 : A statistic T is sufficient if and only if, for $x_1, x_2 \varepsilon \mathcal{X}$, $T(x_1) = T(x_2)$ implies $L(\omega|x_1) \sim L(\omega|x_2)$.

In other words, a statistic $T : \mathcal{X} \to \mathcal{J}$ is sufficient if and only if, for every $t \varepsilon \mathcal{J}$, it is true that all points x on the T-surface $\mathcal{X}_t$ generate equivalent likelihood functions. The following result is then an immediate consequence of the above.

Theorem 2 : For any experiment $\mathcal{E} = (\mathcal{X}, \Omega, p)$ the map (statistic) $x \to \bar{L}(\omega|x)$, from x to a (standardized) likelihood function $\bar{L}$ on Ω, is the minimum sufficient statistic, that is, the above statistic is sufficient and every other sufficient statistic is wider than it.

Definition 8 (Mixture of experiments) **:** Suppose we have a number of experiments $\mathcal{E}_i = (\mathcal{X}_i, \Omega, p_i)$, $i = 1, 2, \ldots$, with the same parameter space Ω, to choose from. And let $\pi_1, \pi_2, \ldots$ be a preassigned set of nonnegative numbers summing to unity. The mixture $\mathcal{E}$ of the experiments $\mathcal{E}_1, \mathcal{E}_2, \ldots$ according to mixture (selection) probabilities $\pi_1, \pi_2, \ldots$ is defined as a two stage experiment that begins with (i) a random selection of one of the experiments $\mathcal{E}_1, \mathcal{E}_2, \ldots$ with selection probabilities $\pi_1, \pi_2, \ldots$, followed by (ii) the performing of the experiment selected in stage (i). Clearly, the sample space $\mathcal{X}$ of the mixture experiment $\mathcal{E} = (\mathcal{X}, \Omega, p)$ is the set of all pairs (i, x_i) with $i = 1, 2, \ldots$ and $x_i \varepsilon \mathcal{X}_i$ (that is, $\mathcal{X}$ is the disjoint union of the sets $\mathcal{X}_1, \mathcal{X}_2, \ldots$). And the probability function $p : \mathcal{X} \times \Omega \to [0, 1]$ is given by

$$P(x|\omega) = \pi_i p_i(x_i|\omega)$$

when $x = (i, x_i)$.

It is important to note our stipulation that the mixture probabilities $\pi_1, \pi_2, \ldots$ are preassigned numbers and, therefore, unrelated to the unknown parameter ω. Given an experiment $\mathcal{E} = (\mathcal{X}, \Omega, p)$ and an ancillary statistic $T : \mathcal{X} \to \mathcal{T}$, we may view $\mathcal{E}$ as a mixture of the family

$$\{\mathcal{E}_t^T : t \varepsilon \mathcal{T}\}$$

of conditional experiments, with mixture probabilities

$$\pi_t = p_T(t|\omega), \quad t \varepsilon \mathcal{T}$$

which do not depend on ω since T is ancillary.

Definition 9 (Similar experiments) **:** The experiments $\mathcal{E}_1 = (\mathcal{X}_1, \Omega, p_1)$ and $\mathcal{E}_2 = (\mathcal{X}_2, \Omega, p_2)$ with the same parameter space Ω are said to be similar or statistically isomorphic if there exists a one to one and onto map $g : \mathcal{X}_1 \to \mathcal{X}_2$ such that

$$p_1(x_1|\omega) = p_2(gx_1|\omega)$$

for all $x_1 \varepsilon \mathcal{X}_1$ and $\omega \varepsilon \Omega$. The function g is then called a similarity map.

We end this section with a definition, due to D. Blackwell (1950), of the sufficiency of an experiment for another experiment and a few related remarks.

Definition 10 (Blackwell sufficiency) : The experiment $\mathcal{E}_1 = (\mathcal{X}_1, \Omega, p_1)$ is sufficient for the experiment $\mathcal{E}_2 = (\mathcal{X}_2, \Omega, p_2)$ if there exists a transtion function $\pi : \mathcal{X}_1 \times \mathcal{X}_2 \to [0, 1]$ (with the usual condition that $\sum_{x_2} \pi(x_1, x_2) = 1$ for all $x_1 \varepsilon \mathcal{X}_1$) which satisfies the additional requirement that

$$p_2(x_2|\omega) = \sum_{x_1} p_1(x_1|\omega)\, \pi(x_1, x_2)$$

for all $\omega \varepsilon \Omega$ and $x_2 \varepsilon \mathcal{X}_2$.

The sufficiency of $\mathcal{E}_1$ for $\mathcal{E}_2$ means exactly this : that the experiment $\mathcal{E}_2$ may be simulated by first performing $\mathcal{E}_1$ and noting its outcome x_1, and then obtaining a point x_2 in $\mathcal{X}_2$ via a secondary randomization process that is defined in terms of the transition function $\pi(x_1, \cdot)$. Note that, for each $x_1 \varepsilon \mathcal{X}_1$, the function $\pi(x_1, \cdot)$ defines a probability distribution on $\mathcal{X}_2$ that is free of the unknown ω. We refer to Blackwell (1950) for an alternative but equivalent formulation of Definition 10 in terms of the average performance characteristics of statistical decision functions.

If for an experiment $\mathcal{E} = (\mathcal{X}, \Omega, p)$ the statistic $T : \mathcal{X} \to \mathcal{J}$ is sufficient (Definition 4), then the marginal experiment $\mathcal{E}_T = (\mathcal{J}, \Omega, p_T)$ is sufficient (Definition 10) for $\mathcal{E}$. The converse proposition is also true. If $\mathcal{E}_1$ and $\mathcal{E}_2$ are similar (Definition 9) experiments with $g : \mathcal{X}_1 \to \mathcal{X}_2$ as a similarity map, then the Kronecker delta function $\delta(gx_1, x_2)$ may be taken as the transition function $\pi(x_1, x_2)$ to prove the sufficiency of $\mathcal{E}_1$ for $\mathcal{E}_2$. In a like manner the similarity map $g^{-1} : \mathcal{X}_2 \to \mathcal{X}_1$ proves the sufficiency of $\mathcal{E}_2$ for $\mathcal{E}_1$. Furthermore, any decision function δ_2 for $\mathcal{E}_2$ can be completely matched (in terms of its average performance characteristics) by the decision δ_1 for $\mathcal{E}_1$ defined as

$$\delta_1(x_1) = \delta_2(gx_1) \quad \text{for all } x_1 \varepsilon \mathcal{X}_1 .$$

3. Some Principles of Inference

Instead of plunging headlong into a controversial definition of Inf($\mathcal{E}$, x), let us follow a path of less resistance and formulate, on the model of A. Birnbaum (1962), some guidelines for the recognition of equivalence of two different bits of statistical information. Each such guideline is stated here as a Principle (of statistical inference).

Looking back on definition 9 of the previous section, it is clear that two similar experiments $\mathcal{E}_1$ and $\mathcal{E}_2$ are identical in all respects excepting in the manner of labelling their sample points. Since the manner of labelling the sample points of an experiment should not have any effect on the actual information obtained in a particular trial, the following principle is almost self evident.

Principle $\mathcal{I}$ (The invariance or similarity principle) : If $\mathcal{E}_1 = (\mathcal{X}_1, \Omega, p_1)$ and $\mathcal{E}_2 = (\mathcal{X}_2, \Omega, p_2)$ are similar experiments with $g : \mathcal{X}_1 \to \mathcal{X}_2$ as a similarity map of $\mathcal{E}_1$ onto $\mathcal{E}_2$, then

$$\text{Inf}(\mathcal{E}_1, x_1) = \text{Inf}(\mathcal{E}_2, x_2)$$

if $gx_1 = x_2$.

Now, suppose the two points x' and x'', in the sample space of an experiment $\mathcal{E} = (\mathcal{X}, \Omega, p)$, give rise to identical likelihood functions, that is, $p(x'|\omega) = p(x''|\omega)$ for all $\omega \in \Omega$. We can then define a similarity map $g : \mathcal{X} \to \mathcal{X}$ of $\mathcal{E}$ onto itself in the following manner :

$$gx = \begin{cases} x & \text{if } x \notin \{x', x''\} \\ x' \text{ or } x'' & \text{acc. as } x = x'' \text{ or } x'. \end{cases}$$

The following is then a specialization of principle $\mathcal{I}$ to the case of a single experiment $\mathcal{E}$.

Principle $\mathcal{I}'$ (A weak version of $\mathcal{I}$) : If $p(x'|\omega) = p(x''|\omega)$ for all $\omega \in \Omega$, then

$$\text{Inf}(\mathcal{E}, x') = \text{Inf}(\mathcal{E}, x'').$$

Principle $\mathcal{I}'$ induces the following equivalence relation on the sample space of an experiment : The points x' and x'' in the sample space $\mathcal{X}$ of an experiment $\mathcal{E}$ are equivalent or equally informative if they generate identical likelihood functions.

Let us look at Definition 2 in Section 2 and reassert the almost self evident proposition : 'No additional information can be generated about a partially known parameter ω by performing a statistically trivial experiment $\mathcal{E}$ '. It follows then that once an experiment $\mathcal{E}_1$ has been carried out resulting in the outcome y, it is not possible to add to the information $\text{Inf}(\mathcal{E}_1, y)$ so obtained by carrying out a further 'postrandomization' exercise — that is, by performing a secondary experiment $\mathcal{E}_{(y)}$ whose randomness structure may depend on the outcome y of $\mathcal{E}_1$ but is completely known to the experimenter. Let us formally rewrite the above in the form

$$\text{Inf}(\mathcal{E}_1, y) = \text{Inf}\{(\mathcal{E}_1 \to \mathcal{E}_{(y)}), (y, z)\}$$

where $(\mathcal{E}_1 \to \mathcal{E}_{(y)})$ stands for the composite experiment '$\mathcal{E}_1$ followed by $\mathcal{E}_{(y)}$' and y, z are the outcomes of $\mathcal{E}_1$ and $\mathcal{E}_{(y)}$ respectively.

Now let $T : \mathcal{X} \to \mathcal{Y}$ be a sufficient statistic for $\mathcal{E} = (\mathcal{X}, \Omega, p)$ and let $\mathcal{E}_T$ and $\{\mathcal{E}_t^T : t \in \mathcal{Y}\}$ be respectively the marginal experiment and the family of conditional experiments as defined in Section 2. Now, we may look upon a performance of $\mathcal{E}$ and the observation of the outcome x as 'a performance of the marginal experiment $\mathcal{E}_T$, observation of its outcome $t = T(x)$, followed by a postrandomization exercise $\mathcal{E}_t^T$ of identifying the exact location of x on the surface $\mathcal{X}_t = \{x' | T(x') = t\}$'. Since T is sufficient, the conditional experiment $\mathcal{E}_t^T$ is statistically trivial for every $t \in \mathcal{Y}$. Looking back on the argument of the previous paragraph, one may now claim that the following principle has been sort of 'proved by analogy'.

Principle $\mathcal{S}$ (The sufficiency principle) : If, in the context of an experiment $\mathcal{E}$, the statistic T is sufficient then, for all $x \in \mathcal{X}$ and $t = T(x)$,

$$\mathrm{Inf}(\mathcal{E}, x) = \mathrm{Inf}(\mathcal{E}_T, t).$$

If T is sufficient and $\mathcal{X}_t$ a particular T-surface, then from Principle $\mathcal{S}$ it follows that $\mathrm{Inf}(\mathcal{E}, x)$ is the same for all $x \in \mathcal{X}_t$. In the literature we often find the sufficiency principle stated in the following alternative (and perhaps a trifle less severe) form :

Principle $\mathcal{S}'$ (Alternative version of $\mathcal{S}$): $\mathrm{Inf}(\mathcal{E}, x') = \mathrm{Inf}(\mathcal{E}, x'')$ if for some sufficient statistic T it is true that $T(x') = T(x'')$.

From Theorems 1 and 2 of Section 2 it follows at once that the following is an equivalent version of $\mathcal{S}'$:

Principle $\mathcal{L}'$: (The weak likelihood principle) : $\mathrm{Inf}(\mathcal{E}, x') = \mathrm{Inf}(\mathcal{E}, x'')$ if the two sample points x' and x'' generate equivalent likelihood functions, that is, if $L(\omega|x') \sim L(\omega|x'')$.

Clearly, $\mathcal{L}'$ implies $\mathcal{I}$. Before we turn our attention to some other guiding principles of statistical inference, let us summarize our findings about the logical relationships among the principles $\mathcal{I}$, $\mathcal{I}'$, $\mathcal{S}$, $\mathcal{S}'$ and $\mathcal{L}'$ in the following :

Theorem 1 : $\mathcal{I} \Rightarrow \mathcal{I}', \quad \mathcal{S} \Rightarrow \mathcal{S}' \Leftrightarrow \mathcal{L}' \Rightarrow \mathcal{I}'$

Whereas the sufficiency principle warns us to be vigilant against any 'postrandomization' in the statistical experiment and advises us to throw away the outcome of any such exercise as irrelevant to the making of inference, the conditionality principle (which we now discuss) concerns itself in a like manner with any 'prerandomization' that may have been built into the structure of an experiment. Consider an experiment $\mathcal{E} = (\mathcal{X}, \Omega, p)$ which is a mixture (Definition 8, Section 2) of the two experiments $\mathcal{E}_i = (\mathcal{X}_i, \Omega, p_i)$, $i = 1, 2$, where the mixture probabilities π and $1-\pi$ are known. A typical outcome of $\mathcal{E}$ may then be represented as $x = (i, x_i)$, where $i = 1,2$ and $x_i \varepsilon \mathcal{X}_i$. Now, having performed the mixture experiment $\mathcal{E}$ and recognizing the sample as $x = (i, x_i)$, the question that naturally arises is whether we should present the data (for analysis) as $(\mathcal{E}, x)$ or the simpler form of $(\mathcal{E}_i, x_i)$. To me it seems almost axiomatic that the second form of data presention should not entail any loss of information and this is precisely the content of the following :

Priciple $\mathcal{C}'$ (The weak conditionality principle): If $\mathcal{E}$ is a mixture of $\mathcal{E}_1$, $\mathcal{E}_2$ as described above, then for any $i \varepsilon \{1, 2\}$ and $x_i \varepsilon \mathcal{X}_i$

$$\text{Inf}(\mathcal{E}, (i, x_i)) = \text{Inf}(\mathcal{E}_i, x_i).$$

In the literature we frequently meet a somewhat stronger version of $\mathcal{C}'$ which may be stated as follows :

Principle $\mathcal{C}$. (The conditionality principle) : If $T : \mathcal{X} \to \mathcal{T}$ is an ancilary statistic (Definition 3, Section 2) associated with the experiment $\mathcal{E} = (\mathcal{X}, \Omega, p)$, then, for all $x \varepsilon \mathcal{X}$ and $t = T(x)$,

$$\text{Inf}(\mathcal{E}, x) = \text{Inf}(\mathcal{E}_t^T, x).$$

We are now ready to state the centrepiece of our discussion in this essay - the likelihood principle. Let $\mathcal{E}_1$, $\mathcal{E}_2$ be any two experiments with the same parameter space and let x_i be a typical outcome of $\mathcal{E}_i (i = 1, 2)$.

Principle $\mathcal{L}$ (The likelihood principle) : If the data $(\mathcal{E}_1, x_1)$ and $(\mathcal{E}_2, x_2)$ generate equivalent likelihood functions on Ω, then $\text{Inf}(\mathcal{E}_1, x_1) = \text{Inf}(\mathcal{E}_2, x_2)$.

Before going into the far-reaching implications of $\mathcal{L}$, let us briefly examine the logical relationships in which $\mathcal{L}$ stands vis a vis the principles stated earlier. That $\mathcal{L} \Rightarrow \mathcal{G}$ follows at once from the definition of similar experiments. From the definition of a sufficient statistic it follows that the likelihood functions $L(\omega|\mathcal{E}, x)$ and $L(\omega|\mathcal{E}_T, t)$ are equivalent, whenever T is sufficient and $t = T(x)$. So $\mathcal{L} \Rightarrow \mathcal{S}$. Likewise, when T is ancillary, the likelihood functions generated by the data $(\mathcal{E}, x)$ and $(\mathcal{E}_t^T, x)$ are equivalent, whenever $t = T(x)$. Therefore, $\mathcal{L} \Rightarrow \mathcal{C}$. The following theorem asserts that the two weak principles $\mathcal{G}'$ and $\mathcal{C}'$ are together equivalent to $\mathcal{L}$.

Theorem 2 : $(\mathcal{G}'$ and $\mathcal{C}') \Rightarrow \mathcal{L}$.

Proof : Suppose the data $(\mathcal{E}_1, x_1)$ and $(\mathcal{E}_2, x_2)$ generate equivalent likelihood functions, that is, there exists $c > 0$ such that

$$L(\omega | \mathcal{E}_1, x_1) = cL(\omega | \mathcal{E}_2, x_2) \qquad \dots (*)$$

for all $\omega \varepsilon \Omega$. Using $\mathcal{I}'$ and $\mathcal{C}'$ we have to prove the equality $\text{Inf}(\mathcal{E}_1, x_1) = \text{Inf}(\mathcal{E}_2, x_2)$. To this end let us contemplate the mixture experiment $\mathcal{E}$ of $\mathcal{E}_1$ and $\mathcal{E}_2$ with mixture probabilities $1/(1+c)$ and $c/1+c$ respectively. Now, $(1, x_1)$ and $(2, x_2)$ are points in the sample space of the mixture experiment $\mathcal{E}$. In view of (*) and our choice of the mixture probabilities, it is clear that the data $(\mathcal{E}, (1, x_1))$ and $(\mathcal{E}, (2, x_2))$ generate identical likelihood functions, and so from $\mathcal{I}'$ it follows that

$$\text{Inf}(\mathcal{E}, (1, x_1)) = \text{Inf}(\mathcal{E}, (2, x_2)).$$

Now, applying $\mathcal{C}'$ to each side of the above equality we arrive at the desired equality.

Since $\mathcal{S}' \Rightarrow \mathcal{I}'$, we immediately arrive at the following corollary which was proved earlier by A. Birnbaum (1962).

Corollary : $(\mathcal{S}'$ and $\mathcal{C}') \Rightarrow \mathcal{L}$.

4. Information as a Function

From our exposition so far it should be amply clear that we are looking upon 'statistical information' as some sort of a function that maps the space D of all conceivably attainable data $d = (\mathcal{E}, x)$ related to ω into an yet undefined range space Λ. For the logical development of any concept it is important to agree in advance upon a 'universe of discourse'. In our case it is the space of all attainable data $d = (\mathcal{E}, x)$, where $\mathcal{E} = (\mathcal{X}, \Omega, p)$ is a typical statistical experiment concerning ω and x a typical outcome that may arise when $\mathcal{E}$ is performed. But what data are attainable, in other words, what triples $(\mathcal{X}, \Omega, p)$ correspond to performable statistical experiments ? The question is a tricky one and has escaped the general attention of statisticians.

Given a state of nature ω, not all conceivable triples $(\mathcal{X}, \Omega, p)$ can be models of performable statistical experiments. The situation is quite different in probability theory where we idealize the notion of a random experiment in terms of a single probability measure P on a measurable space $(\mathcal{X}, \mathcal{A})$. These days, with the help of powerful computers, we can simulate any reasonable random experiment upto almost any desired degree of approximation. That the situation is not quite the same with statistical experiments should be clear from the following.

Example : Let ω be the unknown probability of heads for a particular unsymmetric looking coin. One may argue that no informative (see Definition 2 in Section 2) statistical experiment concerning ω can be performed by anyone who is not in possession of the coin in question. With the coin in possession we can plan a Negative Binomial experiment $\mathcal{E}$ for which $\mathcal{X} = \{1, 2, 3 \ldots\}$ and $p(x|\omega) = \omega(1-\omega)^{x-1}, x \epsilon \mathcal{X}$. It is then not difficult to see how we can plan a (marginal) experiment $\mathcal{E}_1$ for which $\mathcal{X}_1 = \{0, 1\}$ and $p_1(0|\omega) = 1/(2-\omega)$. But can we plan an experiment $\mathcal{E}_2$ for which $\mathcal{X}_2 = \{0, 1\}$ and $p_2(0|\omega) = \sqrt{\omega}$ or $\sin(\frac{1}{2}\pi\omega)$? Intuitively, we feel that such strange looking functions of ω are unlikely to appear as probabilities in 'performable' experiments. They might, and an interesting mathematical problem associated with our coin is to determine the class of functions L that can arise as likelihoods, that is

$$L(\omega) = \text{Prob}\,(A|\omega)$$

where A is an event defined in terms of a 'performable' experiment with the coin. But it is not easy to see how we can give satisfactory mathematical definition of 'performability'.

If we insist on our universe of discourse to be the class of all conceivable triples $(\mathcal{X}, \Omega, p)$, then it is plausible that we shall end up with paradoxes such as those that have arisen in set theory in the past. Without labouring the point any further let us then agree that we are concerned with a rather small class $\mathfrak{E}$ of 'performable' experiments. Let this $\mathfrak{E}$ be our tongue-in-the-cheek definition of performability ! If $\mathcal{E}_1$ and $\mathcal{E}_2$ are performable experiments then it stands to reason to claim that any mixture of $\mathcal{E}_1$, $\mathcal{E}_2$ with known mixture probabilities is also performable. In other words, we may assume that the class $\mathfrak{E}$ is convex, i.e., closed under known mixtures. It also seems reasonable to claim that our class $\mathfrak{E}$ is closed under 'marginalization', that is, if $\mathcal{E} = (\mathcal{X}, \Omega, p)$ is performable then for any statistic $T : \mathcal{X} \to \mathcal{Y}$ the marginal experiment $\mathcal{E}_T = (\mathcal{Y}, \Omega, p_T)$ as defined in Section 2 is also performable. But how secure is the case for the conditional experiment $\mathcal{E}_t^T = (\mathcal{X}_t, \Omega, p_t^T)$? If T is sufficient, then, for every $t \epsilon \mathcal{Y}$, the conditional experiment $\mathcal{E}_t^T$ is non-informative and so is performable in a sense — the experiment can be simulated with the help of a random number table. Now, note that for a description of the general conditionality principle $\mathcal{C}$ we need to assume that for any ancillary statistic T (and every t in the range space of T) the conditional experiment $\mathcal{E}_t^T \epsilon \mathfrak{E}$. [In Chapter 1, I rejected the reasonableness of such an assumption and thereby sought to explain away certain anomalies that arise in an unrestricted use of principle $\mathcal{C}$ in the manner advocated by R. A. Fisher. Those anomalies arose only because I was then trying to reconcile $\mathcal{C}$ with the traditional 'sample space' analysis of data - in terms of the average performance characteristics of some inference procedures.] However, note that our description of the weaker conditionality principle $\mathcal{C}'$ and our derivation of $\mathcal{L}$ from $\mathcal{I}'$ and $\mathcal{C}'$ cannot be faulted on the ground of nonperformability of any experiment. In this connection it is interesting to look back on a derivation of the above implication theorem by Hájek (1967). Not only is Hajek's proof longer and somewhat obscure, but it appears to presuppose (in a quite unacceptable manner) that $\mathfrak{E}$ consists of all triples $(\mathcal{X}, \Omega, p)$.

Having recognized 'information' as a function Inf with its domain as the space D of all data $d = (\mathcal{E}, x)$ with $\mathcal{E} \in \mathfrak{E}$, let us finally turn our attention to the range of Inf. If we accept the likelihood principle, i.e., if we agree that

$$\text{Inf}(\mathcal{E}_1, x_1) = \text{Inf}(\mathcal{E}_2, x_2)$$

whenever $L(\omega | \mathcal{E}_1, x_1) \sim L(\omega | \mathcal{E}_2, x_2)$, then we may as well take a short step further and agree to view Inf as a mapping of the space D of all attainable data $d = (\mathcal{E}, x)$ onto the set Λ of all realizable likelihood functions $L = L(\omega | \mathcal{E}, x)$. Once again we repeat that our definition of equality on Λ is that of proportionality : $L_1 \sim L_2$ if there exists $c > 0$ such that $L_1(\omega) \equiv cL_2(\omega)$.

5. Fisher Information

R.A. Fisher's controversial thesis regarding the logic of statistical inference rests on an unequivocal and complete rejection of the Bayesian point of view. He drew the attention of the statistical community away from the Bayesian 'prior' and 'posterior' and focussed it on the likelihood function. Although we do not find the likelihood principle explicitly stated in the writings of Fisher, yet it is clear that he recognized the truth that statistical inference should be based on the 'whole of the relevant information' supplied by the data and that this information is contained in the likelihood function. However, quite a few of the many ideas formulated by Fisher are not in full accord with the above principal theme of his writings. One such idea is that of 'Fisher Information' which we discuss briefly in this section.

In the situation where the universal parameter is a number θ belonging to an interval subset of the real line, and some regularity conditions are satisfied by $p(x|\theta)$ as a function of θ, the Fisher Information is defined as

$$I(\theta) = E_\theta\{\frac{\partial}{\partial\theta} \log p(X|\theta)\}^2$$

$$= - E_\theta \{\frac{\partial^2}{\partial\theta^2} \log p(X|\theta)\}$$

where X is regarded as a random variable ranging over $\mathcal{X}$. How did Fisher arrive at such a notion of information that does not depend on the sample X = x ? Has I(θ) got anything to do with the kind of information that we are talking about? We speculate here on what might have led Fisher to the above mathematically interesting but statistically rather fruitless notion.

If $\hat{\theta} = \hat{\theta}(x)$ is the maximum likelihood estimate of θ, then the true value of θ ought to lie in some small neighbourhood of $\hat{\theta}$- at least in the large sample situation. Writing $\Lambda(\theta) = \log L(\theta)$ – dealing with log-likelihood was a matter of mathematical convenience with Fisher – we can then say that Λ(θ) is approximately equal to

$$\Lambda(\hat{\theta}) + \frac{1}{2}(\theta-\hat{\theta})^2\,\Lambda''(\hat{\theta})$$

for all θ in a small neighbourhood of $\hat{\theta}$ (where the true θ ought to be). Writing J(θ) for $-\Lambda''(\theta)$, the log-likelihood Λ(θ) may be approximately characterized as

$$\Lambda(\hat{\theta}) - \frac{1}{2}(\theta-\hat{\theta})^2\,J(\hat{\theta})$$

where $J(\hat{\theta})$ is (normally) a positive quantity. Now, the magnitude of the statistic $J(\hat{\theta}) = -\Lambda''(\hat{\theta})$ tells us how rapidly the likelihood function drops away from its maximum value as θ moves away from the maximum likelihood estimate. (Note that $J(\hat{\theta}) = -L''(\hat{\theta})/L(\hat{\theta})$ and this is the reciprocal of the radius of curvature of the likelihood function at its mode). It seems clear that Fisher recognized in $J(\hat{\theta})$ a convenient and reasonable numerical measure for the quantum of information contained in a particular likelihood function. For example, if $x = (x_1, x_2, \dots, x_n)$ is an n-tuple of i.i.d. random variables with x_i distributed as $N(0, \sigma^2)$, then

$$\hat{\sigma}^2 = \Sigma x_i^2/n, \quad J(\hat{\sigma}^2) = 2n/\hat{\sigma}^2$$

and the latter varies from sample to sample (as information usually should).

At some stage of the game Fisher became interested in the notion of average information available from an experiment, that is, in

$$E_\theta(J(\hat{\theta})). \qquad \dots (*)$$

It is not easy to get a neat general expression for the above, and so it seems plausible that Fisher had the inconvenient $\hat{\theta}$ in (*) substituted by θ (the true value, which ought to be near $\hat{\theta}$ anyway) and thus arriving at

$$E_\theta J(\theta) = E_\theta\left\{-\frac{\partial^2}{\partial\theta^2}\log L(\theta|X)\right\}$$

$$= \sum_x \left\{\frac{\partial}{\partial\theta}\log p(x|\theta)\right\}^2 p(x|\theta) \qquad \dots (**)$$

which is the Fisher information $I(\theta)$. At this stage one may well wonder as to whether Fisher ever thought of first rewriting $J(\hat{\theta})$ as $- L''(\hat{\theta})/L(\hat{\theta})$ and then substituting $\hat{\theta}$ by θ before computing its average value as in (**). For, in this case he would have arrived at the number zero as his average information !

6. The Likelihood Principle

If we adopt the Bayesian point of view, then the likelihood principle becomes almost a truism. A Bayesian looks upon the data, or rather its information content $\mathrm{Inf}(\mathcal{E}, x)$, as some sort of an operator that transforms the pattern q of his prior beliefs (about the parameter ω) into a new (posterior) pattern q^*. He formalizes the notion of a 'pattern of beliefs' about ω as a probability distribution on Ω, and postulates that probability as a 'measure of (coherent) belief' obeys the same laws as 'frequency probability' is supposed to obey. The transformation $q \to q^*$ is then effected through a formal use of the Bayes theorem (of conditional probability) as

$$q^*(\omega | \mathcal{E}, x) = L(\omega | \mathcal{E}, x)q(\omega)/\Sigma L(\omega | \mathcal{E}, x)q(\omega)$$

$$\sim L(\omega | \mathcal{E}, x)q(\omega).$$

In view of the above, a Bayesian should not have any qualms about identifying $\mathrm{Inf}(\mathcal{E}, x)$ with the likelihood function $L(\omega | \mathcal{E}, x)$.

Fisher was not the first statistician to look upon the sample x as a variable point in a sample space $\mathcal{X}$, but it was certainly he who made this approach popular. He put forward the notion of 'average performance characteristics' of estimators and sought to justify his method of maximum likelihood on this basis. In the early thirties Neyman and Pearson, and then Wald (in the forties) pushed the idea of 'performance characteristics' to its natural limit. Principle $\mathcal{L}$ is in direct conflict with this neoclassical approach to statistical inference. With $\mathcal{L}$ as the guiding principle of data analysis, it no longer makes any sense to investigate (at the data analysis stage) the 'bias' and 'standard error' of point estimates, the probabilities of the 'two kinds of errors' for a test, the 'confidence-coefficients' associated with interval estimates, or the 'risk functions' associated with rules of decision making.

Principle $\mathcal{L}$ rules out all kinds of post randomization. If, after obtaining the data d, an artificial randomization scheme (using a random number table or a modern computer) generates further data d_1, then the likelihood functions generated by d and (d, d_1) coincide (are equivalent). Since the generation of d_1 does not change the information (i.e., the likelihood function), it should not have any bearing on the inference about ω, or on any assessment of the quality of the inference actually made. Being only a principle of data analysis, $\mathcal{L}$ does not rule out the reasonableness of any prerandomization being incorporated into the planning of experiments. However, it does follow from $\mathcal{L}$ that the exact nature of any such prerandomization scheme is irrelevant at the data analysis stage – what is relevant is the actual outcome of the prerandomization scheme, not its probability. [The latter appears only as a constant factor in the likelihood function eventually obtained.] This last point has a far reaching consequence in the analysis of data produced by survey sampling. If we are not to take into account the sampling plan (the prerandomization scheme choosing the units to be surveyed) at the data analysis stage, then we have to throw overboard a major part of the current theories regarding the analysis of survey data. Recently a great deal has been written on the 'randomization analysis' of experimental data [Curiously, it was again Fisher who initiated this kind of analysis and we sometimes hear it said that this was his most important contribution to statistical theory!] Principle $\mathcal{L}$ rejects this kind of analysis of data.

No wonder then that there is so much resistance to $\mathcal{L}$ among contemporary statisticians. But it is truly remarkable how universal the acceptance of the sufficiency principle ($\mathcal{S}$ and its variant $\mathcal{S}'$) is even though, in the context of a particular experiment, the two principles $\mathcal{L}$ and $\mathcal{S}'$ are indistinguishable. The general acceptance of $\mathcal{S}$ appears to be based on a widespread belief that the reasonableness of the principle has been mathematically justified by the Complete Class Theorem of the Rao-Blackwell vintage. Let us examine the question briefly.

In the context of some point estimation problems, the Rao-Blackwell theorem indeed succeeds in providing a sort of decision-theoretic justification for $\mathcal{S}$. But this success is due to (i) the atypical fact that, in a point estimation problem with a continuous parameter of interest, the action space $\mathcal{A}$ may be regarded as a convex set, and also to (ii) the somewhat arbitrary assumption that the loss function $W = W(\omega,a)$ is convex in a (the action) for each fixed $\omega \in \Omega$. Now, let us formalize the notions of (i) a statistical decision problem as a quintuple

$$\mathcal{D} = (\mathcal{X}, \Omega, p, \mathcal{A}, W),$$

(ii) a nonrandomized decision function as a point map of $\mathcal{X}$ into $\mathcal{A}$ and (iii) a randomized decision function as a transition function mapping points in $\mathcal{X}$ into probability measures on $\mathcal{A}$. Let $T : \mathcal{X} \to \mathcal{T}$ be a sufficient statistic for the experiment $(\mathcal{X},\Omega,p)$. Then, for each decision function δ, we can find an equivalent (in the sense that they generate identical risk functions) decision function δ^* which depends on the sample x only through its T-value T(x). But the snag in this kind of Rao-Blackwellization is that δ^* will typically be a randomized decision function and so its use for decision making will entail a direct violation of $\mathcal{S}$ (which is nothing but a rejection of all post-randomizations). How can a principle be justified by an argument that invokes its violation?!

It is difficult to understand why among contemporary statisticians the support for $\mathcal{S}$ is so overwhelming and unequivocal, and yet that for $\mathcal{L}$ is so lukewarm. In a joint paper with Jenkins and Winsten, it was argued by Barnard(1962) that $\mathcal{S}'$ implies $\mathcal{L}$. Although this attempted deduction of $\mathcal{L}$ from $\mathcal{S}'$ turned out to be fallacious, the fact remains that even as late as in 1962 Barnard found it hard to distinguish between the twin principles of sufficiency and likelihood. [In the writings of Fisher also it is very hard to find an instance where he has stated $\mathcal{L}$ separately from $\mathcal{S}$. It seems that Fisher always meant by a sufficient statistic T the minimum or smallest sufficient statistic and invariably visualized it as that characteristic of the sample knowing which the likelihood function can be determined upto an equivalence.] In view of the many 'unpleasant' consequences of $\mathcal{L}$, Barnard seems to have lost a great deal of his early enthusiasm for $\mathcal{L}$ though his conviction in $\mathcal{S}$ remains unshaken. Birnbaum (1962) deduced $\mathcal{L}$ from $\mathcal{S}'$ and $\mathcal{C}'$ and stated that $\mathcal{S}'$ can be deduced from $\mathcal{C}'$, implying thereby that $\mathcal{C}'$ implies $\mathcal{L}$. In 1962 Birnbaum found in $\mathcal{C}'$ a statistical principle that is almost axiomatic in its import and was, therefore, duly impressed by $\mathcal{L}$ which he (mistakenly) thought to be a logical equivalent of $\mathcal{C}'$. At present Birnbaum too seems to have lost his earlier enthusiasm for $\mathcal{L}$, though it is not clear whether his conviction in $\mathcal{C}'$ has suffered in the process or not.

Let us look back on the simplest (and perhaps the least controversial) of the eight principles stated in Section 3, namely, the invariance principle. To the author, principle $\mathcal{I}$ seems axiomatic in nature. Yet one may argue that $\mathcal{I}$ is far from convincing under the following circumstances. Let $\mathcal{E}_1 = (\mathcal{X}_1, \Omega, p_1)$ and $\mathcal{E}_2 = (\mathcal{X}_2, \Omega, p_2)$ be two statistically isomorphic or similar experiments with $g: \mathcal{X}_1 \to \mathcal{X}_2$ as the similarity map. Principle $\mathcal{I}$ then asserts the equality

$$\mathrm{Inf}(\mathcal{E}_1, x_1) = \mathrm{Inf}(\mathcal{E}_2, x_2)$$

for each $x_1 \varepsilon \mathcal{X}_1$ and $x_2 \varepsilon \mathcal{X}_2$ such that $x_2 = gx_1$. Now, suppose the sample space $\mathcal{X}_1$ is endowed with an order structure that is in some way related to some natural order structure in the parameter space Ω, whereas the sample space $\mathcal{X}_2$ has no such discernable order structure. For example, suppose $\mathcal{X}_1$ consists of the six numbers

1, 2, 3, 4, 5 and 6 whereas $\mathcal{X}_2$ consists of the six qualities R (red), W (white), B (black), G (green), Y (yellow) and V (violet). If the statistician feels that he knows how to 'relate' the points in $\mathcal{X}_1$ with the unknown ω in Ω, and if he also feels that he does not know how to 'relate' the points in $\mathcal{X}_2$ with points in Ω(excepting through what he knows about the similarity map $g : \mathcal{X}_1 \to \mathcal{X}_2$), then he may 'feel more informed' about ω when $\mathcal{E}_1$ is performed resulting in x_1 than when $\mathcal{E}_2$ is performed resulting in $x_2 = gx_1$. When it comes to a matter of feeling, not much can be done about it. It is however difficult to see how one can build up a coherent theory of 'information in the data' that will allow one to discriminate between the data $(\mathcal{E}_1, x_1)$ and its g-image $(\mathcal{E}_2, gx_1)$, where g is a similarity map.

Perhaps the point can be emphasized more forcefully in terms of the weak invariance principle $\mathcal{J}'$. [In Section 3 we recognized $\mathcal{J}'$ as a corollary to both $\mathcal{J}$ and the sufficiency principle.] If the two points x and x' in the sample space of the experiment $\mathcal{E} = (\mathcal{X}, \Omega, p)$ generate identical likelihood functions, i.e., if $p(x|\omega) = p(x'|\omega)$ for all $\omega \varepsilon \Omega$, then $\mathcal{J}'$ asserts the equality of $\text{Inf}(\mathcal{E}, x)$ and $\text{Inf}(\mathcal{E}, x')$. Now, a statistician with a strong intuitive feeling for the relevance of 'related order structures' in $\mathcal{X}$ and Ω will perhaps rebel against principle $\mathcal{J}'$ if he is confronted with the following kind of a situation. Suppose the statistical problem is the traditional one of testing a simple null hypothesis H_0 about the probability distribution of a one dimensional random variable X on the basis of the experiment $\mathcal{E}$ that consists of taking a single observation on X. Let $(\mathcal{X}, \Omega, p)$ be a suitable statistical model (for the experiment $\mathcal{E}$) that subsumes H_0 as the hypothesis $\omega = \omega_0$. Consider now the case where we recognize two points x and x' in $\mathcal{X}$ such that they both generate identical likelihood functions and yet x is near the centre (say, the mean) of the distribution of X under H_0 whereas x' is out at the right tail end(say, the 1% point) of the same distribution. Notwithstanding $\mathcal{J}'$, which asserts that x and x' are equally informative, our statistician (with the strong intuition) may well assert that x (being near the centre of X under H_0) sort of confirms H_0, whereas x' (being out in the tail area) sort of disproves the null hypothesis.

One may take an uncharitable view about the above statistical intuition and lightly dismiss the whole matter as a prejudice that has been nurtured in the classical practice of null hypothesis testing (formulated without any explicit mention of the plausible alternatives). It will, however, be charitable to concede that in great many situations it is true that points in the tail end of the distribution of X under H_0 differ greatly in their information aspects from points in the centre part of the same distribution. We should also concede that our formulation of the equality of statistical information in the data $(\mathcal{E}, x)$ and $(\mathcal{E}, x')$ was made relative to a particular model

$(\mathcal{X}, \Omega, p)$ for the experimental part of the data. It is now plausible to suggest that our statistician (with the strong intuition) is not really rejecting principle $\mathcal{S}'$ in the present instance, but is only doubting the adequacy or appropriateness of the particular statistical model $(\mathcal{X}, \Omega, p)$.

This points to the very heart of the difficulty. All statistical arguments are made relative to some statistical model and there is nothing very sacred and irrevocable about any particular model. When an inference is made about the unknown ω, the fact should never be lost sight of that, with a different statistical model for $\mathcal{E}$, the same data $(\mathcal{E}, x)$ might have warranted a different inference. No particular statistical model is likely to incorporate in itself all the knowledge that the experimenter may have about the 'related order structure' or any other kind of relationship that may exist between the sample space and the parameter space. But if we agree to the proposition that our search for the 'whole of the relevant information in the data' must be limited to within the framework of a particular statistical model, then I cannot find any cogent reason for not identifying the 'information in the data' with the likelihood function generated by it. If in a particular instance the experimenter feels very upset by the look of the likelihood function generated by the data, then he may (and indeed should) reexamine the validity and adequacy of the model itself. A strange looking likelihood function does not necessarily destroy the likelihood principle. [Later on, we shall take up several such cases of apparent likelihood principle paradox.]

On p. 334 of Barnard, Jenkins and Winsten (1962) we find the following astonishing assertion which is in the nature of a blank cheque for all violations against $\mathcal{L}$. [In this and in the following two quotations from Barnard, we have taken the liberty of slightly altering the notations so as to bring them in line with those in this chapter.]

"In general, it is only when the triplet $(\mathcal{X}, \Omega, p)$ can by itself be regarded as specifying all the inferential features of an experimental situation that the likelihood principle applies. If $\mathcal{X}$ and Ω are provided with related ordering structures, or group structures, or perhaps other features, it may be reasonable to apply a form of argument which would not apply if these special features were not present. The onus will, of course, be on anyone violating the likelihood principle to point to the special feature of this experiment and to show that it justifies his special argument."

Does it ever happen that a triple $(\mathcal{X}, \Omega, p)$ specifies 'all the inferential features' of an experimental situation ? Can any experimenter be ever so dumb as not to be able to recognize some 'related order structures or group structures or perhaps other features' connecting $\mathcal{X}$ and Ω ? If we are to take the above assertion at its face value, then we must conclude that under hardly any circumstances is Barnard willing to place his immense authority unequivocally behind the likelihood principle ! As a discussant of Birnbaum (1962, p. 308), Barnard made the point once again as follows :

"The qualification concerns the domain of applicability of the principle of likelihood. To my mind, this applies to those situations, and essentially to only those situations, which are describable in terms which Birnbaum uses — that is, in terms of the sample space $\mathcal{X}$, and the parameter space Ω and a probability function p of x and ω defined for x in $\mathcal{X}$ and ω in Ω . If these elements constitute the whole of the data of a problem, then it seems to me the likelihood principle is valid. But there are many problems of statistical inference in which we have less than this specified, and there are many other problems in which we have more than this specified. In particular, the simple tests of significance arise, it seems to me, in situations where we do not have a parameter space of hypotheses; we have a single hypothesis essentially, and the sample space then is the only space of variables present in the problem. The fact that the likelihood principle is inconsistent with significance test procedures in no way, to my mind, implies that significance tests should be thrown overboard; only that the domain of applicability of these two ideas should be carefully distinguished. We also, on the other hand, have situations where more is given than simply the sample space and the parameter space. We may have properties of invariance, and such things, which enable us to make far wider, firmer assertions of a different type ; for example, assertions that produce a probability when these extra elements are present. And then, of course, there are the decision situations where we have loss functions and other elements given in the problem which may change the character of the answers we give".

If, following Barnard, we set up the test of significance problem in the classical manner of Karl Pearson and R.A. Fisher — with a single probability distribution on the sample space and without any tangible parameter space — then the sample will not produce any likelihood function. Without a likelihood function how can we possibly violate principle $\mathcal{L}$? In the other kind of situations, where we have 'invariance and such other things', Barnard says that we can make assertions that are 'far wider and firmer'. But, wider and firmer than what? What does $\mathcal{L}$ assert that is not sufficiently firm or wide? We must recognize this basic fact that $\mathcal{L}$ does not assert anything that can be measured in terms of its operating characteristics. It appears that in this instance Barnard is confusing principle $\mathcal{L}$ with a set of his favourite likelihood methods of inference (see Section 1 of Chapter 3) and it is this set of likelihood methods that he is now finding to be generally lacking in width and firmness. Before returning to the question of the true implication of $\mathcal{L}$, let us quote once again from Barnard and Sprott (1971, p. 176) :

" $\mathcal{L}$ applies to problems for which the model consists of a sample space $\mathcal{X}$, a parameter space Ω and a family of probability functions $p : \mathcal{X} \times \Omega \to R^+ \ldots$. For two such problems $(\mathcal{X}, \Omega, p)$ and $(\mathcal{X}', \Omega, p')$, principle $\mathcal{L}$ asserts that if $x \epsilon \mathcal{X}$ and $x' \epsilon \mathcal{X}'$ and $p(x|\omega)/p'(x'|\omega)$ is independent of ω, then the inference from x must be the same as the inference from x'. We may distinguish three forms of $\mathcal{L}$.

1. Strongly restricted $\mathcal{L}$: Principle $\mathcal{L}$ applicable only if $(\mathcal{X}, \Omega, p) = (\mathcal{X}', \Omega, p')$. This is equivalent to the sufficiency principle.

2. Weakly restricted $\mathcal{L}$: Principle $\mathcal{L}$ applicable (a) whenever $(\mathcal{X}, \Omega, p) = (\mathcal{X}', \Omega, p')$ and (b) when $(\mathcal{X}, \Omega, p) \neq (\mathcal{X}', \Omega, p')$ but there are no structural features of $(\mathcal{X}, \Omega, p)$ (such as group structures) which have inferential relevance and which are not present in $(\mathcal{X}', \Omega, p')$.

3. Unrestricted $\mathcal{L}$: Principle $\mathcal{L}$ applicable to all situations which can be modelled as above.

4. Totally unrestricted $\mathcal{L}$: As in 3, but, further, all inferential problems are describable in terms of the model given.

As we understand the situation, almost everyone would accept 1, while full Bayesians would accept 4. George Barnard's own position is now, and has been since 1957, 2".

The distinction that Barnard is trying to make above between the two forms (3 and 4) of unrestricted $\mathcal{L}$ is not clear and is perhaps not relevant to our present discussion. In 1 Barnard recognizes the equivalence of $\mathcal{S}$ and $\mathcal{L}$ in the context of a single experiment and appears to have no reservations about $\mathcal{S}$. But in 2 we once again come across the same astonishing blank cheque phrased this time in terms of the all embracing double negatives : 'there are no structural features ... which are not present'.

In Chapter 4, we shall discuss in some detail the two principal sources of Barnard's discomfiture with the unrestricted likelihood principle – the Stein Paradox and the Stopping Rule Paradox. For the moment, let us briefly discuss what we consider to be the real implication of $\mathcal{L}$.

Apart from identifying the information content of the data ($\mathcal{E}$, x) with the likelihood function L(ω | $\mathcal{E}$, x) generated by it, principle $\mathcal{L}$ tells us hardly anything else. It certainly does not tell us how to make an inference (based on the likelihood function) in any particular situation. It is best to look upon $\mathcal{L}$ as a sort of code of conduct that ought to guide us in our inference making behaviour. In this respect it is analogous to the unwritten medical code that requires a doctor to make his diagnosis and treatment of a patient dependent wholly on (i) the case history of and the outcomes of some diagnostic tests carried out on that particular patient, and (ii) all the background information that the doctor (and his consultants) may have on the particular problem at hand. It is this same unwritten code that disallows a doctor to include a symmetric die or a table of random numbers as a part of his diagnostic gadgets. It also forbids him to allow his judgement about a particular patient to be coloured by any speculations on the types and number of patients that he may have later in the week. [Of course, like any other rule the above must also have its exceptions. For instance, if our doctor in a far away Pacific island is running short of a drug that is particularly effective against a prevalent disease, he may then be forgiven for treating a less severely affected patient in an unorthodox manner.]

In the colourful language of J. Neyman, the making of inference is nothing but an 'act of will'. And this act is no more (and no less)objective than that of a medical practitioner making his routine diagnoses. We are all too familiar with the beautiful mathematical theory of Neyman-Pearson-Wald about what is generally recognized as correct inductive behaviour. In principle $\mathcal{L}$ we recognize only a preamble to an antithesis to the currently popular N.P.W. thesis. [For a well-stated version of $\mathcal{L}$ from the Bayesian point of view, refer to Lindley (1965, p. 59) or Savage (1961).]

CHAPTER III

STATISTICAL INFORMATION AND LIKELIHOOD
PART II : METHODS

0. Notes

Author's Note : The year was 1955, the event was a lecture by Professor R. A. Fisher at the Indian Statistical Institute, Calcutta. That was when and how I came to know about the likelihood methods of inference described in this chapter. Professor Fisher began his talk with some disparaging remarks about the Neyman-Pearson methods of Statistics, which he generally characterized as Acceptance Sampling Procedures, and lightly dismissed them as of no relevance in Scientific Inference. And then he proceeded to describe how in some situations the likelihood alone could be the sole (and sound) basis for making inference about the parameter.

At the end of the talk, I was naive enough to ask how we could reconcile the likelihood methods with notions like the bias and the variance of an estimate, the size and the power of a test, etc. Sir Ronald was visibly annoyed with my question. In his typical belligerant manner he wanted to know whether I had been sleeping through the earlier part of his talk! He then explained that those notions belonged to acceptance sampling procedures and were hardly relevant in scientific inference.

In this chapter I established to my own satisfaction that, although Fisher was quite right in his assertion that the likelihood carries all the information in the data about the parameter, all the likelihood methods of inference proposed by him were wrong! Fisher refused to recognize that likelihood was meant to be weighted and added. I argued here that it would be a mistake to look upon the likelihood only as a "point function", as a scale that could be used only for pairwise comparisons of individual parameter values. This issue was hotly debated by George Barnard in the correspondence reported at the end of Chapter V.

The Strong Law of Likelihood mentioned here need not be taken very seriously. It was there only to prove the point that construction of a measure on the parameter space by adding likelihoods might be useful and that it would not lead to a contradiction. The likelihood measure so defined is after all the same as the Bayesian posterior for the uniform prior.

When I met George Barnard at the Aarhus Symposium in the Summer of 1973, the first thing he said was : Basu, you called me a likelihoodwallah, so I would call you Muhammad Tughlak. Why? Because, like the Indian Emperor of the Middle Ages you will destroy everything (in Statistics) by trying to be too logical!

The monograph has three principal themes, the Likelihood Principle, Sir Ronald A. Fisher and counterexamples. All these will appear in a large measure in the next chapter also.

STATISTICAL INFORMATION AND LIKELIHOOD

PART II : METHODS

1. Non-Bayesian Likelihood Methods

In Part I our main concern was with the notion of statistical information in the data, and with some general principles of data analysis. Now we turn our attention from principles to a few methods of data analysis. By a non-Bayesian likelihood method we mean any method of data analysis that neither violates $\mathcal{L}$ - the likelihood principle — nor explicitly incorporates into its inference making process any prior information (that the experimenter may have about the parameter ω) in the form of a prior probability distribution over the parameter space Ω. The origin of most of such methods may be traced back to the writings of R.A. Fisher. In this section we list several such methods. To fix our ideas, let us suppose that Ω is either discrete or an interval subset of the real line. In the latter case, we shall also suppose that the likelihood function $L(\omega)$ is a smooth function and has a single mode (whenever such an assumption is implicit in the method).

(a) Method of maximum likelihood : Estimate the unknown ω by that point $\hat{\omega} = \hat{\omega}(x)$ where the likelihood function $L(\omega)$, generated by the data ($\mathcal{E}$, x), attains its maximum value. Fisher tried very hard to elevate this method of point estimation to the level of a statistical principle. Though it has since fallen from that high pedestal, it is still widely recognized as the principal method of point estimation. Note that this method is in conformity with $\mathcal{L}$ as long as we do not try to understand and evaluate the precision of the maximum likelihood estimate $\hat{\omega} = \hat{\omega}(x)$ in terms of the sampling distribution of the 'estimator' $\hat{\omega}$. However, most users of this method quite happily violate $\mathcal{L}$ in order to do just that.

(b) Likelihood interval estimates : Choose and fix a fairly large number λ (20 or 100 are usually recommended values) and consider the set

$$I_\lambda = \{\omega : L(\hat{\omega})/L(\omega) \leqslant \lambda\}$$

where $\hat{\omega}$ is the maximum likelihood estimate of ω. If the likelihood function is unimodal then the set I_λ is a subinterval of Ω and is intended to be used as a sort of 'likelihood confidence interval' for the parameter ω.

(c) Likelihood test of a null-hypothesis : If the null hypothesis to be tested is H_0 (Hypothesis $\omega = \omega_0$), then the method is : Reject H_0 if and only if ω_0 does not belong to the likelihood interval I_λ defined in (b) above. As before, 20 or 100 are recommended values. [The numbers 20 and 100 correspond roughly to the mystical 5% and 1% of the classical tests of significance.]

(d) Likelihood ratio method : If Ω consists of exactly two points ω_o and ω_1 then $\mathcal{L}$ implies that the likelihood ratio $\rho = L(\omega_1)/L(\omega_o)$ generated by the data ($\mathcal{E}$, x) should provide the sole basis for making judgements about whether the true ω is ω_o or ω_1. The method is : Choose and fix λ (20 or 100 say) and then reject the hypothesis $\omega = \omega_o$ if $\rho \geqslant \lambda$, and accept the hypothesis $\omega = \omega_o$ if $\rho \leqslant \lambda^{-1}$, but do not make any judgement if $\lambda^{-1} < \rho < \lambda$. Wald's method of sequential probability ratio test is really an outgrowth of the above. However, we shall discuss later how principle $\mathcal{L}$ is violated in Wald's analysis of sequentially observed data.

(e) General likelihood ratio method : In a general testing situation with two composite hypotheses

$$H_o : \text{Hypothesis that } \omega \varepsilon \Omega_o \subset \Omega$$

and

$$H_1 : \text{Hypothesis that } \omega \varepsilon \Omega_1 = \Omega - \Omega_o ,$$

the method requires computation of a ratio statistic $\rho = \rho(x)$, defined as the ratio $L(\Omega_1)/L(\Omega_o)$ where

$$L(\Omega_i) = \sup_{\omega \varepsilon \Omega_i} L(\omega) \qquad (i = 0,1)$$

and then rejecting the null hypothesis H_o if and only if the ratio ρ is considered to be too large - greater than a prefixed critical value λ . [This method, along with the methods (b), (c) and (d) given above, draws its inspiration from the maximum likelihood method of point estimation. The method has great practical (computational) advantages when the basic statistical model is that of a multivariate normal distribution (with some unknown parameters). Indeed, a major part of the classical theory of multivariate analysis is nothing but a systematic exploitation of the method in a variety of situations. We should not however lose sight of the fact that in these applications of the method the critical value λ for the ratio ρ is determined with reference to the sampling distribution (under H_o) of the ratio statistic ρ and that this constitutes a violation of $\mathcal{L}$.

(f) Nuisance parameter elimination method : Consider the situation where $\omega = (\theta , \phi)$, θ is the parameter of interest and ϕ is the nuisance parameter. From the data ($\mathcal{E}$, x) we have a likelihood function $L(\theta, \phi)$ that involves the nuisance parameter. The following is a very popular method of eliminating ϕ from L . Maximise $L(\theta,\phi)$ w.r.t. ϕ thus arriving at the eliminated likelihood function

$$L_e(\theta) = \sup_{\phi} L(\theta, \phi)$$

where e denotes the fact of elimination. Having eliminated ϕ from the likelihood function, the method then requires that all inferences about θ should be carried out with the eliminated likelihood function $L_e(\theta)$ along the lines suggested earlier. Method (f) may be looked upon as a natural generalization of method (e).

Let us end this section with a few comments on some common features of these methods.

(i) For going through the motions of any of these methods, it is not necessary to know any details of the sample x other than the likelihood function generated by it. In their pure (that is, uncontaminated by the Neyman-Pearson type arguments) forms, the methods are in conformity with principle $\mathcal{L}$. However, note that none of the above methods can be logically deduced from $\mathcal{L}$ alone.

(ii) In none of the methods we find any mention of the two elements q and Π that we briefly talked about in Section 1 of Chapter II. Let us recall that in q we have incorporated all the background (prior) information that the experimenter has about ω and other related entities. In Π is incorporated all the particular features (such as, the relative hazards of making wrong inferences of various kinds, etc.) of the inferential problem at hand. The likelihood methods of this section differ from standard Bayesian methods mainly in their failure to recognize the relevance of q and Π.

(iii) In their pure forms, these methods do not require the evaluation of the average performance characteristics of anything. This, however, does not mean that we should not speculate about long term characteristics of such methods. Advocates of likelihood methods are surely not averse to the idea of comparing their methods with any other well defined method on the basis of their average performance characteristics in a hypothetical sequence of repeated applications of the methods. [Even Bayesians, who do not usually care for the frequency interpretation of probability, do care very much about one kind (perhaps, the only kind that is relevant) of frequency, namely, the long term success ratio of their methods. After all, the real proof of the pudding lies in the eating.] From our description of the non-Bayesian likelihood methods it is not clear with what kind of average performance characteristics in mind these methods were initially proposed. Indeed, in some later sections we shall give examples of situations where simple-minded applications of these methods will have disastrous long term performance characteristics. Such examples will not, however, disprove $\mathcal{L}$ because the methods do not follow from $\mathcal{L}$.

(iv) The differences between the Bayesian and the (non-Bayesian) Likelihood schools of data analysis may be summarised as follows : Whereas the Bayesian looks upon the likelihood function $L(\omega)$ as an intermediate step – a link between the prior and the posterior – the Likelihoodwallah* looks upon $L(\omega)$ as a sort of an end in itself. Furthermore, the latter looks upon $L(\omega)$ as a point function – $L(\omega)$ is the relative magnitude (or intensity) with which the data supports the point ω – that should never (well, almost never) be looked upon as something that can generate a measure of support (for subsets of Ω that are not single-point sets). In the next section we discuss this point in some detail.

2. Likelihood : Point Function or a Measure ?

It was R.A. Fisher who first thought of likelihood as an alternative measure of rational belief. The following quotation clearly spells out Fisher's own ideas on the subject. [These remarks of Fisher appear to have greatly influenced the thinking processes of many of our contemporary statisticians.] Discussing the likelihood function, Fisher (1930, p. 532) wrote :

"The function of the θ's maximised is not however a probability and does not obe the laws of probability; it involves no differential element $d\theta_1\, d\theta_2\, d\theta_3$ ...; it does none the less afford a rational basis for preferring some values of θ , or combination of values of the θ's, to others. It is, just as much as a probability, a numerical measure of rational belief, and for that reason called the likelihood of $\theta_1, \theta_2, \theta_3$, ... having given values, to distinguish it from the probability that θ_1, θ_2, θ_3, ... lie within assigned limits, since in common speech both terms are loosely used to cover both types of logical situation.

If A and B are mutually exclusive possibilities the probability of "A or B" is the sum of the probabilities of A and of B, but the likelihood of A or B means no more than "the stature of Jackson or Johnson", you do not know what it is until you know which is meant. I stress this because inspite of all the emphasis that I have always laid upon the difference between probability and likelihood there is still a tendency to treat likelihood as though it were a sort of probability.

The first result is that there are two different measures of rational belief appropriate to different cases. Knowing the population we can express our incomplete knowledge of, or expectation of, the sample in terms of probability; knowing the

*'Wallah' in Hindi means a peddler and is a nonderogatory term. The name, Likelihoodwallah, then denotes a peddler of an assortment of non-Bayesian likelihood methods.

sample we can express our incomplete knowledge of the population in terms of likelihood. We can state the relative likelihood that an unknown correlation is + 0.6, but not the probability that it lies in the range .595-.605".

From the above it is clear that Fisher intended his notion of likelihood to be used as some sort of a measure of (the degree of) rational belief. But all the same he was very emphatic in his denial that likelihood is not a measure like probability - it is not a set function but only a point function. It is not however clear why this data induced likelihood measure of rational belief (about various simple hypotheses related to the population) must differ from the other measure of rational belief (namely, probability) in being nonadditive. Why can't we talk of the likelihood of a composite hypothesis in the same way we talk about the probability of a composite event ?

In our quotation we find Fisher lightly dismissing the question with the curious analogy of "the stature of Jackson or Johnson, you do not know what it is until you know which is meant". Twentysix years later we find Fisher (1956, p. 69) still persisting with the same analogy - only this time it was "the income of Peter or Paul". These analogies are particularly inept and misleading. Both stature and income are some kind of measure - the former of size and the latter of earning power. Why can't we talk of the total stature or the total income of a group of people ? It should be noted that when Fisher is talking of 'Jackson or Johnson' he is using the conjunction 'or' in its everyday disjunctive sense of 'either-or'. On the other hand, when we talk about the degree of rational belief (probability or likelihood) in 'A or B' the 'or' is the logical (set-theoretic) connective 'and/or' (union).

Ian Hacking (1965) in his very interesting and informative book, Logic of Statistical Inference, has given a detailed and eminently readable account of how this Fisher-project of building an alternative likelihood framework for a measure of 'rational belief' may be carried out. The expression 'rational belief' sounds a little awkward in the present context as the whole exercise is about a mathematical theory of what 'the data has to tell' rather than about what 'the experimenter ought to believe'. Hacking therefore suggests an alternative expression, 'support-by-data'. About this theory of 'support' Hacking (1965, p. 32) writes :

"The logic of support has been studied under various names by a number of writers. Koopman called it the logic of intuitive probability; Carnap of confirmation. Support seems to be the most general title. ... I shall use only the logic of comparative support, concerned with assertions that one proposition is better or worse supported by one piece of evidence, than another proposition is by other or the same evidence. .... The principles of comparative support have been set out by Koopman; the system of logic which he favours will be called Koopman's logic of support".

The Fisher-project of building an alternative likelihood framework for 'support-by-data' is then carried out by Hacking as follows. Hacking begins with Koopman's postulates of intuitive probability - the logic of support - and enriches it with an additional postulate, which he calls the Law of Likelihood. A rough statement of the law may be given as follows :

Law of Likelihood : Of two hypotheses that are consistent with given data, the better supported (by the data) is the one that has greater likelihood.

In terms of our notations, the Law tells us the following : If $L(\omega_1) > L(\omega_2)$ then the data $(\mathcal{E}, x)$ supports the hypothesis $\omega = \omega_1$ better than the hypothesis $\omega = \omega_2$. The Law sets up a linear order on the parameter space Ω . Any two simple hypotheses $\omega = \omega_1$ and $\omega = \omega_2$ may be compared on the basis of the intensity of their support by the data. But how about composite hypotheses like $\omega = \omega_1$ or ω_2 ? Suppose $A = \{\omega_1, \omega_2\}$ and $B = \{\omega'_1, \omega'_2\}$ and suppose further that $L(\omega_i) > L(\omega'_i), i=1,2$. Would the statistical intuition of Sir Ronald have been outraged by the suggestion that, under the above circustances, it is right to say that the data supports the hypothesis $\omega \in A$ better than the hypothesis $\omega \in B$? The author thinks not.

At the risk of scandalizing some staunch admirers of Sir Ronald, the author now suggests a stronger version of Hacking's law of likelihood.

The Strong Law of Likelihood : For any two subsets A and B of Ω , the data supports the hypothesis $\omega \in A$ better than the hypothesis $\omega \in B$ if

$$\sum_{\omega \in A} L(\omega) > \sum_{\omega \in B} L(\omega) .$$

[Let us recall the assumption (rather, assertion) in Section 2 of Chapter II that all our sets (the sample space, the parameter space etc.) are finite. Because of this we ran into no definition trouble.] Before looking into the possibility of any inconsistencies that may arise out of this Strong Law of Likelihood, let us consider some of its consequences.

With the Strong Law of Likelihood incorporated into Koopman's logic of support we can now identify the notion of 'support-by-data' for the hypothesis $\omega \in A$ with its likelihood $L(A)$ defined as

$$L(A) = \sum_{\omega \in A} L(\omega) .$$

Given a data d, its support for various hypotheses about the population is then a true measure - the likelihood measure of Fisher. Since a scaling factor in the likelihood function does not alter its character, we may as well work with the standardized likelihood function

$$\bar{L}(\omega) = L(\omega)/L(\Omega),$$

and then the corresponding set function $A \to \bar{L}(A)$ gets endowed with all the characteristics of a probability measure.

No Likelihoodwallah can possibly object to our scaling of the likelihood to a total of unity. They can however challenge the Strong Law of Likelihood. But observe that the Strong Law is nothing but the Law of Likelihood (which all Likelihoodwallahs accept) together with an additivity postulate for the logic of support-by-data. [It should be noted that the additivity postulate is not in the set that Hacking (1965, p. 33) borrowed from Koopman's logic of intuitive probability. However, in a later part (Chapter IX) of his book, Hacking introduced this postulate in his logic of support with a view to developing the idea as a sort of "consistent explication of Fisher's hitherto inconsistent theory of fiducial probability". The author had difficulties in following this part of Hacking's arguments.] One may ask : "How can you assume that data support hypotheses in an additive fashion ?" But then the same question may be asked about the other postulates also.

The author is willing to postulate additivity because (i) it is not in conflict with his own intuition on the subject, (ii) it makes the logic of support neat and useful, but mainly because (iii) he does not know how to 'prove' it ! The author is not a logician. The long winded 'proofs' that some subjective probabilists give about the additivity of their measure of 'rational belief' leave the author bewildered and bemused. He finds it a lot easier to accept additivity as a primary postulate for probability. When it comes to likelihood (a measure of support-by-data) he finds it equally easy to accept it as additive. If we can accept that the mind of a rational homosapien ought to work in an additive fashion when it comes to his pattern of belief in various events, why can't we also accept that the inanimate data should lend its support to various hypotheses in a similarly additive manner? Let us not forget that Fisher used the term 'rational belief' and not 'support-by-data'. The 'belief' of what rational mind was he contemplating? Certainly, not that of the statistician (experimenter). Because he is a rational being, the experimenter cannot (and must not) forget all the other (prior) information that he has on the subject. It seems Fisher was contemplating an extremely intelligent being—

a Martian perhaps - who at the same time is totally devoid of any background information about ω other than what is contained in the description of the statistical model ($\mathcal{X}$, Ω, p) for the experiment $\mathcal{E}$ and the data ($\mathcal{E}$, x). Our intelligent Martian objectively weighs all the evidence given by the data and then makes up his/her own mind about the various possibilities related to ω . Fisher wanted to distinguish this posterior pattern of the Martian's 'rational belief' with the ordinary kind of 'rational belief', which we call probability, by calling the former likelihood. But why did he insist so vehemently that likelihood is not additive ?

The answer lies in Fisher's preoccupation with the illusory notions of the infinite and the infinitesimal. Suppose we have formulated in our mind an infinite set of hypotheses H_1, H_2, H_3, ... and suppose our experiment is the trivial one of tossing a symmetric coin once, resulting in the sample H (head). Now, the data equally support each member of our infnite set of hypotheses. There is no difficulty in visualising the likelihood as a nice, flat point function. But how can we convert this into an ordinary kind of a probability measure? Even Hacking, the logician, seems to have been taken in by the force of this argument. On p. 52 of his book Hacking writes : "Likelihood does not obey Kolmogoroff's axioms. There might be continuously many possible hypotheses; say, that P(H) lies anywhere on the continuum between 0 and 1 . On the data of two consecutive heads, each of this continuum of hypotheses (except P(H) = 0) has likelihood greater than zero. Hence the sum of the likelihoods of mutually exclusive hypotheses is not 1, as Kolmogoroff's axioms demand; it is not finite at all".

The author finds the above remark all the more surprising because in the very next paragraph Hacking writes : "..., in any real experimental situation, there are only a finite number of possible outcomes of a measurement of any quantity, and hence a finite number of distinguishable results from a chance setup. Continuous distributions are idealizations." If Hacking is willing to concede that all sample spaces are in reality only finite, why does he not agree to the proposition that the parameter space also is in reality only finite ?

A finite and, therefore, realistic version of the Hacking-idealization of the parameter θ = P(H) lying "anywhere on the continuum between 0 and 1" may be set up as follows : Stipulate that θ varies over some finite and evenly spread out set like $J = \{.00, .01, .02, \ldots .99, 1.00\}$. On the basis of the data (of two consecutive heads in two throws) our Martian then works out his likelihood measure over the set J in terms of the standardized likelihood function $\bar{L} : J \to [0, 1]$ defined as

$$(*) \qquad \bar{L}(\theta) = \theta^2 / \sum_{\theta \in J} \theta^2$$

Now, the above discrete likelihood measure can be reasonably (and rather usefully) approximated by a continuous (likelihood) distribution over the unit interval [0,1] that is defined by the density function

$$(**) \qquad \bar{l}(\theta)d\theta = 3\theta^2 d\theta$$

Note that the (true) likelihood function $\bar{L}(\theta)$ in (*) has no differential element attached to it, whereas its idealized counterpart in (**) has. In order to avoid the logical hazards of the infinitesimal, it is better to look upon the density function $\bar{l}(\theta)$ only as a convenient tool and nothing else.

Now, let us examine how our clever but very ignorant Martian reacts to a restatement of the statistical model in terms of a transformation of the parameter θ. Suppose we write $\phi = \theta^2$ and describe the model in terms of the parameter ϕ. In order to be consistent with our earlier stipulation that $\theta \varepsilon J$, we have to inform the Martian that $\phi \varepsilon J_1$ where $J_1 = \{(.00)^2, (.01)^2, \ldots (.99)^2, (1.00)^2\}$. Looking at the data of two consecutive heads, the Martian will now arrive at his likelihood measure on J_1 on the basis of the standardized likelihood function $\bar{L}_1$ defined as

$$\bar{L}_1(\phi) = \phi / \sum_{\phi \varepsilon J_1} \phi.$$

And this measure on J_1 is entirely consistent with the measure on J obtained earlier in (*). In view of the fact that the set J_1 is not evenly spread out over the interval [0, 1], the idealized limiting version of the above discrete distribution on J_1 is not given by the density $2\phi\, d\phi$ but by the natural progeny of (**) obtained in the usual manner as

$$\bar{l}_1(\phi)d\phi = \bar{l}_1(\theta) \left| \frac{d\theta}{d\phi} \right| d\phi$$

$$= \frac{3}{2} \sqrt{\phi}\, d\phi, \quad 0 \leqslant \phi \leqslant 1.$$

It should be noted that the function $\bar{l}_1(\phi) = \frac{3}{2}\sqrt{\phi}$ has no likelihood interpretation as a point function. However, for reasonable sets A, the integral $\int_A \bar{l}_1(\phi)d\phi$ may be interpreted as the likelihood of the hypothesis $\phi \varepsilon A$ but then only as an approximation.

At this point one may ask the question : "Why is it that the Martian is reacting differently to the two parametrizations of the model in terms of θ and ϕ ?" In the first case we find that the likelihood function $\bar{L}(\theta)$ is proportional to the likelihood density $\bar{l}(\theta)$. But in the second case the two functions $\bar{L}_1(\phi)$ and $\bar{l}_1(\phi)$ are not proportional. The answer lies of course in the fact that the parameter spaces J and J_1 are differently oriented. Suppose, instead of telling the Martian that $\phi \epsilon J_1$, we leave him to his own devices with the vague assertion that ϕ lies somewhere in the continuous interval [0,1]. Now the computer like mind of the Martian will immediately translate our vague statement about ϕ into a statement like $\phi \epsilon J = \{.00, .01, \dots .99, 1.00\}$ and proceed to evaluate the evidence of the data in precisely the same way as he did for θ. His likelihood function $\bar{L}_2 : J \to (0, 1]$ will now be defined as

$$\bar{L}_2(\phi) = \phi / \sum_{\phi \epsilon J} \phi$$

and its idealized continuous version will be described in terms of the density function

$$\bar{l}_2(\phi)d\phi = 2\phi d\phi \ , \quad 0 \leqslant \phi \leqslant 1 \ .$$

The fact that the density function $\bar{l}_2(\phi)d\phi$ is not consistent with the density function $\bar{l}(\theta)d\theta$ was the principal reason why Fisher rejected the idea of likelihood as an additive measure. His mind probably worked in the following fashion : The map $\theta \to \theta^2 = \phi$ sets up a one-one correspondence between the intervals [0, 1] and [0, 1]. The statements $\theta \epsilon [0, 1]$ and $\phi \epsilon [0, 1]$ are therefore equivalent in every way. If on the basis of equivalent background information the Martian is liable to arrive at different (inconsistent) measures of rational belief, then it is clear that we cannot trust his methods for converting the likelihood function into an additive measure. It is therefore safer to regard likelihood only as a point function. This way we cannot have paradoxes of the above kind.

Let us analyse the flaw in the above argument. The assertion that $\theta \to \phi$ is a one-one map is strictly true only in the idealized continuous case. To recognize this we have only to look at a finite (non-infinitesimal) version, say, J of [0, 1]. For each θ (in J) there is a ϕ (in J), which is well defined as $\phi = \theta^2$ correct to its second decimal place. But now the correspondence is many-one and not onto. For example, the statement $\phi = 0$ is the union of the eight statements $\theta = .00, \theta = .01, \dots \theta = .07$, and the statement $\phi = .99$ corresponds to no elementary statement about θ. The assertions $\theta \epsilon J$ and $\phi \epsilon J$ are therefore quite different (both logically and statistically) in nature and our Martian cannot be faulted for reacting differently to two different bits of information.

Let us look back on the passage that we quoted in the beginning of this section from Fisher (1930). Curiously enough, it was in this 1930 paper that Fisher first introduced us to his fiducial probabiity methods for constructing an additive measure of support-by-data, which according to him must be recognized as ordinary frequency probability. It now appears that Fisher was only protesting too much when he so severely deplored the "tendency to treat likelihood as though it were a sort of probability".

There are no compelling arguments in favour of the often repeated assertion that likelihood is only a point function and not a measure. I do not see what inconsistencies can arise from the postulation of the Strong Law of Likelihood in the Koopman-Hacking logic of support-by-data. On the other hand, we shall show later how some of the non-Bayesian likelihood methods get into serious trouble because of their nonrecognition of the additivity of the likelihood measure.

3. Maximum Likelihood

Volumes have been written seeking to justify the maximum likelihood (ML) method of point estimation (and its sister method - the likelihood ratio method for test of hypotheses), and yet I cannot find any logical justification for upholding the method as anything but a simplistic tool that may (with some reservations) be used for routine data analysis in situations where the sample size is not too small and the statistical model not too shaky (unrobust). By definition, the ML estimate $\hat{\omega}$ (of the value of ω that obtains) is the point in the parameter space that is best supported by the data. But what logical compulsions guide us to the maximum likelihood principle: "The best (or most reasonable) estimate of a parameter is that value (of the parameter) which is best supported by the data" ? If we contemplate for a moment our very ignorant Martian, who is trying to make sense of data related to a parameter about which he has absolutely no preconceived notions, then we ought to be more prepared in our mind to accept the reverse proposition : "The most reasonable estimate of a parameter will rarely coincide with the one that has the greatest support from the data".

If Fisher ever thought in terms of the idealization of a Martian, then he must have visualized him/her as a rational being who not only is very ignorant (about the parameter of interest) but is also endowed with very limited capabilities. Fisher's Martian does not know how to add likelihoods, he can only compare them. His recognition of points in the parameter space is only microscopic (pointwise). He compares parameter points pairwise - he can only tell how much more likely a particular point is compared to another. Given two composite hypotheses $\omega \epsilon A$ and $\omega \epsilon B$, the only thing that he can do, in the way of comparing the likelihoods (of the composite hypotheses being true), is to compare the likelihoods of the best supported points $\hat{\omega}_1$ and $\hat{\omega}_2$ in A and B respectively. This is the Martian's Likelihood Ratio method for testing a composite hypothesis against a composite alternative and is analogous to a child's method for picking the winning team in a tug-of-war contest by concentrating his whole attention on the anchors of the two teams ! He has no understanding of any natural topology on the parameter space that may exist. And finally, he does not know anything about the relative hazards of incorrect inferences. The six likelihood methods that we have described in Section 1 are geared to the needs and limitations of such a Martian. It is easy to constuct examples where uncritical uses of such methods will lead to disastrously inaccurate inferences. Here is one such.

Example 1 : An urn contains 1000 tickets, 20 of which are marked θ and the remaining 980 are marked 10θ, where θ is the parameter of interest. A ticket is drawn at random and the number x on the ticket is observed. The ML estimate of θ is then $x/10$. In this case, the ML estimation procedure leads to an exact estimate with a probability of .98. So everything seems to be as it should be. But consider a slight variant of the urn model, where we still have 20 tickets marked θ, but the remaining 980 tickets are now marked $\theta a_1, \theta a_2, \ldots, \theta a_{980}$ respectively, and where the 980 constants $a_1, a_2, \ldots, a_{980}$ are all known, distinct from each other, and all of them lie in the short neighbourhood (9.9,10.1) of the number 10. The situation is not very different from the one considered just before, but now look what happens to our Martian. Noting that the likelihood function is

$$L(\theta|x) = \begin{cases} .02 & \text{for } \theta = x \\ .001 & \text{for } \theta = xa_i^{-1}, \quad i = 1, 2, \ldots \\ 0 & \text{otherwise} \end{cases}$$

the Martian now recognizes x as the ML estimate of θ. He also declares (see method (b) of Section 1) that x is at least 20 times more likely than any other point in the parameter space and, therefore, identifies the single-point set $\{x\}$ as the likelihood interval I_λ with $\lambda = 20$. Irrespective of what the true value of θ is, the ML method now over-estimates it with a factor of nearly 10 and with a probability of .98. As a confidence interval the likelihood interval I_λ (with $\lambda = 20$) has a confidence coefficient of .02.

The source of the Martian's trouble with this example is easy to fathom. If he knew how to add his likelihood measure, then he would have recognized that the likelihood of the true θ lying in the interval $J = (x/(10.1), x/(9.9))$ is .98. Furthermore, if he could recognize that (for medium sized x) the interval J is a narrow one and that small errors in estimation are much less hazardous than an overestimate with a factor of 10, then he would surely have recognized the reasonableness of estimating the true θ by a point like $x/10$ rather than by the ML estimated x.

We all know that under certain circumstances the ML method works rather satisfactorily in an asymptotic sense. But the community of practising statisticians are not always informed of the fact that under the same circumstances the Bayesian method : "Begin with a reasonable prior measure q of your belief in the various possible values of θ , match it with the likelihood function generated by the data, and then estimate θ by the mode of the posterior distribution so obtained", will work as well as the ML method, because the two methods are asymptotically equivalent.

And once we take the final Bayesian step of 'matching the likelihood function with some reasonably formulated prior measure of our personal belief', we can then orient the task of inference making to all the realities -ω , Ω, q, Π, $\mathcal{E}$, x, etc.- of the particular situation. If we look back on the six likelihood methods described in Section 1, it will then appear that, excepting for method (d) - the likelihood ratio method of testing a simple hypothesis against a simple alternative - all the other methods are too simplistic and rather disoriented towards the complex realities of the respective inference making situations.

We end this section with another example to demonstrate how disastrously disoriented the Martian can get (in his efforts to evaluate the likelihood evidence given by the data) because of his inability to add likelihoods. Let us look back on methods (e) and (f) described in Section 1 and then consider the following.

Exapmle 2 : The universal parameter ω is (θ, ϕ), where θ (the parameter of interest) lies in the two-point set $I = \{-1, 1\}$ and the nuisance parameter ϕ lies somewhere in the set $J = \{1, 2, \ldots, 980\}$. Our task is to draw a ticket at random from an urn containing 1000 tickets and then to guess the true value of θ on the basis of the observed characteristics of the sample ticket. About the 1000 tickets in the urn we have the information that (i) the number θ is written in large print on exactly 980 tickets and the number $-\theta$ appears in large print in the remaining 20 tickets, and (ii) the 980 tickets marked θ carry the distinguishing marks 1,2, ..., 980 respectively in microscopic print, whereas, the remaining 20 tickets carry the mark ϕ in microscopic print (where the unknown $\phi \in J$). Let x and y be the numbers in large and small print respectively on our sample ticket. Our sample space then is $I \times J$, which is also our parameter space.

Let us suppose for a moment that either we do not have a magnifying glass to read the small print y or for some reason we consider it right to suppress this part of the data from our Martian. The Martian will then be very pleased to discover that his likelihood function (based on x alone) does not depend on the nuisance parameter and is

$$(*) \qquad L(\theta) = L(\theta,\phi|x) = \begin{cases} .98 & \text{when } \theta = x \\ .02 & \text{when } \theta = -x \end{cases}$$

and so he will come out strongly in support of the guess : 'the true θ is x'. No doubt we should feel proud of our clever Martian because, irrespective of what θ is, the probability of his guessing right in the above circumstances is .98.

But see what happens when we can read y and cannot find any good reason for suppressing this part of the data. With the full sample (x,y) in his possession, the Martian will routinely analyse the data by first setting up the likelihood function as

$$(**) \qquad L(\theta,\phi|x,y) = \begin{cases} .001 & \text{when } \theta = x, \quad \phi \varepsilon J \\ .02 & \text{when } \theta = -x, \quad \phi = y \\ 0 & \text{otherwise} \end{cases}$$

and then eliminating φ from (**) as per method (f) of Section 1 . The eliminated likelihood function is

$$(***) \qquad L_e(\theta) = \sup_{\phi} L(\theta,\phi|x,y) = \begin{cases} .001 & \theta = x \\ .02 & \theta = -x \end{cases}$$

and so this time the Martian comes out strongly in support of the guess $\theta = -x$. With the full data, the performance characteristic of the Martian's method is now 'only 2% probability of success' !

It should be observed that the real source of the Martian's debacle lies in his inability to add likelihoods. Before the data was available, the Martian's ignorance about the parameter $\omega = (\theta,\phi)$ extended over the 2 × 980 points of the set $I \times J$. With the sample reading (x,y), the Martian correctly recognized in (**) that his ignorance about (θ, φ) is cut down to the smaller set $A \cup B$ where

$$A = \{(x, 1), (x, 2), \ldots (x, 980)\}$$

and

$$B = \{(-x, y)\}$$

and that the likelihood of each of the 980 points in A is .001 and that of the single point in B is .02. From the Strong Law of Likelihood (see Section 2) it follows that the likelihood support (by the data) for the composite hypothesis $\omega \varepsilon A$ (that is, $\theta = x$) should be worked out as

$$L(A) = \sum_{(\theta,\phi)\varepsilon A} L(\theta, \phi | x, y) = .98$$

and this compares very favourably with the likelihood support of .02 for the hypothesis $\omega \varepsilon B$ (that is, $\theta = -x$).

This elimination of the nuisance parameter ϕ by the above method of addition (of the likelihood function over the range of ϕ for fixed θ) certainly smacks of Bayesianism, but it appears to be a much more natural thing to do than the Fisher-inspired elimination method by maximization (w.r.t. ϕ for fixed θ). [In the present example, it so happens that the 'addition method' of elimination (of ϕ) leads to the same eliminated likelihood function as was achieved earlier in (*) by the 'marginalization method' of suppressing the y-part of the data. However, the author cannot see how a good case can be made for such a marginalization procedure, even though the distribution of x (as a random variable) depends only on the parameter of interest θ, and that of y depends on the nuisance parameter ϕ alone. Note that, for fixed (θ,ϕ), the statistics x and y are not stochastically independent. It follows that, even when the parameters θ and ϕ are entirely unrelated (independent a-priori), suppression of y may lead to valuable loss of information. In order to see this, suppose that we knew for sure that $\phi = 1$ or 2. Now, the statistic y will give us extra information about θ - if $y > 2$ then we know for sure that $\theta = x$ etc.]

Let us close this section with the remark that, however well suited the 'addition method' of elimination (The Strong Law of Likelihood) may be to the needs and capabilities of our ignorant Martian, the method is not being recommended here as a routine statistical procedure to be adopted by any knowledgeable scientist.

CHAPTER IV

STATISTICAL INFORMATION AND LIKELIHOOD
PART III : PARADOXES

0. Notes

Author's Note: In Mathematics, a single counterexample will disprove a Theorem. In statistics we deal with inductive methods. Here it is often possible to hide behind the maxim: Exception proves the rule. However, I regard it a sacred duty of all statisticians to continually check every statistical method in the glaring light of counterexamples.

In Section 1, a clever Buehler counter argument is used to demonstrate why fiducial probability cannot be interpreted as ordinary frequency probability. In Sections 2 and 3 we take up the challenge of two celebrated counterexamples to the Likelihood Principle. We demonstrate that the counterexamples are really targeted against some familiar statistical methods rather than the Principle itself.

Most non-Bayesian statistical methods get into trouble when we ask how one should modify the method in the presence of ironclad prior knowledge that the parameter of interest lies in an interval (a, b). Sometime in the late fifties Professor R. R. Bahadur had had an appointment to discuss statistics with Sir Ronald who was then visiting the Indian Statistical Institute. Bahadur tagged me along with him. Seeing me Sir Ronald flared up and asked me why I was having so much trouble understanding the fiducial logic. So I asked : With a sample x on a Normal variable with unknown mean μ and known variance σ^2, the fiducial distribution of μ is $N(x,\sigma^2)$. How should we modify the fiducial distribution when we have the sure prior knowledge that μ lies in the closed interval [0, 1]? Fisher replied : All the probability mass of the fiducial distribution $N(x,\sigma^2)$ that lie on the right of 1 should be stacked on 1. Similarly stack all the mass on the left of 0 on 0. I knew what the reply would be and so I was prepared with my next question: Sir, consider the situation where the known variance σ^2 is a very large number. Even before the sample x is observed, we then know for sure that it is going to fall outside the interval [0, 1] and that we are going to put well over 50 percent of the fiducial probability mass at the two end points 0 and 1. Thus the mere knowledge that μ lies in [0, 1] makes us mentally prepared to accept the proposition that $\mu = 0$ or 1. Professor Fisher became terribly angry at my impertinence. When he regained control of himself, he solemnly said : "Basu, you either believe in what I say or you don't, but never ever try to make fun of my theories".

STATISTICAL INFORMATION AND LIKELIHOOD
PART III : PARADOXES

1. A Fallacy of Five Terms

I vividly recall an occasion in late 1955 when Sir Ronald (then visiting the Indian Statistical Institute, Calcutta and giving a series of seminars based on the manuscript of his forthcoming book) got carried away by his own enthusiasm for fiducial probability and tried to put the fiducial argument in the classical form of the Aristotelian syllogism known as Barbara : 'A is B, C is A, therefore C is B'. The context was : A random variable X is known to be normally distributed with unit variance and unknown mean θ about which the only information that we have is, $-\infty < \theta < \infty$. The variable X is observed and the observation is 5. Sir Ronald declared that the following constitutes a 'proof' :

Major premise : Probability that the variable X exceeds θ is 1/2.

Minor premise : The variable X is observed and the observation is 5.

Conclusion : Probability that 5 exceeds θ is 1/2.

We know that in Aristotelian logic an argument of the kind : 'Caesar rules Rome, Cleopatra rules Caesar, therefore, Cleopatra rules Rome', is classified as a 'fallacy of four terms'- the four terms being (i) Caesar, (ii) one who rules Rome, (iii) Cleopatra, and (iv) one who rules Caesar. Sir Ronald is perhaps the only person (in the history of scientific thought) who ever dared (even in a moment of euphoria) to suggest a three line proof involving five different terms - the terms being (i) $\Pr(X > \theta)$, (ii) 1/2, (iii) the observed value of X, (iv) 5, and (v) $\Pr(\theta < 5)$!

About Fisher's fiducial argument Hacking (p. 133) writes : "No branch of statistical writing is more mystifying than that which bears on what he calls the fiducial probabilities reached by the fiducial argument. Apparently the fiducial probability of an hypothesis, given some data, is the degree of trust you can place in the hypothesis if you possess only the given data." The confusion has been further compounded by Fisher's repeated assertions that in those circumstances where he considers it right to talk about fiducial probabilities, the notion should be understood in exactly the same way as a gambler understands his (frequency) probability. Neyman's theory of confidence intervals arose from his efforts to understand the fiducial argument and to reinterpret the concept in terms of frequency probability.

Recently, Fraser, with his structural probability methods, is trying to build a mathematical framework for Fisher's ideas on fiducial probabilities. Whereas Neyman never had had any illusions about his 'confidence coefficients' being the same as ordinary probabilities, it appears that Fraser (like Fisher does not make any logical distinction between ordinary and structural (fiducial) probabilities.

On the surface the fiducial method may appear to be of the true likelihood vintage - an exercise in analysing the mind of the Martian (the particular data at hand). A little reflection (see Anscombe [1957] in this connection) however will prove otherwise. Consider the context where the variable X is known to have a $N(\theta\ ,1)$ distribution, the only background information about θ is that $-\infty < \theta < \infty$, and the observed value of X is x. The fiducial argument leads to the fiducial distribution N(x,1) for θ. The argument has hardly anything to do with the fact that the data generates the likelihood function $\exp\{-(\theta-x)^2/2\}$, but is based on (i) the fortuitous discovery of the pivotal quantity $X-\theta$ with a standard normal distribution, (ii) a reinterpretation of our lack of prior information about θ, and of course (iii) that X is observed as x . The fiducial argument clearly does not respect the likelihood principle.

In the present context we have two unobservable entities - the parameter θ and the (pivotal) quantity $Y = X - \theta$. About θ the statistician (rather, the Martian) is supposed to know nothing other than that the parameter lies in the infinite interval $(-\infty, \infty)$. About Y, on the other hand, he has the very precise information that Y is distributed as N(0,1) irrespective of what value θ takes. In a sense we may then say that the (unobservable) random quantity Y is stochastically independent of the parameter θ. Now, the sum $\theta+Y = X$ is observable and has actually been observed as x. The fiducial argument then somehow justifies the assertion that the observation $\theta+Y = x$ altered the logical status of the parameter θ from that of an unknown quantity lying somewhere in the interval $(-\infty, \infty)$ to that of a random variable with the probability distribution N(x,1). In particular, the argument seeks to prove $\Pr(\theta < x) = 1/2$. Following Neyman, we may interpret the above only to mean that if, under the above kind of situation, we always assert $\theta < x$ then, in a long sequence of (independent) such situations - with the unobservable θ's varying in an arbitrary manner and with varying observations x- we shall be right in approximately 50% of cases. But Fisher (also Fraser) seems to be saying something more than this. In effect he is saying that the observation $X = x$ does not have any effect on the probability distribution of the quantity $Y = X - \theta$ - that is, given $X = x$ the quantity $Y = X - \theta$ is distributed as N(0,1). In other words, Fisher is saying that Y is independent of $X(= \theta+Y)$. Note the inherent contradiction between this assertion of independence and our earlier stipulation that Y is independent of θ . If θ has the character of a random variable and is independent of Y, then Y and $Y+\theta$ can never be independent of each other unless Y is a constant (which it is not). If not, then it is not clear what we are talking about.

Let us try to understand in another way what Fisher really had in mind when he said (in the context of our present X and θ) to the effect : When X is observed as x, we can regard θ as a random variable with $Pr(\theta < x) = 1/2$, and this irrespective of what x is. Furthermore, the statement $Pr(\theta < x) = 1/2$ can be interpreted in the same way as we interpret the statement : "For a fair coin Pr(Head) = 1/2."

In order to do so, let us see if we can distinguish between the following two guessing situations :

Situation I : Every morning Peter confronts Paul with an integral number x that he (Peter) has freshly selected that very morning, and then challenges Paul to hazard a guess (on the basis of the number x) about the outcome Y of a single toss of a fair coin (to be carried out immediately afterwards). Clearly, the number x gives Paul no information whatsoever about Y. And if we are to believe in the fairness of the coin (as the frequency probabilists understand it), then there exists no guessing strategy for Paul that, in the long run, will make him guess correctly in more (or less) than 50% of the mornings on which he chooses to hazard a guess. In the language of Fisher, Paul cannot 'recognize' any subsequence of mornings on which the long run relative frequency of occurance of heads will be different from 1/2.

Now consider

Situation II : Every morning Peter confronts Paul with a bag containing two tickets numbered respectively as $\theta-1$ and $\theta+1$, where the number θ is an integer that has been selected by Peter that very morning. Each morning Paul's task is to draw a ticket at random from the bag, observe the number x on the ticket drawn, and then hazard a guess on whether the number θ (the mean of the two numbers in the bag) is $x - 1$ or $x + 1$.

Clearly, situation II is a simplified (integral) version of the Fisher-problem we started this section with. Let us suppose that Paul has no idea whatsoever about how θ gets selected on any particular morning. He only knows that the unobservable θ can take any value in the infinite set $\{0, \pm 1, \pm 2, \ldots\}$. He also knows that for given θ, the observable X takes only the two values $\theta-1$ and $\theta+1$ with equal probabilities. As before we have the unobservable (pivotal) quantity $Y = X - \theta$ with a well defined probability distribution. In accordance with the Fisher logic, the only thing that the datum $X = x$ tells on any morning about the particular θ that obtains, is simply this : θ is either $x-1$ or $x+1$ with equal probabilities. It seems to me that Fisher would not have recognized any qualitative difference between the two situations. If Paul cannot read the mind of Peter then there is no way he can guess right in more (or less) than 50% of the mornings that he chooses to guess on.

Now, let us look at the following interesting argument given by Buehler (1971), p. 337). That Paul can do better than being right in only 50% of the guesses that he is going to make, is shown by Buehler as follows. Suppose Paul refuses to guess whenever $x < 0$, but always guesses θ as $x-1$ whenever $x \geqslant 0$. Now, let us classify all future mornings of Paul on the basis of the values of θ (that Peter is going to select) as follows :

$$M_1(\theta \leqslant -2), \quad M_2(\theta = -1 \text{ or } 0), \; M_3(\theta \geqslant 1).$$

On M_1-mornings, Paul never guesses and, therefore, is never wrong. Paul makes a guess on 50% of the M_2-mornings and is always right on such occasions. On M_3-mornings Paul always makes a guess and is right in only 50% of such guesses.

No doubt the Buehler argument will be endlessly debated by the advocates of the fiducial and structural probability methods. But let us point out that the argument is in the nature of a broadside against the improper Bayesians also. An improper Bayesian is one who systematically exploits the mathematical advantages of neat improper 'priors' and generally ignores the first requirement of Bayesian data analysis, namely, that the 'prior' ought to be an honest representation of the Bayesian's prior pattern of belief. Observe that in situation II above, an improper Bayesian will note with great relish the fact that the data allow him to assume that the parameter space is the unrestricted set I of all integers and that the likelihood function generated by the observation $X = x$ has the simple form

$$L(\theta|x) = \begin{cases} \frac{1}{2} & \text{when } \theta \in \{x-1, x+1\} \\ 0 & \text{for all other } \theta \text{ in I.} \end{cases}$$

He will now simplify everything by starting with the uniform prior over the infinite set I thus arriving at a posterior distribution which is the same as the uniform fiducial distribution over the two point set $\{x-1, x+1\}$.

2. The Stopping Rule Paradox

The controversy about the relevance of the stopping rule at the data analysis stage is best illustrated by the following simple example :

Example : Suppose 10 tosses of a coin, with an unknown probability θ for landing heads, resulted in the outcome

$$x = \text{THTTHHTHHH}$$

Now, for each of the following four experimental procedures :

E_1 : Toss the coin exactly 10 times;

E_2 : Continue tossing until 6 heads appear;

E_3 : Continue tossing until 3 consecutive heads appear;

E_4 : Continue tossing until the accumulated number of heads exceeds that of tail by exactly 2;

and indeed for any sequential sampling procedure (of the usual kind, with prescience denied) that could have given rise to the above sequence of heads and tails, the likelihood function (under the usual assumption of independence and identity of tosses) is the same, namely,

$$L(\theta|x) = \theta^6(1-\theta)^4 .$$

From the likelihood principle ($\mathcal{L}$) it then follows that at the time of analysing the information contained in the data ($\mathcal{E}$,x) we need not concern ourselves about the exact nature of the experiment $\mathcal{E}$ -our whole attention should be rivetted on the likelihood function $\theta^6(1-\theta)^4$, which does not depend on the stopping rule. In general terms, we may state the following principle due to George Barnard :

Stopping Rule Principle (for a sequential sampling plan) : Ignore the sampling plan at the data analysis stage.

This suggestion will no doubt shock and outrage anyone whose statistical intuition has been developed within the Neyman-Pearson-Wald framework. Even some enthusiastic advocates of $\mathcal{L}$ find the stopping rule principle embarrasingly hard to swallow. It will be quite interesting to make a survey of contemporary practising statisticians with a suitably framed questionnaire based on the above example. However the matter cannot be settled democratically! Dennis Lindley, having seen an earlier draft of this article, wrote to say the following : "You may like to know that in my third year course I have, for many years now, given the class the results of an experiment like you give, and ask them if they need any more information before making an inference. I have never had a student ask what the sample space was. I then point out to them that they could not construct a confidence interval, do a significance test, etc., etc. Although they are not practising statisticians, they have had two years of statistics . They just don't feel that the sample space is relevant. I have tried this out with more experienced audiences and only occasionally had an enquiry about whether it was direct or inverse sampling".

The rest of this section is devoted to a detailed discussion of the famous Stopping Rule Paradox*, which is generally believed to have knocked out the logical basis of principle $\mathcal{L}$. In order to isolate the various issues involved, it will help if we denote by $\mathcal{F}$ the following set of three classical (Fisherian) methods of statistical inference.

The $\mathcal{F}$ methods : The data consist of a prefixed number n of independent observations on a random variable X that is known to be normally distributed with unknown mean $\theta(-\infty < \theta < \infty)$ and known variance 1. The data then generates the information (likelihood function)

(i) $$L(\theta) \sim \exp\{-n(\theta-\bar{x}_{(n)})^2/2\}$$

where $\bar{x}_{(n)} = (x_1+x_2+\dots+x_n)/n$. Under the above circumstances, let $\mathcal{F}$ consist of the trilogy of statistical methods :

$\mathcal{F}$(a) : If $|\bar{x}_{(n)}-\theta_0| > 3/\sqrt{n}$, then reject the null-hypothesis $H_0 : \theta = \theta_0$ and declare that the data is highly significant.

$\mathcal{F}$(b) : The statement $\theta\varepsilon(\bar{x}_{(n)} - 3/\sqrt{n}, \bar{x}_{(n)} + 3/\sqrt{n})$ may be made with a great deal (well over 99%) of 'self-assurance' or 'confidence'.

$\mathcal{F}$(c) : The sample mean $\bar{x}_{(n)}$ is the most 'appropriate' point estimate of θ and the estimate is associated with a 'standard error' of $1/\sqrt{n}$.

Now consider the sequential sampling procedure based on the stopping rule:

$\mathcal{R}$: Continue observing X until the sample mean $\bar{x}_{(n)}$ satisfies the inequality $|\bar{x}_{(n)}| > 3/\sqrt{n}$.

If N is the (random) sample size associated with our rule $\mathcal{R}$, then it is easy to prove that N is finite with probability one if $\theta \neq 0$, and when $\theta = 0$ this conclusion still holds. [The latter may be deduced from the Law of the Iterated Logarithms, but can be proved much more easily directly. It should be noted, however, that $E(N|\theta)$ is finite only when $\theta \neq 0$.] Thus our rule $\mathcal{R}$ is mathematically well defined in the sense that N is finite with probability one for all possible values of θ . Suppose, following the rule $\mathcal{R}$, we generate the sample $x_1, x_2, \dots, x_N$. Our N is now random (not prefixed) but somehow the likelihood function fails to recognize this fact, for it is in the familiar form (see (i) above)

---

*I am unaware of who first formulated this clever paradox.

(ii) $$L(\theta) = (\sqrt{2\pi})^{-N} \exp\{-\Sigma(x_i - \theta)^2/2\}$$
$$\sim \exp\{-N(\theta-\bar{x}_{(N)})^2/2\}.$$

Now, if we combine $\mathcal{L}$ with $\mathcal{F}$, then looking back on (i) and (ii), we shall be forced to admit that, even when the sample $x_1, x_2, \ldots, x_N$ is generated by the sequential sampling rule $\mathcal{R}$, the following two inferences are also appropriate :

(a') The null-hypothesis $H_0 : \theta = 0$ should be rejected, at a very high level of significance (assurance), since $|\bar{x}_{(N)}| > 3/\sqrt{N}$ holds by definition.

(b') We ought to place more than 99% confidence or assurance in the truth of the assertion that the true value of θ lies in the interval $(\bar{x}_{(N)} - 3/\sqrt{N}, \bar{x}_{(N)} + 3/\sqrt{N})$.

The Paradox : The stopping rule paradox lies in the observation that method (a') leads to a sure rejection of hypothesis H_0 (at a high level of significance) even when H_0 is true. Also observe that the confidence interval $\bar{x}_{(N)} \pm 3/\sqrt{N}$ constructed for the unknown θ surely excludes the point $\theta = 0$ even when H_0 is true. Clearly, there must be something very wrong with principle $\mathcal{L}$!

For the moment let us only reverse the charge and claim that the stopping rule paradox, instead of discrediting $\mathcal{L}$, ought to strengthen our faith in the principle by exposing the naivety of certain standard statistical methods that are not truly in accord with the spirit of $\mathcal{L}$. To establish our claim, let us first of all concentrate our attention on the $\mathcal{F}$(a) method of testing the null hypothesis $H_0 : \theta = 0$.

Intuitively, it seems that the sequential sampling rule $\mathcal{R}$ used above is especially well suited to the problem of obtaining information on whether the hypothesis $H_0 : \theta = 0$ is true or not. When θ is appreciably different from zero we do not need too many observations on X before we lose faith in H_0, whereas when θ is nearly zero, we need quite a large sample before we could be reasonably sure that H_0 is false.

Why then should a 'reasonable' sampling plan $\mathcal{R}$, when coupled with $\mathcal{L}$ and the standard method $\mathcal{F}$(a), lead us to a testing procedure (a') with a power function

$$\pi(\theta) = \Pr(\text{Test ends with rejection of } H_0 | \theta)$$

that is uniformly equal to one? Is there any paradox at all?

Could the trouble lie in the fact that our rule $\mathcal{R}$ is not bounded above and, therefore, is perhaps a nonperformable experiment? To see if this might be so, let us define a bounded version $\mathcal{R}_M$ of $\mathcal{R}$ as follows:

$\mathcal{R}_M$: Continue observing X until the sample mean $\bar{x}_{(n)}$ satisfies the inequality $|\bar{x}_{(n)}| > 3/\sqrt{n}$ or $n = M$, whichever happens first. Our M is a fixed but possibly very large integer. With such a 'performable' rule $\mathcal{R}_M$ replacing $\mathcal{R}$, our power function $\pi_M(\theta)$ will now have the familiar U-shape that many of us like so much. Now, one might argue that it is only in the idealized limiting situation ($M \to \infty$) that our test becomes endowed with the (very desirable) property of having maximum power* of discernment against H_0, when the hypothesis is false, coupled with the (rather undesirable !) property of nonrecognition of H_0 when it is true. Let us look at the problem from another angle.

Is it not illogical to talk of a null-hypothesis H_0 that is specified by a particular value of a continuous parameter θ ? Are we not insisting from the beginning that all our realities are finite and therefore discrete? How can a pinpointed hypothesis like $H_0 : \theta = 0$ be classified as anything but an illusory idealization? Surely, such an 'infinitesimal' hypothesis (as H_0) is 'certainly false' to begin with, and ought to be rejected out of hand however large the sample is. How can a testing procedure be faulted for suggesting just that!

In the same spirit that we replaced the unbounded stopping rule $\mathcal{R}$ by a bounded version $\mathcal{R}_M$, let us replace the infinitesimal hypothesis H_0 by a non-infinitesimal version.

$$H_\delta : \text{Hypothesis that } \theta \varepsilon (-\delta, \delta),$$

where δ is some suitable positive number.

Let us see what happens to our paradox when we work with the finite (bounded) stopping rule $\mathcal{R}_M$ and finite (non-infinitesimal) hypothesis H_δ to be tested. If $x = (x_1, x_2, \dots x_N)$ be the sample observations on X that we obtain following rule $\mathcal{R}_M$, then what is the quality and strength of our information Inf($\mathcal{R}_M$,x) regarding the hypothesis H_δ ? Principle $\mathcal{L}$ tells us not to take into account any details of the statistical structure of the experiment performed or of the sample obtained other than the nature of the likelihood function $L(\theta|x)$ generated by the data. Fortunately, $\mathcal{L}$ does not stop us from using any background (prior) information about the parameter θ that we might have had to begin with. However, only a Bayesian knows how to match his 'prior information' with the 'likelihood information' supplied by the data.

*Indeed it was the stopping rule paradox that awakened me (about five years ago) about the possibility of the Darling-Robbins type tests with power one for the hypothesis $\theta \leqslant 0$ against the alternative $\theta > 0$.

[Many valiant and rather desperate attempts have been made by believers in $\mathcal{L}$-like Fisher, Barnard and others- to avoid taking this final Bayesian step, but such efforts have not met with much success.] So let us examine how the Bayesian method works in the present case.

Suppose, for the sake of this argument, that our Bayesian decides upon a uniform distribution over the interval (- 20, 20) as a reasonable approximation to the information (or the general lack of it) that he has about the unknown θ. Looking back on (ii), it is clearly very unlikely that we shall end up with a likelihood function L that does not lie well within (in the obvious sense) the interval (- 20, 20). With L lying well within the interval (-20, 20) the 'posterior density' of θ will be worked out by our Bayesian as roughly proportional to L and so he will evaluate the posterior probability of H_δ as

(iii)
$$\Pr(H_\delta|x) = \int_{-\delta}^{\delta} \frac{\sqrt{N}}{\sqrt{2\pi}} \exp\{-N(\theta-\bar{x}_{(N)})^2/2\}d\theta$$
$$= \Pr\{-\delta\sqrt{N} - \sqrt{N}\,\bar{x}_{(N)} < Z < \delta\sqrt{N} - \sqrt{N}\,\bar{x}_{(N)}\}$$

where Z is a N(0, 1) variable.

The stopping rule $\mathcal{R}_M$ is such that with a fair sized N the sample mean $\bar{x}_{(N)}$ is either roughly equal to $\pm 3/\sqrt{N}$ or is some number in between. Let us consider the situation when $\bar{x}_{(N)}$ is just above $3/\sqrt{N}$ and ignore the overshoot. Formula (iii) now becomes

(iv)
$$\Pr(H_\delta|x) = \Pr(-\delta\sqrt{N} - 3 < Z < \delta\sqrt{N} - 3)$$

and so the 'Bayesian significance' of the data depends entirely on the size of the statistic N. In order to see this let us suppose that $\delta = 1/10$. When N = 100, the right hand side in (iv) becomes $\Pr(-4 < Z < -2)$ which is less than 0.025. Whereas, when N = 10,000, the expression in (iv) becomes $\Pr(-13 < Z < 7)$ which is far in excess of 0.999!!

The point is clear : It is naive to propose $\mathcal{F}$(a) as a realistic statistical method. It simply does not make good statistical sense to set up a pinpoint (infinitesimal) null-hypothesis like $H_0 : \theta = 0$ and then to recommend its rejection whenever $|\bar{x}_{(n)}| > 3/\sqrt{n}$, where $\bar{x}_{(n)}$ is the observed mean of n (prefixed) independent observation on an X distributed as $N(\theta, 1)$ with $-\infty < \theta < \infty$. It should be recognized that the level of significance of the data vis a vis the hypothesis H_0 does not depend on the magnitude of $|\sqrt{n}\,\bar{x}_{(n)}|$ alone. It also depends, in a very crucial manner, on the magnitude of the sample size n. A Fisherian will perhaps feel quite satisfied with the information that $\sqrt{n}\,\bar{x}_{(n)} = 3$, and will, in any case, confidently reject the hypothesis H_0. But a Bayesian will surely enquire about the size of n (even though he may be quite uninterested at the data analysis stage to know whether n was prefixed or not). And, as we have just seen, the Bayesian's reactions to the two situations, n = 100 and n = 10,000, will be entirely different. In the first case he will consider it very unlikely that the true θ lies in the interval (- 0.1, 0.1), whereas in the second case he will have an enormous amount of confidence in the same hypothesis.

The stopping rule paradox should really be recognized as just another paradox of the infinitesimal. To emphasize this once again, let us briefly return to that part of the paradox that refers to (b') that is, to the fact that, with $\mathcal{R}$ as the stopping rule, the 3σ -interval $\bar{x}_{(N)} \pm 3/\sqrt{N}$ will always exclude the point 0 even when $\theta = 0$. This should not worry the planner of the experiment $\mathcal{R}$ if he bears in mind the fact that, in an hypothetically infinite sequence of repeated trials with θ fixed at 0, the variable N will usually take extremely large values, since $E(N|\theta = 0) = \infty$. For then he will recognize that the 3σ-interval $\bar{x}_{(N)} \pm 3/\sqrt{N}$ will in general be extremely short and will have its centre exceedingly near the point 0. In other words, the 3σ-interval will, with a great deal of probability, overlap very largely with the experimenter's indifference zone $(-\delta, \delta)$ around the point $\theta = 0$. Let us repeat once again that the pinpoint hypothesis $\theta = 0$ is only a convenient idealization and should never be mistaken for a reality.

3. The Stein Paradox

In 1961 L. J. Savage wrote : "The likelihood principle, with its at first surprising conclusions, has been subject to much oral discussion in many quarters. If the principle were untenable, clearcut counter-examples would by now have come forward. But such examples seem, rather, to illuminate, strengthen, and confirm the principle". In the following year, Charles Stein (1962) took up the challenge and came up with his famous paradoxical counter-example. It is popularly believed that the Stein paradox demolishes principle $\mathcal{L}$. We propose to show here why the paradox should really be regarded as something that illuminates, strengthens and confirms the likelihood principle.

The counter example is based on the function

$$f(y) = y^{-1}\exp\{-50(1 - y^{-1})^2\}, \quad 0 < y < \infty$$

defined over the positive halfline. Note that $\lim f(y) = 0$ both when $y \to 0$ and $y \to \infty$, and that in the latter case the rate of convergence (to zero) is slow enough to make the integral $\int_0^\infty f(y)dy$ diverge. We can therefore choose a and b such that

(i) $$\int_0^b af(y)dy = 1 \quad \text{and} \quad \int_{10}^b af(y)dy = 0.99$$

In fact the number b is exceedingly large - larger than 10^{1000}.

Now suppose that the probability distribution of the observable Y involves the unknown θ as a scale parameter in the following manner. The probability density function of Y is given by

(ii) $$p(y|\theta) = \begin{cases} a\theta^{-1}f(y\theta^{-1}), & 0 < y < b\theta \\ 0 & y \geqslant b\theta \end{cases}$$

Let us also suppose that our only prior knowledge about θ is $0 < \theta < \infty$.
With a single observation y on Y we end up with the likelihood function

(iii) $$L(\theta|y) \sim \begin{cases} \exp\{-100(\theta-y)^2/2y^2\} & yb^{-1} < \theta < \infty \\ 0 & 0 < \theta \leqslant yb^{-1} \end{cases}$$

Note that the maximum likelihood (ML) estimate of θ is y itself. But from (i) and (ii) we have

(iv) $$\Pr(Y > 10\theta|\theta) = \int_{10\theta}^{b\theta} p(y|\theta)dy$$

$$= \int_{10}^{b} f(y)dy = 0.99.$$

In other words, we have a situation where the ML estimator overestimates the true θ by a factor in excess of 10 and with a degree of certainty that is 99% ! The force of this criticism is, however, not directed against principle $\mathcal{L}$. We have seen earlier in Section 3 of Chapter III that simpleminded, unquestioning applications of the ML method can lead us into serious trouble. The Stein example is another such signpost warning us against uncritical use of the ML method. In this respect it is analogous to the following variant of an urn model that we considered earlier in Section 3, Chapter III .

Example : Suppose $0 < \theta < \infty$ and that an urn contains 1000 tickets out of which 10 are numbered θ and the remaining 990 are marked respectively as θa_1, θa_2, ..., θa_{990}, where the a_i's are known numbers all greater than 10. The random variable Y is the number on a ticket that is to be drawn at random from the urn. Here $\Pr(Y > 10\theta|\theta) = 0.99$; and when Y is observed as y, the unknown θ becomes 10 times more 'likely' to be equal to y than any one of the other 990 possible values, namely, ya_i^{-1} (i = 1,2 ...,990).

Stein's ingenious arguments against principle $\mathcal{L}$ run along the following lines : If Y were distributed as $N(\theta, \sigma^2)$ with $-\infty < \theta < \infty$ and σ known, then an observation y on Y would have generated the 'normal' likelihood function

(v) $$\exp\{-(\theta - y)^2/2\sigma^2\}, \quad -\infty < \theta < \infty$$

and in such a case it would have been clearly correct (method $\mathcal{F}$(b) of Section 2) to make an assertion like

(vi) $$y - 3\sigma < \theta < y + 3\sigma$$

with an associated level of assurance (confidence) that is at least 99% . Now, if we look back on L in (iii) and remember that $b > 10^{1000}$, then we have to admit that, for all practical purposes and irrespective of what y is, the likelihood function L in (iii) is indistinguishable from the one in (v) above with $\sigma = y/10$. Invoking principle $\mathcal{L}$ together with the 3σ-interval method $\mathcal{F}$(b), Stein concludes that it must then be appropriate to associate at least 99% confidence in the truth of the proposition

(vii) $$(0.7)y < \theta < (1.3)y$$

where y is the observed value of the random variable Y distributed as in (ii) and θ is the value of the unknown parameter that obtains. But from (iv) it follows that, having observed Y = y, we are also entitled to make the assertion

(viii) $$\theta < (0.1)y$$

with a 99% degree of confidence.

The Stein paradox then lies in the observation that the two statements (vii) and (viii) are mutually exclusive and, therefore, in no meaningful sense can they both be associated with degrees of confidence that are as high as 99%. According to Stein, this paradox clearly proves the untenability of principle $\mathcal{L}$, and a great many contemporary statisticians seem to be in wholehearted agreement with him.

A reexamination of the Stein argument will make it clear how the anomaly was forged out of the union of $\mathcal{L}$ with method $\mathcal{F}$(b)–the 3σ interval-estimation method based on an observation y on $Y \sim N(\theta,\sigma^2)$, with $-\infty < \theta < \infty$ and σ known. But what is the logical status of method $\mathcal{F}$(b)? And then, how compatible is $\mathcal{F}$(b) with principle $\mathcal{L}$? We know all too well how the 3σ-interval is justified in the Neyman-Pearson theory in terms of the 'coverage probability' of the corresponding (random) interval estimator $(Y-3\sigma, Y+3\sigma)$. We are also aware of the Fisher/Fraser efforts of justifying the same interval in terms of fiducial/structural probability. But such 'sample space' arguments are not compatible with $\mathcal{L}$, nor are they applicable to the present case.

There are two well known likelihood routes following which one may seek to arrive at method $\mathcal{I}$(b) from principle $\mathcal{L}$. The first route is briefly charted out in our description of method (b) in Section 1, Chapter III — the LR (likelihood ratio) method of interval estimation. Following this route, one first recognizes the 3σ-interval in (vi) and (vii) as the LR interval

$$I_\lambda = \{\theta : L(\hat{\theta})/L(\theta) < \lambda\}$$

where $\hat{\theta}$ (= y) is the ML estimate of θ and $\lambda = e^{4.5}$, and then the argument is allowed to rest on the largeness of the number $\lambda(= e^{4.5})$. However, observe that the Stein paradox does not relent a bit even when one increases the λ to the staggering level of $e^{40.5}$ - that is, replaces the 3σ-interval by the 9σ-interval. In Sections 2 and 3 of Chapter III we have argued at length against likelihood methods that are based solely on pointwise comparisons of likelihood ratios. The Stein paradox ought to be recognized as just another signpost of warning against uncritical uses of the ML and the LR methods.

The other slippery route that will generate the 3σ-intervals (vi) and (vii) from $\mathcal{L}$ is of course the way of the improper Bayesians. Looking at the likelihood function (v), an improper Bayesian will immediately recognize the enormous mathematical advantages of beginning his Bayesian data analysis rituals with the uniform prior over the infinite parameter space. This will allow him to claim that, given Y = y, the posterior distribution of θ is $N(y,\sigma^2)$. And then he will arrive at the 3σ-interval $(y-3\sigma, y+3\sigma)$ in the approved manner and associate the interval with more than 99% posterior probability. In a moment of euphoria an improper Bayesian may even put down the following as a fundamental statistical principle :

Principle $\mathcal{IB}$: If the likelihood function L generated by the data is indistinguishable from the normal likelihood (v) above, and if our prior knowledge about the parameter θ is very diffuse, then it is right to associate over 99% confidence (probability) in the truth of the proposition that the true θ lies in the 3σ-interval (vi).

Stein's denunciation of the likelihood principle is apparently based on the supposition that $\mathcal{IB}$ is a corollary to $\mathcal{L}$. In his example, the L in (iii) is truly indistinguishable from (v) and this is so irrespective of the magnitude of the observed y. It is $\mathcal{IB}$ (and not $\mathcal{L}$) then that justifies a posterior probability measure in excess of 99% for the interval in (vii), and this for all possible observed values y for Y. Written formally as a conditional probability statement, the above will look like : If θ is uniformly distributed over the parameter space $(0, \infty)$ and if Y, given θ, is distributed as in (ii), then

(a) $$\Pr(A \mid Y = y) > 0.99 \qquad \text{for all } y \varepsilon (0, \infty),$$

where the event A is defined by the inequality $(0.7)Y < \theta < (1.3)\,Y$. But from (iv) we know that

(b) $$\Pr(A \mid \theta) < 0.01 \qquad \text{for all } \theta \varepsilon (0, \infty).$$

Of course, all our probabilistic intuitions will rebel against the suggestion that there can exist a random event A whose conditional probability is either uniformly greater than 0.99 or uniformly smaller than 0.01 depending on whether we choose the conditioning variable as Y or θ. But it should be realized that the improper Bayesian has lifted the subject matter to the rarefied, metaphysical plane of infinite (improper) probabilities and so no mathematical contradictions are involved, since both θ and Y are (marginally) improper random variables and the unconditional probability of A is infinite.

To a proper Bayesian, the Stein paradox is merely another paradox of the infinite. In order to see this, let us see what happens if we couple a proper prior density function q to the likelihood function in (iii) and then obtain the shortest 99% confidence interval (in the approved Bayesian manner) as the interval $I_q(y) = (m(y), M(y))$. We now have

$$\Pr(\theta \varepsilon I_q(y) \mid Y = y, q) = 0.99.$$

And if we consider θ as fixed and speculate about the 'coverage probability' of the (random) interval-estimator $I_q(Y)$, then we arrive at the performance characteristic

$$\pi(\theta) = \Pr\{\theta \varepsilon I_q(Y) \mid \theta\} = \Pr\{m(Y) < \theta < M(Y) \mid \theta\}.$$

Since q is a prior, we now recognize (thanks to Fubini) that

$$\int_0^\infty \pi(\theta) q(\theta) d\theta = 0.99$$

and we are saved from an embarrassment of the kind that the improper Bayesian suffered in (b) above - his $\pi(\theta)$ was uniformly smaller than 0.01 !

All of us have our favourite paradoxes of the infinite and the infinitesimal. I cannot resist the temptation of setting down here my favourite paradox of the infinite.

Example : Peter and Paul are playing a sequence of even money games of chance in which the odds are heavily stacked against Paul– the games are identical and independent, and in each game Paul's chance of winning is only 0.01. Paul, however, has the choice of stakes and can decide when to stop playing. Paul considers the situation to be highly favourable to himself, but bemoans the fact that his chance of winning in a single game is not low enough - he would have much preferred it to be, say, one in a million. Simple ! Paul trebles the stakes after each loss, and continues to play until his first (or the n-th) win. Observe that we have opened our windows to three infinities : Paul's capital, Peter's capital and the playing time - all are supposed to be unbounded.

What then is the real status of the 3σ -interval in (vii) ? Principle $\mathcal{SB}$ notwithstanding, it is certainly wrong to say : "No matter how large or small y is, the interval $J(y) = (0.7y, 1.3y)$ should be associated with a high degree of confidence/likelihood/probability for containing the true θ" . Only a Bayesian, working with an honest (and, therefore, proper) measure of prior belief, is able to give a reasonable answer to the question : "Under what circumstances is it plausible to associate a 3σ -likelihood interval like (vii) with a posterior measure of belief that is in excess of 99%". His answer will be something like : "When the prior distribution is found to be nearly uniform (with a positive density) over the 3σ-interval". Suppose, for the sake of the argument, that the Bayesian regards a uniform probability distribution over the interval (0, C) as a fair representation of the state of knowledge that he started with about the parameter θ. This means, in particular, that he has about 90% prior belief in the proposition $\theta > (0.1)C$. So when he plans to take an observation on the Stein variable Y he is already very confident that the observation y will fall well outside the interval (0,C). He will not be at all surprised to find the 3σ-likelihood interval J(y) to be disjoint with his parameter space (0,C) and will naturally allot a zero measure of (posterior) belief to the 3σ-interval then.

Mathematics is a game of idealizations. We must however recognize that some idealizations can be relatively more monstrous than others. The idea of a uniform prior over a finite interval (0, C) as a measure of belief is a monstrous one indeed. But the super idealization of a uniform prior over the infinite half-line $(0, \infty)$ is really terrifying in its monstrosity. Can anyone be ever so ignorant to begin with about a positive parameter θ that he/she is (infinitely) more certain that

θ lies in the interval (C, ∞) than in the interval $(0, C)$—and this for all finite C however large? ! Naturally, everything goes completely haywire when such a person, with his mystical all consuming belief in $\theta > C$ for any finite C, is asked to make an inference about θ by observing a variable Y which is almost sure to be at least 10 times larger than θ itself !

According to my monstrosity scale for mathematical idealizations, the uniform prior over the halfline $(0, \infty)$ is rated as only half as monstrous as the prior distribution defined in terms of the improper density function $d\theta/\theta$. Stein cleverly exploited the logical vulnerability of the former at the infinite end. The latter is vulnerable at the zero end also. Anyone endowed with this latter kind of prior knowledge about θ must regard each of the two statements $0 < \theta < \varepsilon$ and $C < \theta < \infty$ as infinitely more probable than any statement of the kind $\varepsilon < \theta < C$ – and this for all $\varepsilon > 0$ and $C < \infty$!

However, one point in 'favour' of the measure Q on $(0, \infty)$ defined by the density $d\theta/\theta$ is that it is a (multiplicative) Haar measure on the (multiplicative) group of positive numbers - the measure is invariant for all changes of scale (transformations like $\theta \to a\theta$, with $a > 0$, of $(0, \infty)$ onto itself). This, together with the fact that θ enters into the model (for Y) as a scale parameter, make Q almost irresistible to many improper Bayesians who will somehow convince themselves of the necessity of taking Q as a prior measure of rational belief. The rest of their arguments will then follow the standard Bayesian line ending in the 99% posterior probability interval $J_Q(y)$ for θ.

With Q as the Bayesian prior, the posterior distribution of the scale parameter θ is defined in terms of the density function

$$q(\theta|y) = \begin{cases} a\theta^{-1} \exp\{-50(\frac{\theta}{y} - 1)^2\}, & b^{-1}y \leqslant \theta < \infty \\ 0 & 0 < \theta < b^{-1}y \end{cases}$$

and is the same as the fiducial/structural probability distribution of θ that is obtained in the usual manner from the pivotal quantity Y/θ. In view of the fact that the above density function is bimodal (with modes at $b^{-1}y$ and at a point roughly equal to $99y/100$), the usual 99% posterior probability set $J_Q(y)$ will in fact be the union of two intervals and, therefore, different from the 99% confidence interval $J_S(y) = (b^{-1}y, 10^{-1}y)$ suggested by Stein. It should however be noted that the improper Bayesian will evaluate the posterior probability of the interval $J_S(y)$ as 99% and hence the two intervals J_Q and J_S must have an overlap with at least 98% posterior probability.

At this point let us take note of the fact that any recommendation for the use of the prior Q (for weighting the likelihood function) on the score of θ being a scale parameter is contrary to the spirit of principle $\mathcal{L}$. This is because the information that θ is a scale parameter cannot be deciphered from a description of the likelihood function alone. Curiously enough, of all persons George Barnard also has a lot to do with the logical monstrosity of Q. In Barnard (1962) we have a discription of how he proposes to use the posterior (fiducial) distribution q above in conjunction with the likelihood function L to arrive at a confidence interval $J_B(y)$. The interval $J_B(y)$ looks startlingly different from $J_S(y)$ but has the same 99% 'coverage probability' as that of the latter.

Let us close this section by asserting once again that the Stein paradox illuminates the likelihood principle by focussing our attention on the true Bayesian profile of the principle. It also strengthens principle $\mathcal{L}$ by demonstrating the logical inadequacies of some so called likelihood methods/principles like ML, LR, $\mathcal{IB}$, etc.

ACKNOWLEDGEMENT

In the Fall of 1972 I gave a series of lectures on Likelihood at the University of Sheffield. This three part eassy is the rewritten version of part of the lecture notes circulated at that time. I wish to thank Terry Speed and other participants in this seminar series whose unflagging interest in the subject persuaded me to do this rewriting.

My attention has been drawn to a short note by A. Birnbaum in the December 1972 issue of JASA. There is a certain amount of overlap between Birnbaum's note and part one of this essay.

CHAPTER V

STATISTICAL INFORMATION AND LIKELIHOOD

DISCUSSIONS

0. Notes

Author's Note : The likelihood principle antithesis to the sample space based data analysis thesis of Pearson-Fisher-Neyman is indeed very strange. It is like stealing the big stick of the bully and hitting him over the head with it!

But unlike the Bayesian coherency argument, the likelihood principle argument is very persuasive indeed. I know quite a few statisticians (including myself) who grew up in the Neyman-Pearson tradition and then were suddenly dazzled by the light of likelihood. As if the (sample space) scales fell from their eyes and they suddenly recognized the sample, the unique one.

At the Aarhus conference none of the discussants raised any serious objection to the likelihood principle. In the Summer of 1977, I gave three lectures on the topic at the University of California, Berkeley but heard no voice in dissent. [Only Lucien Le Cam told me outside the lecture hall that he could not agree with me.] However, in 1976, when I presented the likelihood thesis at the Iowa State University, my good friend Oscar Kempthorne fought tooth and nail against the Principle. A couple of years later, Kemp sent me a copy of a very polemical writing of his in which he proved to his own satisfaction that my proof in Chapter II that $\mathcal{S}'$ plus $\mathcal{C}'$ implies $\mathcal{L}$ is a bunch of baloney! [This article appeared in an Australian Journal of Statistics, but I cannot recall the exact reference.] Readers of this monograph will find Kemp's discussion in Chapter XVI and my reply to it pretty hillarious.

The Berger-Wolpert (1984) monograph, The Likelihood Principle, is an excellent one and very useful.

STATISTICAL INFORMATION AND LIKELIHOOD

DISCUSSIONS

Editor's Note : Chapters II, III & IV were presented as a three part essay at the Conference on Foundational Questions in Statistical Inference held at the Institute of Mathematics of Aarhus University, Denmark in the Summer of 1973. The essay was read in two instalments and was followed by discussions on each occasion. The following is a consolidated account of the discussions that took place. The discussants were A. W. F. Edwards (in the chair), G. A. Barnard, A. P. Dempster, G. Rasch, D. R. Cox, S. L. Lauritzen, O. Barndorff-Nielsen, P. Martin-Löf and J.D. Kalbfleisch.

Edwards : Professor Basu raised the question of why Fisher felt he had to justify the method of maximum likelihood in repeated sampling terms. I believe he did so in response to an invitation by Karl Pearson : 'If you will write me a defence of the Gaussian method [as Pearson termed maximum likelihood], I will certainly consider its publication'. Thus, ten years after he had originally proposed the method, Fisher examined its repeated sampling properties (1922). But by 1938 he was writing 'A worker with more intuitive insight than I might perhaps have recognized that likelihood must play in inductive reasoning a part analogous to that of probability in deductive problems' (see Jeffreys (1938)).

Barnard : Concerning Fisher's 1912 paper, the justification given for maximum likelihood was to some extent its "absolute" character, in being, unlike χ^2, independent of any arbitrary grouping of the observations, or of any arbitrary choice of variables for fitting moments.

The Bayesian position cannot be reckoned as having been fully stated until they specify how the prior factor q, in the posterior Lq, is to be determined. The last posthumous paper by Jimmie Savage was a serious attempt to do this; but its very length and complexity (and that of a related paper by, I think Winckler, in JASA) show how much has yet to be done here. Sometimes the non-Bayesian position is attacked as leading sometimes to arbitrary conclusions; but any limited degree of arbitrariness there may be is negligible compared with the much greater arbitrariness represented by q.

It is important to realize that the L factor is capable of verification, by repeated experiments; but the q factor is not. This does not mean that the L factor must necessarily be given an oversimplified "frequency" interpretation.

Dempster : Professor Barnard appears to set up a ridiculously strict double standard by requiring that the Bayesian shall say exactly where his prior distribution comes from while assuming that the likelihood is known beyond question. In fact, it is often unclear which of the two sources of uncertainty in the model is the more dangerous.

Rasch : While, of course, admitting the benefit of prior knowledge, if available, I am disinclined to transforming "pure belief" - whether superstitious or not - into a "measure", whether "probabilistic" in some sense or not. Instead I shall ask two questions : In what does the prior information consist ? and : Just where does it come from ?

There seems to be two sources.

One is the insight - direct or indirect - in the field of inquiry of the data, such as it may have accummulated until the actual investigation.

As regards such "insight" I may be a bit more explicit : As "direct" I take, for one thing, knowledge about the conditions under which the data were in fact collected (planned experiment, survey, responses to questionnaires, routine records on the part of the Central Statistical Bureau, regular astronomical observations, or what not). For another thing it includes available theory about the subject matter in question. By "indirect" I am partly thinking of inspired analogies from related fields - more or less distant - partly of general views, e.g. philosophical and technical, both of which may influence the mathematical formalization.

As a case in point I may refer to my realizing the common structure of data on misreadings by schoolchildren exposed to two or more reading tests, and accidents occurring to the population of drivers, when they are riding on different road categories at different days. This gave rise to using the same model in the two cases (the Multiplicative Poisson Model).

However, both direct and indirect insight should, I think, enter into the construction of the model that is going to form the basis for the analysis.

The other source is experience with same or related sorts of data, whether it be from previous studies - whoever made them - or from parallel studies in different places (such as serological analyses of the same substances carried out at different laboratories, as organized by WHO).

But in such cases the available data, or the results of analyzing them, might simply be handled parallel to the actual data, on the basis of models expressed in ordinary probabilistic terms - elaborated, of course, with due respect to differences in conditions.

In principle, this point of view removes the difference between data collected in the past and in future, in one place or another. It aims at giving a model, once (tentatively) established, as broad a background as at all feasible for checking it.

As a case in point I may mention an investigation of the death rates in Denmark through 50 years which disclosed a certain structure in their dependence on age, in spite of relatively strong changes in living conditions. Afterwards the same structure was found in Sweden, and again, some years later, in United Nations data from numerous countries all over the world.

Barnard : Some notion of repeatability is involved in any form of scientific inference. We would not be interested in the behaviour of Nile floods if we knew that the Nile would disappear tomorrow, and, along with it, the area of Abyssinia and other parts of Africa whose weather conditions largely determine the Nile floods.

A repetition need not be an exact replication. Thus a measurement of length to 1 mm may be "repeated" by a test whether the length is > or < 100 cms. And a measurement of rainfall around the Blue Nile may indirectly "repeat" a measurement of the height of a Nile flood. The essential feature is the accumulation of independent pieces of evidence bearing on a given topic. And the meaning of "independence" here is not mere statistical independence (cf. my 1949 paper, pp. 119-120).

Cox : Dr. Basu has talked of analysis not involving a sample space. Yet the start of his treatment is that a parameter ω is given. Quite apart from the issue that the formulation of an appropriate ω is often a key point, how can ω be given a physical meaning without some notion of repetition, even if hypothetical, and hence how can consideration of some sample space be avoided ?

Lauritzen : It seems difficult to me to give any meaning to the parameter ω without referring to outcomes of other experiments.

Rasch : Although agreeing with the view, expressed by Steffen Lauritzen, that assigning a probability distribution to a parameter in general would seem artificial, I may add that there are cases, albeit few in my own experience, where such a superstructure is warranted.

By way of an example I may mention measuring the diameters of 500 red blood corpuscles in each of a number of blood samples, taken in quick succession from the same normal person. Each sample shows a most beautiful normal distribution and the estimated standard deviations lie quite close to each other, but the average diameters varied much more than allowed for by the standard error. The reason for this discrepancy was, however, quite clear : During the technical preparation of a blood sample it is exposed to a certain pressure, exerted by hand – therefore sometimes a bit harder than at other times, thus influencing the sizes of all of the blood cells, but not noticeably the differeces between them.

This, of course, does not turn the problem into a proper Bayesian one. In the instances of repeated sampling the model applied was : the distribution $N(\xi_i \ \sigma^2)$ for diameters within sample no. i and $N(\xi, \tau^2)$ for the variation of mean values ξ_i between samples, which leaves us with an ordinary estimation problem.

Barndorff-Nielsen : In relation to Professor Barnard's remark concerning repeatability of experiments, may I make the following comment? It seems to me that there exist experiments – in the broad sense of the word – which are not repeatable in any real sense, but which do properly belong to the province of science. I am, inter alia, thinking of data pertaining to the geological history of the earth or to the theory of evolution.

Barnard : The current revival of interest in geology is due in large measure to the fact that 1) we have at last another body the moon - which is in some sense a "repetition" of the Earth, and we are beginning to obtain "geological" information about Mars; 2) we have theories of geological processes (continental drift, etc.) which are still going on and which seem likely to enable us eventually to predict earthquakes, etc; 3) experimental work on the behaviour of materials under ultra-light pressures, though difficult, is approaching relevance to geological processes. Thus, although the specific history of the earth is not replicated, the processes involved can be, at least to some extent.

Martin-löf : In response to Barnard, I would like to stress that even when an experiment cannot be repeated (except in our thought as done by Gibbs and von Mises with their ensembles and Kollektivs, respectively) it may be amenable to a statistical analysis. A typical example is Lauritzen's (1973) treatment of the gravitational field of the earth as one observation of a certain Gaussian random field. It is quite enough that we can draw verifiable conclusions from the probabilistic assumptions by means of the interpretation clause which allows us to neglect events of small probability.

Barnard : Professor Basu's claim that the Bayesian will more often be right assumes that the Bayesian's prior will correspond with the actual frequencies arising in the sequence of problems dealt with. But there seems no reason to suppose this will be so. Thus the Bayesian may well be less often right.

Edwards : A measure of the unsatisfactory nature of the confidence estimate is its sensitivity to variation in b, a somewhat hypothetical quantity. I suspect that the likelihood interval is not so sensitive.

Dempster : I wish only to record that the Stein and Stopping Rule paradoxes no longer seem to me to deserve the name paradox. There is no mathematical reason to expect Bayesian and confidence probability levels to agree, and their predictive and post-dictive interpretations are, in any case, incommensurable. The Bayesian approach is right in principle, but may be difficult in practice. If the required prior knowledge is too weak for any reasonably objective Bayesian inference to be allowed, I would back off and use a sampling-rule dependent confidence method, carefully pointing out the tricky and weak associated meaning.

Barnard : I may be wrong, but I believe Fisher did not assert any frequency-covering properties for likelihood intervals. He simply asserted that any specific θ_1 outside the interval

$$\{\theta : L(\hat{\theta})/L(\theta) \leqslant 100\}$$

would have plausibility, relative to the maximum likelihood value $\hat{\theta}$,less than 1/100. Whenever one wishes to make frequency statements concerning a single parameter value θ_1, considered by itrself, one must consider sampling distributions in some way (unless, of course, one is prepared to assume a distribution of θ("prior" distribution) as true of the set of cases with reference to which the frequency is asserted.)

Dempster : I feel that the non-Bayesians in this discussion have not yet been sufficiently nudged to face the difficulties their position. I propose therefore that we consider a game which can actually be played, and which I believe goes to the heart of the issue. Imagine N pairs of statisticians (A_i, B_i) for i = 1,2, ...,N where A_i is non-Bayesian and B_i is Bayesian. Each pair engages an agent C_i to determine a parameter value θ_i where A_i and B_i have some common understanding of how the determination is to be made (e.g., asking a random man in the street for a random number) but neither A_i nor B_i are given the value θ_i . Instead, an experiment is performed, say a sequential experiment, which allows θ_i to be estimated. Both A_i and B_i have a common access to the results of the experiment. A_i then creates a 95% confidence interval I_i for θ_i, which necessarily depends on the sampling rule as well as the likelihood. B_i is then offered the choice of sides in a wager over $\theta_i \varepsilon I_i$ and $\theta_i \notin I_i$ at odds of 19 to 1. A referee totals the net gain or loss of the A team from or to the B team over the N wagers, and declares the winning team accordingly.

There is of course no guarantee that either team will win, even for very large N. The defining property of the confidence intervals undeniably holds when the experimental model specification holds, but this property is inadequate to render the above game fair unless each B_i chooses his side of the wager according to a rule free from both prior knowledge and experimental data. In the real world, every scrap of available information will be used, hence the confidence interval property is inadequate for much of statistical practice. A simplistic Bayesian property also holds, namely, that the Bayesian can quite generally expect positive long run gain under his assumed probability models. But this property is also inadequate since no realistic Bayesian would expect all his model specifications to hold up in a longrun practice.

Where do we stand?! My own view is to distrust non-Bayesian decision theory since it fails to model the free choice aspect of decision-making. While there is no carte blanche in favour of Bayes, I do believe thast the B-team will very often win in the real world precisely because it can reflect real prior knowledge, at least sufficiently well to stay in the black. This is a matter of judgement, not proof.

Kalbfleisch : Professor Dempster has raised the question as to why the many adherents to the frequentist theories of inference have raised no specific objections to this paper. For my part, I find that the paradoxes outlined in this paper are forceful and do lead me to the conclusion that $\mathcal{J}(a)$ and $\mathcal{J}(b)$ cannot be viewed as solutions to all problems. But, the arguments leading to this conclusion are themselves frequentist in nature and there is the feeling that this strengthens rather than weakens the frequentist position. The justifications for accepting the likelihood principle that Professor Basu gives are not essentially different from those given by Birnbaum, and as I have pointed out there are objections which can be raised to these arguments.

The fact that the likelihood function alone is not enough, as Basu's exposition suggests, leads us to try to supplement it — either with the prior information q or with various frequentist arguments — for the solution of certain problems. I think much is to be said for a weaker sequence of principles (like those I have suggested) which allow for many different approaches such as tests of significance, confidence procedures, procedures of the type $\mathcal{F}$(a) and $\mathcal{F}$(b) and Bayesian methods, each applying to certain problems and not to others.

Edwards : Extreme paradoxes such as Stein's are intended to provide us with results so conflicting that we are bound to vote one way or the other. In practice they leave us bemused, and it may be better to focus on less extreme but more realistic examples which similarly contrast likelihood and confidence principles by making use of distributions with unusually long tails.

Consider the case in which a theoretical physicist predicts the value of a fundamental parameter to be $\mu = 0$. After many years' work practical physicists have made just two measurements, 11.5 and 13.5, and then their apparatus blew up. It is agreed thast these measurements may be regarded as a random sample from a normal distribution with unknown variance. Forming the statistic t on one degree of freedom, it is 12.5, not significant at the 5% two-tailed point. But on a support test (see Table 6 of Edwards (1972)) the increase in support available is $\ln(1+(12.5)^2)$, a likelihood ratio of 157.25, an impressive amount.

Barnard : Concerning Professor Basu's example about adding likelihoods, I said that the Bayesian consider it is always possible to add them, i.e. to find λ such that "α or β" = $\lambda\alpha + (1 - \lambda)\beta$.

Dempster : Only "always-Bayesians" think it always possible.

Barnard: I agree.

I said it was only sometimes possible to add likelihoods. So long as we are considering only small sample sizes, Basu's nearly identical hypotheses give the same likelihood orderings and so they clearly can be combined. But larger samples could show up differences between the hypotheses, which could become important, and then one could not add them. Thus, in my view, one cannot always add.

Dempster (note added in written version) : What I had in mind is that some Bayesians may feel comfortable switching over to a significance testing mode to provide checks on their assumed models. Such Bayesians, including myself, are "sometimes Bayesians" (so B's in Barnard's abbreviation) rather than "always Bayesians".

Barnard : I believe the stoping rule paradox was first brought up by Bartlett in a letter to me in the middle 50's. Armitage independently raised it in the discussion initiated by Savage. Although my views on it have not always been the same, I now think it simply serves to show that likelihoods are relevant to comparisons of pairs of (simple) hypotheses; they cannot apply to statements involving a single hypothesis, considered on its own. For the case stated, with n fixed, and x being the variable

$$\frac{|\bar{x}|}{\sqrt{n}} \geqslant 3$$

rejects the hypothesis $\mu = 0$. But if $|\bar{x}|/\sqrt{n}$ is fixed, and n is variable, the test criterion becomes n; low values of n will tend to reject the hypothesis.

Author's reply : We are talking about statistical data — data equipped with statistical models. We are debating about the basic statistical question of how a given data $d = (\mathcal{E}, x)$, where $\mathcal{E} = (\mathcal{X}, \Omega, p)$ is the model and x is the sample, ought to be analysed. My submission to you is that the likelihood principle of data analysis is unexceptionable. The principle simply asserts that if our intention is not to question the validity of the model $\mathcal{E}$ but to make relative (to the model) judgements about some parameters in the model, then we should not pay attention to any characteristics of the data other than the likelihood function generated by it. From the discussions it would appear that very few amongst us is in full agreement with the above proposition. The Neyman-Pearson-Wald antithesis to the likelihood principle is what we may call the principle of performance characteristics which requires us to evaluate the data in full perspective of the sample space. Few, if any, amongst us seem to have any conviction in this unconditional 'sample space' approach to data analysis.

What I am saying is that, for one who truly believes in the likelihood principle, there is hardly any choice left but to act as a Bayesian. If L is the 'whole of the relevant information contained in the data' then we ought to match L with 'all other information' q on the subject. In point of fact we usually have a lot of other information. How can we ignore q ? It seems to me that only an honest Bayesian can give a sensible answer (however clumsy and incompetent it may appear to non-Bayesians) to the basic question : How to analyse a given data ?

Professor Barnard likes the likelihood factor L but does not care for the Bayesian's prior q. He is arguing that the former is verifiable but the latter is not. Our concern here is not with the verification of assumed models but with the question of data analysis relative to such models. In any event, the kind of experiments that we come across in scientific inference can hardly be called repeatable in any meaningful sense of the term. Who has ever heard of a scientific experiment being repeated a number of times with the purpose of checking on the authenticity of an assumed likelihood function ? The likelihood L is no less subjective and hardly any more verifiable than the prior q.

Discussions

Irrespective of whether we believe in repeatability of experiments and frequency interpretation of probability or not, we are all immensely concerned with one kind of frequency, namely, the long run relative frequency of success in our inference making efforts. Whether the Bayesian method of data analysis is superior to any other well defined method cannot be proved mathematically. The long run success of an individual Bayesian will surely depend on his ability to come up with realistic q's and L's. Professor Barnard remarked that a Bayesian can well be less often right if his specification of the prior q is off the mark. He is apparently visualizing a sequence of identical experiments in which the model and, therefore, the L factor is always right but the same off key q is being used again and again. If the Bayesian is allowed to update his prior q for each experiment in the light of his past accumulated experience, then there is no reason to believe that he will fare badly in the long run even in such an unrealistic hypothetical sequence.

In real life, a practising statistician faces a sequence of different inferential problems about different parameters. If in each case he really applies his mind to the task of constructing a realistic likelihood scale L and carefully goes about the task of quantifying the prior information q then it seems entirely believable to me that our Bayesian will fare much better than a traditional 'sample space' data analyst. For one thing, the 'sample space' analyst has to work with a plethora of likelihood functions — one for each point in his sample space. Naturally he can work with only rather simplistic (and, therefore, unrealistic) statistical models. The Bayesian is never inhibited by such constraints. Since he has to work with only one likelihood function — the one that corresponds to the observed sample — he can boldly reach for more sophisticated (and, therefore, more meaningful) statistical models.

I am certainly not averse to the idea of sample space. As Professor Cox pointed out, in some cases even the parameter (say, the true weight of the chalk stick that I am holding in my hand) cannot be defined without the idea of repeated measurements. At the time of planning a statistical experiment we of course need to speculate about its sample space. But with an experiment already planned and performed, with the sample x already before us, I do not see any point in speculating about all the other samples that might have been.

The Bayesian and the Neyman-Pearson-Wald theories of data analysis are the two poles in current statistical thought. To day, I find assembled before me a number of eminent statisticians who are looking for a via media between the two poles. I can only wish you success in an endeavour in which the redoubtable R.A. Fisher failed.

BARNARD-BASU CORRESPONDENCE

Editor's Note : After the Aarhus Conference, Professors Barnard and Basu corresponded on some issues raised by Barnard. This correspondence was published in the conference proceedings and is reproduced here in full.

Brightlingsea, 18th May, 1973

Dear Dev,

1. It was good to see you in Aarhus, and I hope we meet again soon. I liked your paper, especially the first part, which was a very clear account of issues around the Likelihood Principle. But, as I said, I think in Part II you are not wholly fair to Fisher — and having checked with my own papers, which I could not do in Aarhus, I think you are not wholly fair to me.

2. First, in III,1,(a) you say "Fisher tried very hard to elevate maximum likelihood to the level of a statistical principle. Though it has since fallen ...". I don't think this is true. The matter is not easy to discuss in a precise way without specifying precisely what we understand by the problem of point estimation. Nowadays there are many people who seem to identify this with the decision problem, to find a function of the observations which will minimize the mean square deviation from the true value. This is certainly not the sense in which Fisher understood the problem. But my understanding of Fisher is that he pointed to the advantages of the maximum likelihood method, in regular situations, but never claimed it as a matter of principle. For instance, the passage beginning "A realistic consideration of the problem of estimation..." on p.157 (1st Ed.) or p. 160 (2nd Ed.) in Statistical Methods and Scientific Inference shows what I mean.

3. At the same time, I venture the following assertion about ML : Let us call a method of estimation "algorithmic" if, given the specification of the density function of the observations (i.e. given the model), the estimate derived can be obtained by a standard mathematical process such as solution of an equation, maximisation of a given function, etc. I assert that no algorithmic method of estimation is known which is superior to ML.

4. Can you produce a counter example ? In case you should refer to Bayes, I will accept integration of a given function as an algorithmic process; but you must also give an algorithm for determining a (reasonable ?)"prior".

5. Next, in III, 1, (b) you refer to "likelihood intervals" as "likelihood confidence intervals". This would suggest that covering frequency properties

are claimed for them, when in fact this is not so, except in specific cases when additional conditions are satisfied. It was, so far as I can remember, always clear both to Fisher and to me that an interval defined as your I_λ would not necessarily cover the true value with any particular frequency. Your subsequent examples which bring this out in a very strong way should therefore, I think, make clear only that these intervals do not possess a property which was never claimed for them.

6. On III, 2 :I think the statistical intuition of Sir Ronald would have been outraged by the suggestion you make, since the data specified are not inconsistent with the following numerical values :

$L(\omega_1) = 0.011$, $L(\omega'_1) = 0.01$, $L(\omega_2) = 0.101$, $L(\omega'_2) = 0.10$ and the prior probabilities 0.25, 0.005, 0.25, 0.495 respectively. A priori, the hypotheses $\omega \varepsilon A$ and $\omega \varepsilon B$ are equally probable, but given the data, their probabilities are proportional to 0.028 and 0.04955. Thus the data support B better than A in this case. We certainly could not say, in general, the opposite.

7. More generally, your supposition about adding likelihoods amounts to an assumption that all hypotheses are equally probable a priori. This can be made self-consistent; but I do not accept it as true, any more than Fisher did.

8. I find Fisher's analogy of "the height of Peter or Paul" a good analogy. If we were told that this was to mean "Choose Peter or Paul with equal probability, and then measure the chosen one's height", the phrase would acquire a definite meaning, as a random variable.

9. Your Example 1 in III I find unconvincing, because if your Martian were prepared to regard the range (9.9, 10.1) as of negligible width, he would do this in the first place, and so reduce your second case to the first. But if (as might be), he was interested in being exactly right, with "a miss" being "as good as a mile" (as the saying goes), then in the second case his best bet really would be $\theta = x$.

10. In your discussion of the fiducial argument in IV, 1 , I think you should say, to begin with, that $X - \theta$ is $N(0, 1)$, and then proceed to discuss θ and X on a symmetrical footing. There is no particular reason to suppose that either is unobservable.

11. With Buehler's argument, in IV, 1 , I think you should point out that, unless the M_2 mornings have positive density in the long run — and there is nothing

to guarantee this — then Paul will, in the long run, be right no more often than 50% of the time.

12. A small point on IV, 2 . I enclose an offprint which indicates that I was considering the stopping rule paradox before 1964, and the associated idea of tests of power 1. The priority over Darling-Robbins is unimportant, but since you have been referring to me, it should perhaps be made clear that, presumably, I have some way of dealing with the problem.

13. Finally, on IV,3, you say my likelihood interval fails the Stein test "miserably". I think you will find it meets the frequency test exactly.

Best regards,

George Barnard

Manchester, 29th May, 1973

Dear George,

1. Many thanks for your letter of May 18 which I find very interesting and informative. My views on the various issues raised by you are recorded below. Please note that the paragraphs of this letter correspond to those of yours.

2. I am reassured to learn that you regard ML only as a method of point estimation. I am however not so sure about Sir Ronald's own views on the subject. In any case, hardly anything can be said about Sir Ronald's views on Statistical Inference that cannot be denied. In paragraph 3 of p. 49 in Hacking's book you will find a reference to the Fisher Principle of ML.

3. I am somewhat bewildered by your challenge about producing an "algorithmic" method of point estimation that is "superior" to ML. Superior in what sense ? If you are asking for a method B that is universally (i.e., for all models) and uniformly (i.e., for all parameter values in each particular model) superior to ML in the usual sense of some average performance characteristics then I am affraid I have nothing tangible to offer. But then I can as easily counter your challenge by producing a method B and then asking you to produce something "superior" to that. In Section 3, Part II of my eassy I have elaborated at length on my objections to ML as a method. My objections stem mainly from the fact that the method has nothing to do with the two essential ingredients of inference making that are always present in some measure in every realistic situation and which I have denoted in my essay by the symbols q and Π .

4. Regarding your remark in paragraph 4, I do not know how a (honest) Bayesian's prior can be characterized in terms of the mathematical description of the model. I have made it amply clear why any such attempted characterization will violate the likelihood principle and, therefore, the very essence of Bayesianism.

5. Without disagreeing with your comments in paragraph 5, I have only to say that when I use the word "confidence" I tend to associate it with the elusive notion of a "measure of belief" rather than with that of "frequency probability". With my examples I have been trying to establish this simple fact that there exists no logical (coherent) basis for supposing that a likelihood interval I_λ with a sufficiently large λ has a claim to a large measure of assurance about the true θ lying in that interval. My examples underline the crucial (and to me self-evident) fact that the "information" contained in the likelihood function can be analysed only in the context of the background knowledge q and the inferential problem Π . I consider it utterly self-defeating to try to build a theory of inference on likelihood alone.

6. Your remarks in paragraph 6 made me happy in the knowledge that you are not averse to prior probabilities. It seems to me that we are talking on slightly different wave lengths but essentially about the same thing. We are agreed then there are two sources of information — the prior knowledge q and the likelihood measure L of support-by-data. You have produced an example where the L-support for the composite hypothesis A is greater than that for B, the q-support for A is the same as that for B, but the (L+q)-support for A is less than that for B. Where is the contradiction ?

7. Regarding your remark in paragraph 7, I shall readily concede that the supposition that the likelihood support-by-data is an additive measure is tantamount to the supposition that to the data (or the ignorant Martian) all simple hypotheses are equally probable a priori. Of course, I do not believe in the Martian's "equally distributed ignorance" any more than you do or Fisher did. That is why this insistance about the meaninglessness of L by itself and about the necessity of matching it with an honest prior of the scientist.

8. As regards "the height of Peter or Paul", I still fail to see why it is a better analogy to "the likelihood of A and/or B" than the natural analogy of "the probability of A and/or B". With this (false) analogy Fisher dismissed the Bayesian insight about the likelihood being something that is meant to be weighted and then accumulated. How does your random choice between Peter or Paul make the analogy a better one ?

9. My Example 1 in III, 3 was constructed to demonstrate the fact that methods like ML, LR, etc. are disoriented to the task of inference making. Apart from the fact that such methods do not make any use of q and Π , they are also based on the popular misconception that likelihood is a point function and as such can be interpreted only by maximization and by ratio comparisons.

10. I must admit that I can never cease to be mystified by the fiducial/ structural probability arguments of Fisher/Fraser. How can I "proceed to discuss X and θ on a symmetrical footing" when they are not? I have observed $X = x$ and am trying to make an inference about θ. I also have some preconceived ideas about θ. Where is the symmetry ?

11. Paul ought to be able to recognize some event like $M_2(\theta = -1 \text{ or } 0)$ that has "positive density in the long run". Otherwise, his ignorance about Peter's θ is of such a monstrously all consuming kind (a uniform prior over all integers?) that I refuse to speculate about it.

12. It was very interesting to read through the offprint you sent me. It shows that the stopping rule paradox led you to the idea of tests with power one. I mean to find out from Professor Robbins as to how he was led to the same idea.

13. I am sorry for the error in IV, 3 where I said that the interval $J_B(y)$ fails the Stein test. Please accept my apologies.

With all the best,

Yours sincerely,

Dev Basu

Brithtlingsea, 5th June 1973

Dear Dev,

1. I had better begin by confessing I write this under the selfimposed handicap that I lost my copy of my letter; so please forgive any resulting deviations from logical order. My comments are numbered to yours.

2. I wish I could remember what I said that can have led you to the first sentence. ML is primarily a method of point estimation, and as I read Fisher, this is how he understood it. On referring to Hacking, I find I have marked the passage you mention as being in error. And as far as Fisher's views on inference are concerned, I would have thought we can take his "Contributions to Mathematical Statistics", and "Statistical Methods and Scientific Inference" as representing his views,

and it is reasonable to ask, if someone says Fisher took a certain view, that he should be asked to support the statement by some reference to Fisher's works - not to Hacking's, or anyone else's.

The nearest, I think, you could come to a quotation to justify your statement about Fisher is to be found on p. 100 of Anthony Edwards' book, last paragraph. But I think this clearly, in fact, shows your statement to be unjustified.

3. My challenge about producing a better algorithm than ML stands--and you can determine the sense of "superior" in any reasonable way you like, so long as you say what it is. Of course I am not asking for something that is universally and uniformly superior.

4. I agree. But since the mathematics of Bayes Theorem are very simple, within the scope of any mathematician who can integrate, acceptance of the Bayesian position means that statistics texts will need to concentrate on the very difficult task of enabling people to assess for themselves their prior distributions and their loss functions. I say very difficult because for many of us, we are unaware often of the existence of these things (and, indeed, unpersuaded).

5. I have now checked what Fisher said about likelihood intervals, and it is clear that he, no more than I, did not think that a likelihood interval I_λ would have (except in regular asymptotic cases) any particular probability of containing the true value. Thus, in arguing as you do, I think you are flogging a dead horse. But of course, the fact that I_λ has no particular probability of containing the true value do not justify your "crucial (and to you self-evident) fact". I agree with your last sentence, but nonetheless think it worthwhile to see how far we can go with a theory based on likelihood alone; and clearly I think one can go further than you suggest.

6. Considering that I advocated the use of prior probabilities in 1946 when such a point of view was far from popular I think it clear that I am not averse to them, when they exist, in the sense that they can be subjected at least in principle to some sort of objective verification. And of course I agree that there are two sources of information. But the question is, can the prior knowledge always be expressed in terms of prior distribution ?

As to the example of course there is no contradiction; but there is a paradox. If one piece of information is neutral as between A and B and the other piece favours A, it surely is odd that the two together should favour B. Such a thing cannot happen with simple likelihoods.

7. I think we agree here.

8. You think Fisher's analogy false because you, unlike him, take a Bayesian view.

With regard to your "and/or" what you say in Section 2 of Ch. III is, I think, false. Because A will denote a different parameter value from B, and this will imply that A and B are incompatible. Thus the "or" really is the disjunctive "or". If it were "and/or" one could say that the likelihood of "A or A or A" was 3 times the likelihood of A, which is absurd.

9. I agree with what you say. But I do not find your demonstration convincing.

10. The symmetry is, that I might have observed θ and be trying to make an inference about x. As to preconceived ideas, I may also have such ideas about x. It is part of the argument that I have no knowledge about θ (or, respectively x), other than that specified.

11. A uniform prior for θ over all integers is not required. I do not follow the sense of your "all consuming". The information given by the observations is not "consumed" by the prior ignorance.

12. I would prefer the term "test with power one" to the terms "Darling-Robbins type tests", seeing that Barnard published and used such a test in practice three years before Darling or Robbins.

The term "Darling-Robbins type tests" should, I think, be used for tests whose power function is discontinuous.

13. Many thanks,

Best regards,
George Barnard

Manchester, 12th June, 1973

Dear George,

Many thanks for your letter of June 5. Excepting for two points, I must concede you the last word on all the other issues.

I find your remarks on "and/or" in paragraph 8 very confusing. May be the difficulty is only a matter of semantics. In every introductory course on probability theory, don't we always carefully explain why the expression Pr(A or B) must not be understood to mean "probability of either A or that of B" ? We then explain that the "or" is not to be used in its usual disjunctive sense of "either-or" but in the "accumulative" sense of the set-theoretic/logical connective union/and-or. After that we have a hard time (especially if we take the subjectivist point of view) explaining why Pr(A or B) = Pr(A) + Pr(B) when A and B are exclusive events. In Pt.II of my essay I only suggested that the trouble with the Fisherian analogy of "the height of Peter or Paul" for "the likelihood of A or B" lies in the fact that the "or" in the former is the disjunctive "either-or", whereas the "or" in the latter ought to be understood in the same accumulative sense as we understand it in Pr(A or B). Why not ? After all Fisher wanted us to look upon likelihood as an "alternative measure of rational belief".

Regarding your comments in paragraph 11, the "uniform prior over the infinite set of all integers" was cited by me only as an example of a "monstrously all consuming" (if you do not like the phrase "all consuming", please read it as "all pervading") state of Paul's prior ignorance about the integral parameter θ that makes him allot zero (relatively, that is) prior probability to every finite set of integers. That sensible looking posterior distributions (or knowledge about θ) can often be (mathematically) derived from such a monstrous lack of prior information, is nothing but a piece of mathematical curiosity to me.

With all the best,
Yours sincerely,
Dev Basu

Brightlingsea, 18th June, 1973

Dear Dev,

Thanks for your letter and for the copies of mine. I now have them all clipped together with your paper, so if I lose one I lose the lot.

About the "or" and "and/or", I guess I should try a different approach, along lines I gave in my second talk in Aarhus. Let us agree that a simple statistical hypothesis H is one which specifies completely a probability distribution P(x : H) on a sample space S (finite, for simplicity; x is a point in S). Since x is a point, it specifies completely a possible result of the experiment to which H relates; it can therefore be called a simple event.

Now it is a property of experiments that we can always imagine them modified in such a way that the sample space S becomes S', where the points of S' correspond to the sets of a partition of S. Thus, for example, in throwing a dice,

$S = \{1, 2, 3, 4, 5, 6\}$; we can imagine ourselves incapable of counting the spots, but only capable of seeing whether there is an even or an odd number of them, in which case $S' = \{E, O\}$ corresponds to the partition $S = \{2, 4, 6\} \quad \{1, 3, 5,\}$ of S . It is reasonable to require that the hypothesis H should specify the probability distribution on S' as well as that on S. Evidently this can be done if we use the addition rule, so that, e.g. $P(E:H) = P(2:H) + P(4:H) + P(6:H)$. This is, essentially, what leads us to add probabilities.

You will find distinctions such as those I have indicated in any careful treatment of the foundations of probability. Thus, for example, Renyi (in Foundations of Probability) distinguishes between the outcome of an experiment (my simple event) and an event. An outcome is a point in the sample space, an event is a set of points. For Renyi, an experiment ξ is a non-empty set $\mathcal{X}$ of elements x called outcomes of the experiment and a σ-algebra $\mathcal{A}$ of subsets of $\mathcal{X}$ called observable events. He writes $\xi = (\mathcal{X}, \mathcal{A})$.

In Renyi's terminology, what I am saying is that given any experiment $\xi = (\mathcal{X}, \mathcal{A})$, and any sub-$\sigma$-algebra $\mathcal{A}'$ of $\mathcal{A}$, there exists an experiment $\xi' = (\mathcal{X}, \mathcal{A}')$. It is this fact that gives importance to the addition rule for probabilities, in applications to experiments.

Now given a family Φ of simple hypotheses, with $H \varepsilon \Phi$, what general logical process is there that corresponds to going from $\mathcal{A}$ to $\mathcal{A}'$? I assert that in general there is no such process, although in special cases there may be.

Specifically, given the experiment $\xi = (\mathcal{X}, \mathcal{A})$, and a family Φ of (simple) hypotheses (completely) specifying probability distributions on $\mathcal{X}$, I say that a subset of Φ is a disjunctive subset iff there exists a subalgebra $\mathcal{A}'$ of $\mathcal{A}$ such that every H in the subset assigns the same probability to every member of $\mathcal{A}'$. In the absence of a prior distribution over Φ, the disjunction of a set of hypotheses H can be considered to exist only if the set is a disjunctive set. For only then can the disjunction itself be regarded as a simple hypothesis (about the experiment $(\mathcal{X}, \mathcal{A}')$).

I fear you may find this all too muddling. I'll send you a copy of my second Aarhus paper when I have written it out. Briefly, I am pointing to the fact that the disjunction of simple events can be regarded as a simple event in another experiment; but the disjunction of simple hypotheses can not in general be regarded as a simple hypothesis, because an arbitrary set of Φ will not necessarily be disjunctive.

Incidentally, I have referred to Renyi because I have it handy; there is a similar distinction made by Kolmogoroff, though I don't remember just how he does it.

Regarding the "uniform prior", I guess we should agree to differ. All our analyses of real situations are to some extent approximations. Whether such "complete ignorance" is a useful approximation in any situation will be to some extent a matter of taste.

Yours,
George

CHAPTER VI

PARTIAL SUFFICIENCY

0. Notes

Editor's Note : This chapter is based on an invited paper presented by Basu at the 1976 annual meeting of the Institute of Mathematical Statistics held in New Haven, Connecticut. The paper was published in the Journal of Statistical Planning and Inference, 2, 1978.

Author's Note : The notion of a sufficient statistic is generally recognized as perhaps the most significant contribution of Sir Ronald to Theoretical Statistics. A great deal of mathematical literature has grown around the theme of sufficiency. However, the related concept of a partially sufficient statistic – a statistic that summarizes in itself all the relevant and usable information about a sub-parameter – can be very elusive if, following Fisher, we try to put the question in sample space terms. On the other hand, the Bayesian perspective on the question is straightforward and easy to comprehend.

Suppose (μ, σ) is the universal parameter and let σ be the parameter of interest, with x as the data, $p(x|\mu, \sigma)$ as the likelihood, and $q(\mu, \sigma) = q(\sigma)\, q(\mu|\sigma)$ as the prior. Suppose, also, that the posterior marginal distribution of σ depends on the data x only through the statistic T(x). We may then call T partially sufficient for σ, because with T in hand we can isolate all the information in the data [about the parameter σ] relative to the particular prior opinion q.

Indeed this is how Kolmogorov (1942) sought to define partial sufficiency. [Can you believe that Kolmogorov — one of the founders of the axiomatic theory of probability – has all along been a closet Bayesian?!] Kolmogorov, however, got a little carried away and called T partially sufficient for σ if the posterior (marginal) opinion on σ depends on the data x only through T and this for all possible prior opinion q on (μ, σ).

Hájek (1965) pointed out that if T is partially sufficient for σ in the sense of Kolmogorov, then T must be sufficient for the universal parameter (μ, σ). The Hájek argument is laid out in Sec. 8 of Ch. VII – there is a minor snag in the argument but the conclusion is essentially correct. In this and the following chapter we have discussed at length an alternative Hájek definition of partial sufficiency.

PARTIAL SUFFICIENCY

1. Introduction

In the beginning we have a parameter of interest — an unknown state of nature θ. With a view to gaining additional information on θ, we plan and then perform a statistical experiment $\mathcal{E}$ and thus generate the sample x. The problem of data analysis is to extract all the relevant information in the data $(\mathcal{E}, x)$ about the parameter of interest θ.

The notion of partial sufficiency arises in the context where the statistical model

$$\{(\mathcal{X}, \mathcal{A}, P_\omega) : \omega \in \Omega\}$$

of the experiment $\mathcal{E}$ involves the universal parameter ω and where $\theta = \theta(\omega)$ is a sub-parameter. In this case it is natural to ask :

Question A : What is the whole of the relevant information about θ that is available in the data $(\mathcal{E}, x)$?

It is not easy for a non-Bayesian to face up to this question. Most of us would feel more at ease when the question is rephrased in the following familiar form :

Question B : What statistic T summarizes in itself the whole of the relevant information about θ that is available in the sample x?

Let us understand that the two questions A and B, though similarly phrased, are very different in their orientations. Question A is clearly addressed to the particular data $(\mathcal{E}, x)$. But in B we are searching for a principle of data reduction. We may rephrase B in the following nearly equivalent form :

Question B* : Does there exist a statistic T such that, in some meaningful sense, there is no loss of information on θ in the reduction of the data $(\mathcal{E}, x)$ to $(\mathcal{E}_T, t)$, where $\mathcal{E}_T$ is the marginal experiment — perform $\mathcal{E}$ but record only $t = T(x)$ — corresponding to the statistic T?

Questions B or B*, when asked in the context of the universal parameter ω, led Fisher to the important notion of a sufficient statistic. But the same question, when asked in the context of a sub-parameter θ, turns out to be surprisingly resistant to a neat solution. The notion of partial sufficiency is indeed shrouded in a lot of mystery.

It is interesting to note that Sir Ronald introduced the notion of sufficiency into statistical literature (Fisher, 1920) first in the context of partial sufficiency. With a sample $x = (x_1, x_2, \ldots, x_n)$ from a normal population with unknown μ and σ, Fisher (1920) was concerned with the relative precisions of the two estimators

$$s_1 = (\tfrac{1}{2}\pi)^{1/2} \Sigma|x_i - \bar{x}|/n, \qquad s = [\Sigma(x_i - \bar{x})^2/n]^{1/2}$$

of the standard deviation σ. [Fisher had used the notations σ_1 and σ_2 for the above estimators, but we have opted for the more familiar s.] Introducing this paper in Fisher (1950), Sir Ronald described the main thrust of his 1920 argument in the following terms:

> "..., but the more general point is established that, for a given value of s, the (conditional) distribution of s_1 is independent of σ. Consequently, when s, the estimate based on the mean square, is known, a value of s_1, the estimate based on the mean deviation, gives no additional information as to the true value (of σ). It is shown that the same proposition is true if any other estimate is substituted for s_1, and consequently the whole of the relevant information respecting the variance which a sample provides is summed up in the single estimate s".

[Author's note : The proposition stated in the final sentence of the above quoted paragraph was not proved in Fisher (1920). Indeed, the proposition is not true unless we limit the discussion to location invariant statistics.]

In Fisher (1922), p. 316 we find the first mention of the now famous :

Criterion of Sufficiency : That the statistic chosen should summarize the whole of the relevant information supplied by the sample.

On the same page we find it suggested that, in the case of a sample $x_1, x_2, \ldots, x_n$ from $N(\mu, \sigma)$, the statistic s fully satisfies the criterion of sufficiency. It is thus clear that from the very beginning Sir Ronald had been grappling with the notion of partial sufficiency.

In this article we shall be examining several definitions of partial sufficiency that have been proposed from time to time. In every case we shall look back on this original problem of Fisher and ask ourselves the question : "Does this definition make s partially sufficient for σ ?"

[Author's note : The name "sufficient" is, of course, very misleading. We should never have allowed an expression like "T is sufficient for θ" to creep into any statistical text. It is less misleading to use expressions like "T is sufficient for the sample x" or "T isolates and exhausts all the information in x about θ". Perhaps we should agree to substitute the name "sufficient" by the more descriptive characterization "exhaustive", which also comes from Fisher. Having said all these, we are nevertheless going to use the expression "partially sufficient for θ" in the rest of this essay!]

2. Specific Sufficient Statistics

In Neyman and Pearson (1936) we find one of the earliest attempts at making some sense of the elusive notion of partial sufficiency. Let us suppose that the parameter of interest θ has a "variation independent" complement ϕ–that is, the universal parameter ω may be represented as $\omega = (\theta , \phi)$ with the domain of variation Ω of ω being the Cartesian product $\Theta \times \Phi$ of the respective domains of θ and ϕ. In this case, we have (from Neyman - Pearson) the following :

Definition (specific sufficiency). The statistic $T : \mathcal{X} \rightarrow \mathcal{T}$ is specific sufficient for the parameter θ if, for every fixed $\phi \varepsilon \Phi$, the statistic T is sufficient in the usual sense–that is, T is sufficient with respect to the restricted model

$$\{(\mathcal{X}, \mathcal{A}, P_{\theta,\phi}) : \theta \varepsilon \Theta , \phi \text{ fixed}\}$$

for the experiment $\mathcal{E}$.

With a sample $x = (x_1, x_2, \ldots, x_n)$ of fixed size n from $N(\mu, \sigma)$, the sample mean $\bar{x}$ is specific sufficient for μ. The sample standard deviation s is, however, not specific sufficient for σ. Even though $\bar{x}$ is specific sufficient for μ, in no meaningful sense of the terms can we suggest that $\bar{x}$ exhaustively isolates all the relevant information in the sample x about the parameter μ. Surely, we also need to know s in order to be able to speculate about, say, the precision of $\bar{x}$ as an estimate of μ. Clearly, we are looking for something more than specific sufficiency.

The fact of T being specific sufficient for θ may be characterized in terms of the following factorization of the frequency (or density) function p on the sample space $\mathcal{X}$:

$$p(x|\theta,\phi) = G(T(x), \theta,\phi)\, H(x,\phi).$$

Alternatively, we may characterize the specific sufficiency of T (for θ) by saying that the conditional distribution of any other statistic T_1, given T and (θ, ϕ), depends on (θ,ϕ) only through ϕ.

Before going on to other notions of partial sufficiency, it will be useful to state the following :

Definition (θ-oriented statistics). The statistic $T : \mathcal{X} \to \mathcal{T}$ is θ-oriented if the marginal (or sampling) distribution of T — that is, the measure $P_\omega T^{-1}$ on $\mathcal{T}$ —depends on ω only through $\theta = \theta(\omega)$. In other words, $\theta(\omega_1) = \theta(\omega_2)$ implies

$$P_{\omega_1}(T^{-1} B) = P_{\omega_2}(T^{-1}B)$$

for all 'measurable' sets $B \subset \mathcal{T}$.

It should be noted that the notion of θ-orientedness does not rest on the existence of a variation independent complementary parameter ϕ. In our basic example of a sample from $N(\mu, \sigma)$, observe that $\bar{x}$ is not μ-oriented but that s is σ-oriented.

3. Partial Sufficiency

If we put together the two definitions of the previous section, then we have the following definition of partial sufficiency that is usually attributed to Fraser (1956).

Definition. The statistic T is partially sufficient for θ if it is specific sufficient for θ and is also θ-oriented.

See Basu (1977) for a number of examples of partially sufficient statistics. In the example of a sample $(x_1, x_2, \ldots, x_n)$ from $N(\mu, \sigma)$, the statistic $\bar{x}$ is not partially sufficient for μ as it is not μ-oriented and the statistic s is not partially sufficient for σ as it is not specific sufficient for σ. In view of the specific sufficiency part of the above definition, it is necessary that the parameter θ has a variation independent complement ϕ. The requirement of θ-orientedness brings in the unpleasant consequence that T may be partially sufficient for θ but a wider statistic T_1 need not be. Indeed, the whole sample x is never partially sufficient for θ.

The notion of partial sufficiency may be characterized in terms of the following factorization criterion :

$$p(x|\theta,\phi) = g(T|\theta)\, h(x|T,\phi)$$

where g and h denote respectively the marginal probability function of T and the conditional probability function of x given T. Note that the marginal distribution is θ-oriented and the conditional distribution is ϕ-oriented.

The interest in the Fraser definition of partial sufficiency stems from the following generalization (Fraser, 1956) of the Rao-Blackwell argument. Let $a(\theta)$ be an arbitrary real valued function of θ and let $W(y,\theta)$ denote the loss sustained when $a(\theta)$ is estimated by y. Let us suppose that, for each $\theta \varepsilon \Theta$, the loss function $W(y,\theta)$ is convex in y. Finally, let $\mathcal{U}$ be the class of all estimators U such that the risk function

$$r_U(\theta) = r_U(\theta,\phi) = E[W(U,\theta)|\theta,\phi]$$

is finite and depends on (θ,ϕ) only through θ.

Theorem (Fraser). If T is partially sufficient for θ, then for any $U \varepsilon \mathcal{U}$ there exists an estimator $U_0 = U_0(T) \varepsilon \mathcal{U}$ such that $r_{U_0}(\theta) \leqslant r_U(\theta)$ for all θ Θ.

The proof of the theorem consists of choosing and fixing a particular value ϕ_0 of ϕ and then considering the statistic $U_0 = U_0(T) = E(U|T, \theta, \phi_0)$ as an estimator of $a(\theta)$. That U_0 does not involve the parameter θ follows from the supposition that T is sufficient for θ when ϕ is fixed at ϕ_0. That $U_0 \varepsilon \mathcal{U}$ follows from the supposition that T is θ-oriented. The rest follows at once from Jensen's inequality.

The above theorem may be generalized along the lines suggested by Hájek (1967). Let $\mathcal{U}'$ be the class of all estimators U for which the risk function $r_U(\theta,\phi)$ is finite (but not necessarily free of ϕ). Let $R_U(\theta) = \sup_\phi r_U(\theta,\phi)$ be the maximum risk associated with U for a particular θ.

Theorem (Hájek). If T is partially sufficient for θ, then for any $U \varepsilon \mathcal{U}'$ there exists an $U_0 = U_0(T)$ such that $R_{U_0}(\theta) \leqslant R_U(\theta)$ for all θ.

The definition of U_0 is the same as in the previous theorem. The rest of the proof follows from the following chain of relations

$$R_U(\theta) \geqslant r_U(\theta,\phi_0) \geqslant r_{U_0}(\theta,\phi_0) = R_{U_0}(\theta).$$

If $U \varepsilon \mathcal{U}$, that is, if the risk function for U is free of ϕ, then $r_U(\theta) \equiv R_U(\theta)$ and so the above theorem is a generalization of the Fraser theorem.

Let us take note of the fact that the proofs of the previous two theorems rest heavily on the supposition that T is θ-oriented but make very little use of the supposition that T is specific sufficient for θ. What is needed is the sufficiency of T (for θ) for just one specified value ϕ_0 of ϕ. Consider the following example.

Example. Let $x = (x_1, x_2, \ldots, x_m; y_1, y_2, \ldots, y_n)$ be m+n independent normal variables with unit variances and with $E(x_i) = \theta$ $(i = 1,2, \ldots, m)$ and $E(y_j) = \theta\phi$ $(j = 1,2, \ldots, n)$, where $\theta \varepsilon [a,b]$ is the parameter of interest and $\phi \varepsilon \{0,1\}$ is the nuisance parameter. The likelihood function factors as

$$p(x|\theta,\phi) = A(x)\exp\left[-\tfrac{1}{2}m(\bar{x}-\theta)^2\right]\exp\left[-\tfrac{1}{2}n(\bar{y}-\theta\phi)^2\right].$$

The pair $(\bar{x}, \bar{y})$ constitute the minimal sufficient statistic. The statistic $\bar{x}$ is θ-oriented and is sufficient for θ when $\phi = 0$. Therefore, we can invoke either the Fraser or the

Hájek complete class theorem and suggest a reduction of the data x to the statistic $\bar{x}$. However, such a data reduction will clearly result in a substantial loss of information in the event $\phi = 1$. Looking at the full data we should usually be able to make a good guess of the true value of ϕ. For instance, if $m = 2$, $n = 200$, $\bar{x} = 16.02$ and $\bar{y} = 17.45$ then we know for (almost) sure that $\phi = 1$ and should naturally rebel against the idea of reducing the data to $\bar{x}$.

This example highlights the inherent weakness of the Fraser-Hájek argument. Fraser limited his discussion to the class $\mathcal{U}$ of estimators U whose risk functions involve only θ. It is not at all clear why we have to limit our universe of discourse to such a limited class. [It is true that the statistical literature is so full of Fraser-type limited complete class theorems. Familiar examples of such theorems abound in the theories of best unbiased estimates, best similar region tests, best invariant procedures, etc.] In this example, the class of estimators of θ that are functions of $(\bar{x}, \bar{y})$ is complete in the class $\mathcal{U}'$ of all estimators, provided the loss function is convex. But the only functions of $(\bar{x}, \bar{y})$ that belong to $\mathcal{U}$ are those that do not involve $\bar{y}$. Thus, Fraser's requirement that we limit the discussion to $\mathcal{U}$ sort of forces $\bar{y}$ out of the picture even though it contains a lot of information on θ.

Hájek considered the wider class $\mathcal{U}'$ but eliminated the nuisance parameter from the argument by redefining the risk function as

$$R_U(\theta) = \sup_{\phi} r_U(\theta, \phi).$$

This method of eliminating the nuisance parameter from the risk function has been made popular by Lehmann (1959) in his famous text on tests of statistical hypotheses. A generalized version of the Minimax Principle is being invoked in this elimination argument. The author is not at all clear in his mind about the statistical content of this generalized principle. The example above is clearly in conflict with the principle.

4. H-sufficiency

Hájek (1967) pushed Fraser's notion of partial sufficiency to its natural boundary in the following manner. For each $\theta \in \Theta$ let $\Omega_\theta = \{\omega : \theta(\omega) = \theta\}$ and let $\bar{\mathcal{P}}_\theta$ be the convex hull of the family $\mathcal{P}_\theta = \{P_\omega : \omega \in \Omega_\theta\}$ of the probability measures on the sample space $\mathcal{X}$. The class $\bar{\mathcal{P}}_\theta$ is the class of all probability measures Q_θ on $\mathcal{X}$ that has the representation

$$Q_\theta(A) = \int_{\Omega_\theta} P_\omega(A) d\zeta_\theta(\omega)$$

for all measurable sets A, where ζ_θ is some 'mixing' probability measure on Ω_θ.

[Note that we are riding slipshod over the usual measurability requirements.]

Definition (H-Sufficiency). The statistic T is H-sufficient (partially sufficient in the sense of Hájek) for θ, if, for each $\theta \varepsilon \Theta$, there exists a choice of a $Q_\theta \varepsilon \bar{\mathcal{P}}_\theta$ such that

(i) T is sufficient with respect to the model $\{(\mathcal{X}, \mathcal{A}, Q_\theta) : \theta \varepsilon \Theta\}$ and

(ii) T is θ-oriented in the model $\{(\mathcal{X}, \mathcal{A}, P_\omega) : \omega \varepsilon \Omega\}$.

Observe that the notion of H-sufficiency (unlike the Fraser definition of partial sufficiency) does not require θ to have a variation independent complement ϕ. If T is partially sufficient in the sense of Fraser, then it is also H-sufficient. In order to see this, we have only to choose and fix $\phi_o \varepsilon \Phi$ and then take $Q_\theta = P_{\theta,\phi_o}$ which is a mixture probability corresponding to a degenerate mixing measure.

Also observe that the requirement of θ-orientedness in the definition of H-sufficiency has the same unfortunate consequence (as in the case of Fraser's definition) that T may be H-sufficient but a wider statistic (e.g., the whole sample x) need not be so. Hájek (1967) sought to remedy this fault in his definition by putting in the additional clause (almost as an afterthought) that any statistic T_1 wider than an H-sufficient T should be regarded as H-sufficient. But such a wide definition of partial sufficiency cannot be admitted when we are concerned with the problem of isolating the whole of the relevant information about a sub-parameter.

The two theorems of the previous section may now be consolidated in the following complete class theorem. [For a proof refer to p. 361 of Basu (1977).]

Theorem (Hájek). If T is H-sufficient for θ, then, for any $U \varepsilon \mathcal{U}$, there exists a $U_o = U_o(T)$ such that $r_{U_o}(\theta) \leq r_U(\theta)$ for all θ. Furthermore for any $U \varepsilon \mathcal{U}'$ it is true that $R_{U_o}(\theta) \leq R_U(\theta)$ for all θ.

Let us look back on the classical problem of a sample $x = (x_1, x_2, \ldots, x_n)$ of fixed size n from $N(\mu,\sigma)$. No statistic T can be H-sufficient for μ. This is because T can be μ-oriented only if it is an ancillary statistic, in which case it cannot, of course, be partially sufficient for μ. [This remark holds true for a general location-scale parameter set-up with μ as the location parameter.] On the other hand the statistic s is σ-oriented. Let us examine whether s is H-sufficient for σ.

The density (or likelihood) function factors as

$$p(x|\mu,\sigma) = A(\sigma)\exp\left[-\frac{ns^2}{2\sigma^2}\right]\exp\left[-\frac{n(\bar{x}-\mu)^2}{2\sigma^2}\right]$$

where $A(\sigma) = (\sqrt{2\pi}\,\sigma)^{-n}$.

For each $\sigma \varepsilon (0, \infty)$, let ζ_σ be our choice of the mixing measure on the range space R_1 of the nuisance parameter μ. The corresponding family $\{Q_\sigma : 0 < \sigma < \infty\}$ of mixture measures on the sample space R_n will have the density function

$$\bar{p}(x|\sigma) = \int_{-\infty}^{\infty} p(x|\mu, \sigma) d\zeta_\sigma(\mu)$$

$$= A(\sigma) \exp\left[-\frac{ns^2}{2\sigma^2}\right] \int_{-\infty}^{\infty} \exp\left[-\frac{n(\bar{x}-\mu)^2}{2\sigma^2}\right] d\zeta_\sigma(\mu).$$

We shall recognize s as H-sufficient for σ provided we can find a family $\{\zeta_\sigma\}$ of mixing measures such that

$$\int_{-\infty}^{\infty} \exp\left[-\frac{n(\bar{x}-\mu)^2}{2\sigma^2}\right] d\zeta_\sigma(\mu) = B(\bar{x})C(\sigma) \tag{1}$$

because in that case $\bar{p}(x|\sigma)$ will factor as

$$\bar{p}(x|\sigma) = A(\sigma)\exp\left[-\frac{ns^2}{2\sigma^2}\right] B(\bar{x})C(\sigma)$$

establishing condition (i) of the definition of H-sufficiency.

One way to ensure (1) is to choose for ζ_σ the uniform distribution over the whole of R_1. But, with a family $\{Q_\sigma\}$ of improper mixtures, the proof of the Hájek theorem will break down. If the range of σ is the whole of the positive half line, then it can be shown that the factorization (1) can be achieved with no proper mixing. However, if we are willing to set a finite upper bound K for the parameter σ – from a practical point of view this is hardly a restriction – then it is easy to check that the choice of ζ_σ as the normal distribution with mean zero and variance $(K^2 - \sigma^2)/n$ will achieve the desired factorization (1). The above argument of Hájek (1967) establishing the H-sufficiency of s for $\sigma (0 < \sigma < K)$ is very intriguing. At this point we like to contrast the approaches of Fisher and Hájek to the question of partial sufficiency of s for σ. First, let us look at the question from the :

Fisher Angle : The pair $(\bar{x}, s)$, being jointly sufficient for (μ, σ), contains the whole of the available information on the parameter of interest σ. Furthermore, the two statistics $\bar{x}$ and s, being stochastically independent, yield independent (additive, that is) bits of information on σ. If μ were known, then we have n 'degrees of freedom' worth of information on σ. Of these, the statistic s summarizes in itself n-1 'degrees of freedom' worth of information on σ. If the only (prior) information about μ that we have is $-\infty < \mu < \infty$, then there is no way that we can recover any part of the (at most one 'degree of freedom' worth of) information contained in $\bar{x}$ about μ. It is in this situation of no (prior) information on μ that Fisher would label s as exhaustive of all available and usable information on σ. And in the event of no (prior) information on σ either (other than $0 < \sigma < \infty$) Fisher would invoke his celebrated fiducial argument to declare that the status of the parameter σ has been altered from that of an unknown constant to that of a random variable with (fiducial) probability distribution $\sqrt{n}s/\chi_{n-1}$. Observe that the fiducial distribution of σ depends on the sample only through the statistic s.

A sort of improper Bayesian justification for the Fisher intuition on the problem at hand can be given by suggesting that, for every prior $q(\mu,\sigma)$ for the parameter (μ,σ) that is of the form

$$q(\mu,\sigma)\, d\mu\, d\sigma = g(\sigma) d\mu d\sigma$$

[μ and σ are independent a-priori and μ has the (improper) uniform distribution over the whole real line], the posterior marginal distribution of σ depends on the sample x only through the statistic s. Furthermore, the fiducial distribution of σ corresponds to the case where $g(\sigma) = 1/\sigma (0 < \sigma < \infty)$. Although Fisher never put his arguments in the above straightforward Bayesian framework, the fact remains that Fisher's thinking on the problem of inference had a distinct Bayesian orientation.

Hájek Angle : On the surface, Hájek's partial sufficiency argument carries a distinct Bayesian flavour. His mixing measure ζ_σ —normal with zero mean and $(K^2 - \sigma^2)/n$ as variance — for μ may be interpreted as the prior conditional distribution of μ given σ. With any prior $q(\mu,\sigma)$ of the form

$$q(\mu,\sigma)\; d\mu\, d\sigma \;= [d\zeta_\sigma(\mu)]\; g(\sigma)\, d\sigma$$

the posterior marginal distribution of σ will depend on the sample x only through the statistic s. It will, however, be very hard to make any Bayesian interested in a prior $q(\mu,\sigma)$ of the above form. Apart from the fact that q depends on the sample size (which it should not), it is not possible to make any sense of q as a measure of prior belief pattern. The main thrust of the Hájek argument is, however, not Bayesian at all. He was using the Bayesian device (of averaging over the parameter space) only as a mathematical artifact to prove a complete class theorem in the fashion of Abraham Wald.

We have already pinpointed the flaw in Hájek's definition of partial sufficiency through our example of the previous section. In that example, $\bar{x}$ is H-sufficient for θ even though marginalization to $\bar{x}$ will entail a substantial loss of information on θ in the case of the (easily discernable) event $\phi = 1$.

5. Invariantly Sufficient Statistics

In this section we briefly review George Barnard's thoughts on the knotty question of partial sufficiency of s for σ. The following quotation is from p. 113 of Barnard (1963).

> "The definition of sufficiency which has become universally accepted required that the distribution of any function of the observations, conditional on a fixed value of the sufficient statistic, should be independent of the parameter in question, and there is no doubt that with this definition, s fails to be sufficient for σ. However, as was usual for him, Fisher's definition of sufficiency was designed to embody a logical notion, that of providing the whole of the available relevant information for a given parameter and the definition just referred to does not altogether succeed in this object.
>
> The availability or otherwise of information is critically dependent on knowledge or lack of knowledge. Obviously if σ is already known, s provides us with no information whatsoever. The failure of s to satisfy the definition given above for sufficiency arises from the fact that the distribution of $\bar{x} - \mu$ (with the usual notations) depends also on σ. However, ... μ is given as unknown, and so the information in $\bar{x} - \mu$ is unavailable.
>
> As already remarked, Fisher was very much concerned, up to the end of his life, with the difficulty of expressing in precise mathematical form, the notions corresponding to 'known' and 'unknown'. The present writer several times suggested to him, in connection with parameters such as μ in the case of the normal distribution, ..., that these parameters correspond to groups under which the problems considered are invariant, and the notion of ignorance of μ can be represented in terms of group invariance properties".

Barnard's thoughts on the problem are best understood in the context of the simple example of a sample $x = (x_1, x_2, \ldots, x_n)$ of fixed size n from $N(\mu, \sigma)$. The group $G = \{ g_a : a \in R_1 \}$ of transformations

$$g_a(x_1, x_2, \ldots, x_n) = (x_1 + a, x_2 + a, \ldots, x_n + a)$$

of the sample space R_n onto itself is associated with the group $\bar{G} = \{ \bar{g}_a : a \in R_1 \}$ of transformations

$$\bar{g}_a(\mu, \sigma) = (\mu + a, \sigma)$$

of the parameter space onto itself. The group $\bar{G}$ leaves the parameter of interest σ invariant but acts transitively on (traces a single orbit on the domain of) the nuisance parameter μ.

The problem of estimating the parameter σ is invariant with respect to the group G of transformations $g_a : \mathcal{X} \to \mathcal{X}$. The maximal invariant is the difference statistic

$$D = (x_2 - x_1, x_3 - x_1, \ldots, x_n - x_1).$$

The statistic s is invariantly sufficient for σ in the sense that

(i) s is a function of D and is, therefore, σ-oriented, and

(ii) the conditional distribution of any other invariant statistics $s_1 = s_1(D)$, given s, is the same for all possible values of σ (and, of course, of μ as well).

[The notion of invariantly sufficient statistic is due to Charles Stein. See Hall, Wijsman and Ghosh (1965), and Basu (1969a)for further discussion on the subject.]

We are now ready for the following

Question. What is the logical necessity for restricting our attention to only G-invariant estimators of σ ?

The standard argument for restricting attention to only such T that satisfies the identity

$$T(x_1 + a, x_2 + a, \ldots, x_n + a) = T(x_1, x_2, \ldots, x_n)$$

for all samples $x \in R_n$ and all $a \in R_1$ — that is, to measurable functions of the maximal invariant $D = (x_2 - x_1, x_3 - x_1, \ldots, x_n - x_1)$ — runs along the following lines :

Argument. The sample $(x_1, x_2, \ldots, x_n)$ consist of n i.i.d. $N(\mu, \sigma)$'s with $\mu(-\infty<\mu<\infty)$ 'unknown' and with σ as the parameter of interest. If we shift the origin of measurement to -a, then the sample will take on the new look $(x_1 + a, x_2 + a, \ldots, x_n+a)$. The new model for the new-look sample will then correspond to n i.i.d. $N(\mu+a,\sigma)$'s.

Note that the new mean $\mu + a$ is 'equally unknown' as μ and that σ remains unaltered. The problem of estimating σ (with μ unknown), therefore, remains invariant with any shift in the origin of measurement. Now, an estimator T is a formula for arriving at an estimate $T(x_1, x_2, \ldots, x_n)$ based on the sample $x = (x_1, x_2, \ldots, x_n)$.

With the same sample represented differently as $(x_1 + a, x_2 + a, \ldots, x_n + a)$, but with the problem (of estimating σ) unaltered, the same formula T will yield the estimate $T(x_1 + a, x_2 + a, \ldots, x_n + a)$. Clearly, the formula T will look rather ridiculous if $T(x_1 + a, x_2 + a, \ldots, x_n + a)$ is not equal to $T(x_1, x_2, \ldots, x_n)$ for some x and a .

The above invariance argument of Pitman-Stein-Lehmann has been sold in many different packages to a vast community of statisticians. However, a close look at the present package will immediately reveal the fact that the argument does not really add up to anything that is logically compelling.

For one thing, the part of the argument that asserts that the problem remains invariant with any shift of the origin of measurement is questionable. The argument rests heavily on the supposition that $\mu + a$ is 'equally unknown' as μ. Only an improper Bayesian with uniform prior (over the whole real line) for μ can make a case for such a statement.

Secondly, implicit in the argument lies the supposition that the choice of the estimator (estimating formula) T as a function on the sample space may depend on the statistical model (which, in this case, does not change with any shift in the origin of measurement) and the kind of 'average performance characteristics' that we find satisfactory but must not (repeat not) depend on any pre-conceived notions that we may have on the parameters in the model. This, of course, is not a tenable supposition (as all Bayesians will readily agree).

Let T_q be a typical Bayes estimator of σ that corresponds to the prior distribution q for (μ,σ) – for the sake of this argument let us imagine $T_q(x)$ to be the posterior mean of σ for a given sample x and the prior q. In T_q we thus have a well-defined formula for estimating σ. Every such formula T_q is invariant for every shift in the origin of measurement. This is because when the origin is shifted to - a, the sample $(x_1, x_2, \ldots, x_n)$ shifts to $(x_1 + a, x_2 + a, \ldots, x_n + a)$, the parameters $(\mu, \sigma$ move to $(\mu + a, \sigma)$ and the prior q changes itself to the corresponding prior q_a for $(\mu + a, \sigma)$. It is easy to see then that

$$T_q(x_1, x_2, \ldots, x_n) = T_{q_a}(x_1 + a, x_2 + a, \ldots, x_n + a)$$

for all q, x and a. Thus, no Bayes rule violates the essence of the invariance argument.

However, if for a fixed q we look upon $T_q(x)$ as a function on the sample space, we shall then find that the function typically depends on x through both $\bar{x}$ and s. [As we have noted in the previous section, for all (improper) priors q of the form $q(\mu,\sigma)\,d\mu d\sigma = g(\sigma)\,d\mu\, d\sigma$ and also for some curious looking proper priors of the Hájek kind, the posterior marginal distribution of σ will depend on x only through s and so with such a choice of the prior q, the Bayes estimator $T_q(x)$ for σ will be G-invariant as a function on the sample space.]

There is no logical necessity for restricting our attention to only G-invariant estimators as long as we take care to avoid using estimating procedures that do not recognize the arbitrariness that is inherent in the choice of the origin of measurement, etc. As we just noted, all Bayes estimation procedures are invariant in this sense.

6. Final Remarks

Sir Ronald was deeply concerned with the notion of information (about a parameter) in the data, but never directly faced up to such basic questions as : What is information? How informative is this data? Have we obtained enough information on the parameter of interest? etc.

The mathematical definition of information that we got from Fisher is a most curious one. The definition does not relate to the concept of information in the data but is supposed to bring out the notion of information in (the statistical model of) an experiment and the associated family of marginal experiments. Even then, the Fisher information $I(\omega)$ can hardly be interpreted in terms of the average (or expected) amount of knowledge gained (or uncertainties removed) about the universal parameter ω when the experiment is performed. And we get no prescription from Sir Ronald about how to 'marginalize' his information function (or matrix) to a sub-parameter. We must reject the notion of Fisher information on the ground of irrelevance in the present context.

The Fisher criterion of sufficiency — that the statistic chosen should summarize the whole of the relevant information supplied by the data — should be looked upon only as a principle of data reduction relative to a particular statistical model of the experiment. The earliest thoughts of Fisher on the subject of sufficiency crystalized around the following two propositions that are stated here relative to a fixed experiment $\mathcal{E}$ that is already endowed with an assumed statistical model.

Proposition 1. To reduce (or marginalize) the data x to the statistic $T = T(x)$ will entail a total loss of all available information on the (universal) parameter ω if the marginal distribution of T is the same for all possible values of ω. Any such statistic T may be regarded as 'marginally uninformative' about ω.

Proposition 2. To reduce the data x to the statistic T will entail no loss of available information on ω if the conditional distribution of every other statistic T_1 given T is the same for all possible values of ω. Such a statistic T may be called sufficient, fully informative, or exhaustive of all available information on ω.

It is not remarkable that we now have the notions of 'no information' and 'full information' (meaning, exhaustive of all available information) without ever mentioning what we mean by information?! If by information we mean the state of our knowledge about the parameter ω, then should we not speculate about it in terms of the parameter space Ω rather than in terms of the sample space $\mathcal{X}$?!

It so happens that Fisher's 'sample space' definition of sufficient (information-full, that is) statistic agrees with the following Bayesian definition of sufficiency due to A.N. Kolmogorov (1942):

Definition. The statistic T is sufficient if, for every prior q(.) on Ω, the posterior distribution $q(\,.|x)$ on Ω depends on x only through T(x).

It is to the lasting credit of Sir Ronald that, having discovered the 'sample space' definition of sufficiency, he was able to put the notion in the correct perspective by characterizing a sufficient statistic as that characteristic of the sample knowing which we can determine the likelihood function up to a multiplicative factor. Fisher recognized that, relative to a given model, the whole of the relevant information in the data is summarized in the corresponding likelihood function. This is only a short step away from the Bayesian insight on the knowledge business.

The 'sample space' definition of sufficiency for the universal parameter ω is all right. But the weakness and inadequacy of this approach becomes apparent when we try the sample space way to 'isolate' all the 'available' relevant information on a sub-parameter. Note that we now have to deal with the new term 'isolate' and that the term 'available' suddenly springs to life with a new meaning. Fraser, Hájek and Barnard all seem to have tacitly assumed that T can isolate information on θ only if it is θ-oriented. This sample space requirement of θ-orientedness for the partially sufficient T has been a major source of our trouble with the notion of partial sufficiency. The statistical insight that leads to θ-orientedness as a prime requirement for partial sufficiency, cannot be reconciled with any Bayesian insight on the subject. What if there are no non-trivial θ-oriented statistic? Can't we then isolate the information on θ ? What is information on θ? How can we isolate something that we have not even cared to define?

Barnard (1963) said "... , the notion of ignorance on μ can be represented in terms of group invariance properties." What is ignorance? Lack of prior information? How can we talk about lack of information when we have not even attempted to define what we mean by information? In any case, how can we possibly characterize ignorance on μ in terms of group invariance properties of the model? Who is ignorant? The scientist or the model?!

In September 1967 the author had asked the late Professor Renyi the question :"Why are you a Bayesian ?" Promptly came back the answer : "Because I am interested in the notion of information. I can make sense of the notion in no other way".

CHAPTER VII

ELIMINATION OF NUISANCE PARAMETERS

0. Notes

Editor's Note : The material covered in this chapter was first presented at a Symposium held at the Carleton University, Ottawa, Canada in October, 1974 and published in Proc. Symp. on Statistics and Related Topics, Md. E. Saleh (ed.), Carleton Math. Lecture Notes No. 52, 1975. The article was republished in a revised form in Jl. Am. Statist. Asssoc., 72, 1977, 355-366.

Author's Note : Eliminating the nuisance parameters from a model is a major problem of non-Bayesian statistical theories. A surprisingly large number of research articles were published on the subject. In this chapter we discuss in depth some particular aspects of the marginalizing and the conditioning methods of nuisance parameter elimination. The methods are then evaluated from a Bayesian viewpoint.

This article was supposed to be the first one of a two-part essay. In the second part I proposed to take a hard look at the Fisher phrase "The statistic Y contains no relevant or usable bit of information about the parameter of interest in the absence of any prior knowledge on the nuisance parameter". For instance, in his 2×2 test of independence [see discussion in Ch. XVIII] Fisher conditioned the cell frequencies by holding the marginals fixed on the score that the marginals possess no usable bit of information, etc. Cox (1958) tried to give a mathematical definition of the Fisher concept of no-information, Sprott (I forget the exact reference) defined the term in a very different fashion, Barndorff-Nielsen (1973) introduced the extremely curious notion of M-ancillarity and later coined the phrase "nonformation" to denote his interpretation of what Sir Ronald really had in mind. There was a time when I used to spend a lot of time reading these (and other) obscure writings and constructing little counter-examples to knock down many a red herring. But then I suddenly realized what an exercise in futility it would be to seek for a meaning of "nonformation" (or "allformation") about a parameter in sample space terms. The original mistake was Sir Ronald's.

It was a mistake to rephrase the question "What is all the information in the data?" by the question "What statistic (a function on the sample space) carries all the information in the sample?" Fisher seems to have recognized that the (standardized) likelihood function is the minimum sufficient statistic. But we do not have an unequivocal statement from him saying that the particular likelihood function generated by the data (the unique one) is the information.

ELIMINATION OF NUISANCE PARAMETERS

1. The Elimination Problem and Methods

The problem begins with an unknown state of nature represented by the parameter of interest θ . We have some information about θ to begin with – e.g., we know that θ is a member of some well-defined parameter space Θ– but we are seeking more. Toward this end, a statistical experiment $\mathcal{E}$ is planned and performed and this generates the sample observation x. Further information about θ is then obtained by a careful analysis of the data ($\mathcal{E}$, x) in the light of all our prior information about θ and in the context of the particular inference problem related to θ . For going through the rituals of the traditional sample-space analysis of data, we must begin with the invocation of a trinity of abstractions ($\mathcal{X}$, $\mathcal{A}$, $\mathcal{P}$), where $\mathcal{X}$ is the sample space, $\mathcal{A}$ is a σ- algebra of events (subsets of $\mathcal{X}$), and $\mathcal{P}$ is a family of probability measures on $\mathcal{A}$. If the model ($\mathcal{X}$, $\mathcal{A}$, $\mathcal{P}$) is such that we can represent the family $\mathcal{P}$ as $\{P_\theta : \theta \in \Theta\}$, where the correspondence $\theta \to P_\theta$ is one-one and (preferably) smooth, then we go about analyzing the data according to our own light and are thankful for not having to contend with any nuisance parameters.

However, instances of statistical models with $\mathcal{P}$ indexed by θ alone are very rare. Typically, we have to work with a family $\mathcal{P}$ that is indexed as

$$\mathcal{P} = \{P_{\theta,\phi} : \theta \in \Theta \ , \ \phi \in \Phi\} \ ,$$

where ϕ is an additional unknown parameter. If the inference problem at hand relates only to θ and if information gained on ϕ is of no direct relevance to the problem, then we classify ϕ as the nuisance parameter.

The big question in statistics is : How can we eliminate the nuisance parameter from the argument? During the past seven decades an astonishingly large amount of effort and ingenuity has gone into the search for reasonable answers to this question. Broadly speaking, this collective endeavor of the community of statisticians may be classified into the following overlapping categories:

1. To plan the experiment $\mathcal{E}$ in such a fashion that the model is related to the parameter of interest and is relatively free of disturbing nuisance parameters. In this article we are not concerned with the important problems of planning experiments. Our concern is with the problem of data analysis. However, a few elimination methods, such as randomization and sequential sampling, may well be classified under this heading.

2. To justify a replacement of the basic model ($\mathcal{X}$, $\mathcal{A}$, $\mathcal{P}$) by a related θ-oriented model ($\mathcal{T}$, $\mathcal{B}$, $\mathcal{Q}$), the family $\mathcal{Q}$ is indexed by θ alone. The marginalization and the conditionality arguments that we shall examine in this article belong to this category.

3. To estimate the nuisance parameter away; that is, to substitute the unknown nuisance parameter ϕ by an estimated value $\hat{\phi}$. This classical method of elimination is used repeatedly in the large sample theory of statistics.

4. To Studentize in the manner of W.S. Gossett with the idea in mind to construct a reasonable looking pivotal quantity involving the sample x and the parameter of interest θ.

5. To invoke the invariance argument of Pitman-Stein-Lehmann and marginalize the data to a suitable invariant statistic T whose sampling distribution is free of the nuisance parameter.

5. To delimit the argument to a small class of decision procedures, e.g., unbiased estimators, fixed size confidence intervals, similar tests, etc., whose average performance characteristics are, at least in part, free of the nuisance parameter. Mathematicians love this argument. See, e.g., Linnik (1965, 1968).

7. To eliminate the nuisance parameter from the risk function $r_\delta(\theta , \phi)$ of the decision procedure δ by the invocation of a so-called maximization (or minimax) principle. The recommendation for the choice of δ is then made on the basis of the eliminated risk function

$$R_\delta(\theta) = \sup_\phi r_\delta(\theta , \phi).$$

In Lehmann (1959) we find this argument used quite frequently. For example, the size of a test is always understood as the maximum probability of committing an error of the first kind.

8. To invoke the fiducial argument of R.A. Fisher. With the departure of Sir Ronald from our midst, we seem to have lost our zest for this novel elimination argument.

9. To justify an elimination of the nuisance parameter directly from the likelihood function $L(\theta, \phi | x)$ generated by the particular data ($\mathcal{E}$, x). The idea is to construct a new scale $L_e(\theta, x)$ (the suffix e denotes

the process of elimination of the nuisance parameter) for a direct comparison of the amount of support that the data lends to various values of θ. The maximization of likelihood with respect to ϕ is the classic example of this kind of elimination.

10. To act like a Bayesian; that is, to fix a prior, compute the posterior, integrate out the nuisance parameter from the posterior to arrive at the posterior marginal distribution of the parameter of interest, and then to let the statistical argument rest on the posterior marginal distribution.

In addition, we have the choice of a fairly large number of specialized elimination methods : the two-stage sampling plan of Stein (1945), the randomization method of Durbin (1961), the characterization argument of Prohorov (1967), the partial sufficiency argument of Hájek (1967), the M-ancillarity argument of Barndorff-Nielsen (1973), etc.

After this introduction to the problem and methods of elimination, we plunge headlong into the depths of the marginalization and the conditionality arguments and try to sort out a number of ideas related to partial sufficiency and partial ancillarity.

2. Marginalization and Conditioning

The marginalization method of elimination consists of : Choosing a suitable statistic $T : (\mathcal{X}, \mathcal{A}) \to (\mathcal{T}, \mathcal{B})$, such that the family

$$\mathcal{P}_T = \{P_{\theta,\phi} T^{-1} : \theta \in \Theta, \phi \in \Phi\}$$

of probability measures on $(\mathcal{T}, \mathcal{B})$ is θ-oriented, i.e., the family $\mathcal{P}_T$ is indexed by θ alone, and then recommending that the model $(\mathcal{X}, \mathcal{A}, \mathcal{P})$ be given up in favour of the model $(\mathcal{T}, \mathcal{B}, \mathcal{P}_T)$.

In effect, the method replaces the data $(\mathcal{E}, x)$ by its reduction $(\mathcal{E}_T, t)$, where $T(x) = t$. By $\mathcal{E}_T$ we mean the marginal experiment that may be operationally defined as "perform $\mathcal{E}$ but record only $T(x)$." It is not easy to justify data reduction of the above kind. A great deal of thought and mathematical expertise have gone into the many efforts made so far at such justification. Two distinct major lines of thought in this general direction are : (a) the invariance argument and (b) the partial sufficiency argument. In this article, we shall be concerned with the partial sufficiency argument only.

The conditioning method of elimination consists of : Choosing a suitable statistic $Y : (\mathcal{X}, \mathcal{A}) \rightarrow (\mathcal{Y}, \mathcal{C})$ such that the conditional distribution of the sample x, given Y = y, is θ-oriented (it depends on (θ, ϕ) only through θ) for all $y \in \mathcal{Y}$; and recommending that the data $(\mathcal{E}, x)$ be analyzed by looking at the sample x, not as a random variable with the unconditional distribution model $(\mathcal{X}, \mathcal{A}, \mathcal{P})$ but as a random variable with the θ-oriented conditional distribution model that corresponds to the condition Y = y, where y is the observed value of the statistic Y. In effect, the method aims at replacing the data $(\mathcal{E}, x)$ by the conditioned data $(\mathcal{E}_y^Y, x)$ where $\mathcal{E}_y^Y$ is a conceptual conditional experiment that corresponds to the observed value y of a suitable statistic Y.

For the marginalization argument, the statistic T not only needs to be θ-oriented but also has to be one that, in some sense, summarizes in itself all the relevant and usable information about θ that is contained in the data. Similarly, for the conditionality argument, it is not enough to choose just any statistic Y that will do the elimination job. The statistic Y needs to be such that, in some meaningful sense, we can assert that referring the observed sample x to the reference set of all possible samples x' with Y(x') fixed at the present observed value y = Y(x) entails no loss of information on the parameter of interest θ. The statistical literature is strewn with logicians' nightmares of the above kind. Let us see what sense we can make of such nightmares.

3. Partial Sufficiency and Partial Ancillarity

In this section we put together a number of mathematical definitions.

Definition 1 (Model) : By the model (or statistical structure) of an experiment $\mathcal{E}$ we mean the usual trinity of abstractions $(\mathcal{X}, \mathcal{A}, \mathcal{P})$.

We suppose that the family $\mathcal{P}$ is indexed as $\mathcal{P} = \{P_\omega : \omega \in \Omega\}$ and call ω the universal parameter. Let $\theta = \theta(\omega)$ be the parameter of interest. By a statistic T we mean a measurable map of $(\mathcal{X}, \mathcal{A})$ into another measurable space $(\mathcal{T}, \mathcal{B})$.

Definition 2 (Ancillarity): The statistic T is ancillary if the marginal (or sampling) distribution of T is the same for all $\omega \in \Omega$—i.e., for all $B \in \mathcal{B}$, the function $P_\omega(T^{-1}B)$ is a constant in ω.

Definition 3 (θ-Oriented Statistic) : The statistic T is θ-oriented if the marginal distribution of T depends on ω only through $\theta = \theta(\omega)$. That is, $\theta(\omega_1) = \theta(\omega_2)$ implies $P_{\omega_1}(T^{-1}B) = P_{\omega_2}(T^{-1}B)$ for all $B \in \mathcal{B}$.

Observe that every ancillary statistic is θ-oriented irrespective of what θ is.

Example 1 : Let $x = (x_1, x_2, \ldots, x_n)$, with n fixed in advance, be a sample of n independent observations on a $N(\mu, \sigma)$. Let $D = (x_2 - x_1, x_3 - x_1, \ldots, x_n - x_1)$ be the difference statistic. Clearly, D is σ-oriented and, therefore, so is every measurable function h(D) of D. That the class $\{h(D)\}$ of measurable functions of the difference statistic does not exhaust the family of σ-oriented statistics is seen as follows. Choose and fix two function $h_1(D)$ and $h_2(D)$ that are identically distributed and also a Borel set E in R_1. Since $\bar{x}$ is stochastically independent of D for all (μ, σ), it now follows that the statistic T_E defined as

$$T_E(\bar{x}, D) = \begin{cases} h_1(D) & \text{if } \bar{x} \in E \\ h_2(D) & \text{if } \bar{x} \notin E, \end{cases}$$

is σ-oriented — indeed, T_E is identically distributed as $h_1(D)$ and $h_2(D)$. It is thus clear that D is not the maximum σ-oriented statistic. In fact no maximum σ-oriented statistic exists. (See Basu (1959, 1965b) for more information on this kind of problem.) In this case we have a plentiful supply of σ-oriented statistics. However, the notion of μ-orientedness is vacuous in the sense that no nontrivial (nonancillary) statistic can be μ-oriented. This remark is generally true for the location parameter μ in a location-scale parameter setup.

Definition 4 (Variation Independence) : The two functions $\omega \to a(\omega)$ and $\omega \to b(\omega)$ on the space Ω with respective ranges A and B are said to be variation independent if the range of the function $\omega \to (a(\omega), b(\omega))$ is the Cartesian product $A \times B$.

If the universal parameter ω can be represented as $\omega = (\theta, \phi)$, where θ and ϕ are variation independent in the preceding sense — that is, $\Omega = \Theta \times \Phi$ where Θ and Φ are the respective ranges of θ and ϕ — then we call ϕ a variation independent complement of θ. With θ as the parameter of interest, we may then call ϕ the nuisance parameter.

We have not come across a satisfactory definition of the notion of a nuisance parameter. It is only hoped that the above working definition will meet with little resistance. (See Barndorff-Nielsen (1973) for further details on the notion of variation independence.)

By a sufficient statistic, we mean a statistic that is sufficient in the usual sense with respect to the full model $(\mathcal{X}, \mathcal{A}, \mathcal{P})$. The following definition of a specific sufficient statistic appears in Neyman and Pearson (1936). Let ϕ be a variation independent complement of θ.

Definition 5 (Specific Sufficiency) : The statistic T is specific sufficient for θ if, for each fixed $\phi \in \Phi$, the statistic T is sufficient with respect to the restricted model $(\mathcal{X}, \mathcal{A}, \mathcal{P}_\phi)$, where $\mathcal{P}_\phi = \{P_{\theta,\phi} : \theta \in \Theta, \phi \text{ fixed}\}$.

In Example 1, the sample mean $\bar{x}$ is specific sufficient for μ. In fact, $\bar{x}$ is a minimum specific sufficient statistic for μ. The sample standard deviation s is, however, not sufficient for σ for any specified value of μ. Indeed, a statistic can be specific sufficient for σ only if it is sufficient.

In the spirit of Definition 5, we then define the notion of specific ancillarity in the following terms. As before, let ϕ be a variation independent complement of θ.

Definition 6 (Specific Ancillarity) : The statistic T is specific ancillary for θ if, for each fixed $\phi \in \Phi$, it is ancillary with respect to the restricted model $(\mathcal{X}, \mathcal{A}, \mathcal{P}_\phi)$.

In other words, T is specific ancillary for θ if it is ϕ-oriented, where ϕ is a variation independent complement of θ. It should be noted that the definition of θ-orientedness does not presuppose the existence of a variation independent complement ϕ, but the definitions of specific sufficiency and specific ancillarity (for θ) do.

In Example 1, with σ as the parameter of interest, it is tempting to marginalize to the statistic s. But can we logically justify such a marginalization? In what sense can we say that s summarizes in itself all the relevant and available information about σ in the absence of any information on μ ? We shall return to the question later.

Suppose μ is the parameter of interest in Example 1. Marginalization to the statistic $\bar{x}$, which is specific sufficient for μ, will not eliminate σ as $\bar{x}$ is not μ-oriented. We shall also lose valuable information on μ if we throw away the s-part of the sufficient statistic $(\bar{x}, s)$ and record only $\bar{x}$. For one thing, we shall no longer be able to speculate about the accuracy of $\bar{x}$ as a point estimate of μ. The marginalization method is of no use for the purpose of eliminating the scale parameter σ. As we have noted earlier, if T is μ-oriented then it has to be an ancillary statistic. Surely, we do not want to marginalize to something that has nothing to do with μ! The conditionality argument is also of no use for eliminating σ. If conditioning with respect to Y eliminates σ, then Y has to be specific sufficient for σ. But as we have stated earlier, every such Y has to be sufficient for (μ, σ). Hence, conditioning with respect to Y will eliminate μ as well! The problem of eliminating the scale parameter σ is not an easy one. Student's t-test and Stein's two-stage sampling plan are classical examples of statistical methodology that were developed to solve the problem in non-Bayesian terms.

The following definition of partial sufficiency is usually attributed to Fraser (1956). But we find the definition clearly laid out in Olshevsky (1940), who attributed it to Neyman (1935 a).

Definition 7 (p-Sufficiency) : The statistic T is partially sufficient (denoted by p-sufficient) for θ if T is specific sufficient for θ and T is θ-oriented. From this it is clear that the notion of p-sufficiency for θ presupposes the existence of a variation independent complement ϕ for θ. With the same presupposition, Sandved (1967) defined a notion of partial ancillarity in the following terms.

Definition 8 (S-Ancillarity) : The statistic Y is partial ancillary (S-ancillary) for θ if Y is specific ancillary for θ (Y is ϕ-oriented and Y is specific sufficient for ϕ). It should be noted that in Definitions 7 and 8, we are looking at the same concept but from two different angles. The statistic Y is S-ancillary for θ if and only if it is p-sufficient for ϕ.

The name S-ancillary (ancillary in the sense of Sandved) is due to Barndorff-Nielsen (1973) whose terminology for p-sufficiency is S-sufficiency. Barndorff-Nielsen's mathematical formalization of the twin notions of p-sufficiency and S-ancillarity as a "cut" may be defined as follows.

Definition 9 (Barndorff-Cut) : A statistic $T : (\mathcal{X}, \mathcal{A}) \to (\mathcal{T}, \mathcal{B})$ defines a Barndorff-cut of an experiment

$$\mathcal{E} = \{(\mathcal{X}, \mathcal{A}, P_\omega) : \omega \in \Omega\},$$

if there exist two variation independent and complementary subparameters $\theta = \theta(\omega)$ and $\phi = \phi(\omega)$, such that the marginal experiment $\mathcal{E}_T = \{(\mathcal{T}, \mathcal{B}, P_\omega T^{-1}) : \omega \in \Omega\}$ is θ-oriented ($P_\omega T^{-1}$ depends on ω only through $\theta(\omega)$) and that each one of the family $\{\mathcal{E}_t^T : t \in T\}$ of conditional experiments is ϕ-oriented.

The statistic T is then p-sufficient for θ and S-ancillary for ϕ. Observe that every sufficient statistic defines a Barndorff-cut and so also does every ancillary statistic. In the former case $\theta(\omega) = \omega$ and $\phi(\omega)$ is a known constant, and in the latter case it is the other way around.

In Example 1 there exists no Barndorff-cut that separates μ and σ. The following are a few other examples where the definition yields something.

Example 2 : Let the random variables $x_i (i = 1, 2, \ldots, m)$ be iid $N(\theta, 1)$, and let $y_j (j = 1, 2, \ldots, n)$ be an independent set of iid $N(\phi, 1)$. Clearly, $\bar{x}$ is p-sufficient and $\bar{y}$ is S-ancillary for θ.

Example 3 : Let x and y be independent Poisson variables with means μ and υ, respectively. With the reparametrization $\theta = \mu/(\mu + \upsilon)$ and $\phi = \mu + \upsilon$, it can be checked that $Y = x + y$ is S-ancillary for θ. There does not exist a statistic T that is p-sufficient for θ.

Example 4 : Let $x = (y, z, w)$ have a multinomial distribution with $y + z + w = n$ and p, q, r, $(p + q + r = 1)$ as probabilities (parameters). The parameters p and q are not variation independent. However, when we reparametrize as $\theta = p$, $\phi = q/(1-p)$, it is easy to check that the statistic y becomes p-sufficient for θ (S-ancillary for ϕ).

Example 5 : Let $0 < \theta < 1$ and $0 < \phi < \infty$. Let X be a random variable with pdf

$$p(x \mid \theta, \phi) = \begin{cases} (1 - \theta)\phi e^{\phi x} & \text{for } x \leqslant 0 \\ \theta\phi e^{-\phi x} & \text{for } x > 0. \end{cases}$$

Let $x_1, x_2, \ldots, x_n$ be n independent observations on X. Let T be the number of positive x_i's, and let $Y = \Sigma|x_i|$. Then T and Y are respectively p-sufficient and S-ancillary for the parameter θ.

Note the similarities between Examples 2 and 5. In either case, we have for the parameter of interest θ a statistic T that is p-sufficient and a statistic Y that is S-ancillary. In each case, however, the two statistics are stochastically independent for all possible values of the universal parameter. The fact that this is not generally true is going to bother us in due course.

It will be useful to review the various definitions in terms of the corresponding factorizations of the likelihood functions. To this end let us suppose that the family $\mathcal{P} = \{P_{\theta,\phi} : \theta \varepsilon \Theta, \phi \varepsilon \Phi\}$ is dominated by a σ-finite measure μ and let $\{p(. \mid \theta, \phi)\}$ be the corresponding family of probability density functions. To fix our ideas and to avoid all measure-theoretic difficulties let us pretend for the time being that $\mathcal{X}$ is a countable set and that μ is the counting measure on $\mathcal{X}$. Corresponding to any statistic $T : \mathcal{X} \to \mathcal{J}$ we have a factorization (of p) of the form

$$p(x \mid \theta, \phi) = g(T \mid \theta, \phi)\, f(x \mid T, \theta, \phi),$$

where g defines the marginal distribution of T and f defines the conditional distribution of x given T. (Our notations are admittedly rather sloppy, but there should be no difficulty in following our meaning.) Consider now the following particular cases of the above general factorization.

Case I : $p = g(T|\,\theta\,,\,\phi)\, f(x|\,T)$: this corresponds to the case where T is sufficient.

Case II : $p = g(T)f(x|\,T,\theta\,,\,\phi)$: the statistic T is ancillary.

Case III : $p = g(T|\,\theta)\, f(x|\,T,\theta\,,\,\phi)$: the statistic T is θ-oriented. The case where T is ϕ-oriented is similar.

In the situation where θ and ϕ are variation independent parameters, the notion of θ-orientedness is the same as the notion of specific ancillarity for ϕ. Case III, therefore, also corresponds to the case where T is specific ancillary for ϕ. With θ and ϕ variation independent, we have the next case.

Case IV : $p = g(T|\,\theta\,,\,\phi)\, f(x|\,T,\phi)$: the statistic T is specific sufficient for θ. The case where T is specific sufficient for ϕ is similar.

Case V : $p = g(T|\,\theta)\, f(x|\,T,\phi)$: The statistic T is p-sufficient for θ and is S-ancillary for ϕ.

Case Va : $p = g(T|\,\phi)\, f(x|\,T,\theta)$: the statistic T is S-ancillary for θ and is p-sufficient for ϕ.

Instead of looking at factorizations in terms of marginal and conditional frequencies, suppose we consider factorizations of the more general form

$$p(x|\,\theta\,,\phi) = G(x,\theta\,,\phi)F(x,\theta\,,\phi).$$

The very familiar

Case VI : $p = G(T,\theta\,,\phi\,)F(x),$

when proved equivalent to Case I, constitutes the well-known factorization theorem for sufficiency. Similarly, the factorization

Case VII : $p = G(T,\theta\,,\phi)\, F(x,\phi)$

can be shown to be equivalent to Case IV (the case of specific sufficiency for θ). Now consider

Case VIII : $p = G(T,\theta)\, F(x,\phi).$

Is Case VIII equivalent to Case V? It is important to recognize that the answer is in the negative. The examples in Section 9 will clarify the matter. Finally, we have factorizations of the form

Case IX : $p = G(x,\theta)\, F(x,\phi).$

It will turn out later that we really should be after factorizations of this form. Clearly, p factors in the manner of Case IX whenever we have a Barndorff-cut separating θ from ϕ (as in Cases V or Va). That the converse is not true will be variously exemplified in Section 9.

4. Generalized Sufficiency and Conditionality Principles

To understand the logic of the generalized sufficiency and conditionality principles $\mathcal{S}^*$ and $\mathcal{C}^*$, it is useful to consider a few hypothetical situations. (For a comprehensive discussion on the sufficiency, conditionality, invariance, and the likelihood principles refer to Basu (1973).)

(i) We have two experimental setups $\mathcal{E}$ and $\mathcal{E}'$, where the former provides information only on the parameter of interest θ while the latter is informative about an unrelated parameter ϕ alone – the parameter ϕ is unrelated to θ in the sense that we do not recognize the relevance of any information on ϕ for the purpose of inference making on θ. Faced with data such as $\{(\mathcal{E}, x), (\mathcal{E}', x')\}$, it makes good statistical sense to ignore the second part of the data and concentrate our attention on the relevant part $(\mathcal{E}, x)$.

(ii) Let $\mathcal{E}$ be an experiment whose randomness (probabilistic) characteristics depend only on θ. Having obtained the data $(\mathcal{E}, x)$, suppose we choose to perform a randomization exercise $\mathcal{E}_{(x)}$ thus arriving at the additional data $(\mathcal{E}_{(x)}, y)$. If all the randomness characteristics of $\mathcal{E}_{(x)}$ (possibly influenced by x) are known to us, then the secondary data $(\mathcal{E}_{(x)}, y)$ cannot give us any additional information on θ, or on anything for that matter. It makes good statistical sense then to suggest that the analysis of the data $\{(\mathcal{E}, x), (\mathcal{E}_{(x)}, y)\}$ ought to proceed on a total nonrecognition of the randomization exercise $\mathcal{E}_{(x)}$ and the resulting outcome y. Indeed, this is one way of looking at the sufficiency principle $\mathcal{S}$.

(iii) If in (ii) we find that the randomness characteristics of $\mathcal{E}_{(x)}$ are fully known except for a nuisasnce parameter ϕ that is unrelated to θ, then we are in a situation quite analogous to (i). Conforming to the statistical intuition that told us to ignore $(\mathcal{E}', x')$ in (i), the generalized sufficiency principle $\mathcal{S}^*$ tells us to ignore $(\mathcal{E}_{(x)}, y)$ in this situation.

(iv) We have a choice of k experiments $\mathcal{E}_{(1)}, \mathcal{E}_{(2)}, \ldots, \mathcal{E}_{(k)}$. The randomness structure of $\mathcal{E}_{(y)}$, $y = 1, 2, \ldots, k$, is related only to the parameter of interest. Let $\mathcal{E}$ stand for a randomization exercise that selects one of the k experiments with known (predetermined) selection probabilities $\pi_1, \pi_2, \ldots, \pi_k$. The experiment $\mathcal{E}_{(y)}$ selected by $\mathcal{E}$ is then performed resulting in the outcome x. The full data is $\{(\mathcal{E}, y), (\mathcal{E}_{(y)}, x)\}$. Since the part $(\mathcal{E}, y)$ of the data is totally uninformative, it makes good statistical sense to disregard this part of the data and focus our attention on the relevant part, i.e., $(\mathcal{E}_{(y)}, x)$. This is a version of the conditionality principle.

(v) Now, suppose in (iv) above the selection probabilities $\pi_1, \pi_2, \ldots, \pi_k$ are not fully known but depend on (are functions of) an unrelated nuisance parameter ϕ. We are now in a situation that is very similar to (i). The generalized conditionality principle $\mathcal{C}^*$ tells us to analyze the data by concentrating our whole attention on that part of the data – namely $(\mathcal{E}_{(y)}, x)$ – that is related to θ.

We are now ready to state formally the generalized principles of sufficiency and conditionality.

Principle $\mathcal{S}^*$ (Generalized Sufficiency Principle) **:** If, in terms of the model $(\mathcal{X}, \mathcal{A}, \mathcal{P})$ for the data $(\mathcal{E}, x)$, we recognize the statistic T as p-sufficient (partially sufficient in the sense of Definition 7) for the parameter of interest θ, then the data $(\mathcal{E}, x)$ should be reduced by marginalization to $(\mathcal{E}_T, t)$, where $\mathcal{E}_T$ is the marginal experiment corresponding to T and $t = T(x)$.

Principle $\mathcal{S}^*$ may be stated in a less severe form in the following terms.

Principle $\mathcal{S}^{**}$ **:** If T is p-sufficient for θ, then $T(x') = T(x'')$ implies that the information content (the evidential meaning) of the data $(\mathcal{E}, x')$ and $(\mathcal{E}, x'')$ relative to the parameter θ are identical in all respects. In other words, the data $(\mathcal{E}, x')$ warrants the same inference on θ as does the data $(\mathcal{E}, x'')$.

Principle $\mathcal{C}^*$ (Generalized Conditionality Principle)**:** If Y is S-ancillary (Definition 8) for θ, then the data $(\mathcal{E}, x)$ should be analyzed by reinterpreting it as $(\mathcal{E}_{(y)}, x)$, where $\mathcal{E}_{(y)} (= \mathcal{E}^Y_y)$ is the conditional experiment that corresponds to the observed value $y = Y(x)$ of Y.

As we have said before, corresponding to any statistic T we can conceive of a decomposition of the experiment $\mathcal{E}$ into a two-stage experimental setup in which the marginal experiment $\mathcal{E}_T$ is followed by the conditional experiment $\mathcal{E}^T_t$ that corresponds to the observed value $t = T(x)$ of T. The original data $(\mathcal{E}, x)$ may then be viewed as $\{(\mathcal{E}_T, t), (\mathcal{E}^T_t, x)\}$. If T is p-sufficient for θ then, by definition, the experiment $\mathcal{E}_T$ is θ-oriented, and the experiment $\mathcal{E}_t^T$ is ϕ-oriented. So, in view of (i) and (iii), it makes good statistical sense to invoke principle $\mathcal{S}^*$ and marginalize the data to $(\mathcal{E}_T, t)$. Conversely, if T is S-ancillary for θ then, by definition, $\mathcal{E}_T$ is ϕ-oriented and $\mathcal{E}_t^T$ is θ-oriented. So, in view of (i) and (v), it appears logical that we ought to ignore the $(\mathcal{E}_T, t)$ part of the data and analyze it as $(\mathcal{E}_t^T, x)$. This is the generalized conditionality principle $\mathcal{C}^*$.

5. A Choice Dilemma

In the writings of R. A. Fisher we find the conditionality argument used in three different ways: to recover the ancillary information in the data when it is found that the maximum likelihood estimator is not sufficient; to eliminate the nuisance parameter as in the case of the celebrated test of independence with a 2×2 multinomial data; and to generalize the fiducial argument as in the case of multiple observations on a random variable with a location parameter in its distribution.

In Basu (1964), while studying Fisher's recovery of information argument, I discovered a disturbing inherent difficulty in the conditionality argument. The difficulty flows from the fact that, in general, there does not exist a largest ancillary statistic in the sense of the usual partial order on statistics. Even in the simplest of situations we may have two ancillary statistics Y and U such that the statistic (Y, U) is not ancillary. Indeed, the pair (Y, U) may be fully informative, i.e., sufficient. In such a situation, the conflict between which of the two ancillaries to choose for the purpose of conditioning the data remains unresolved, despite some valiant efforts by Barnard and Sprott (1971) and Cox (1971), in non-Bayesian terms. The generalized conditionality (S-ancillarity) argument founders on the same non-uniqueness rock. We reproduce here an example from Basu (1964) that has attracted a lot of attention from non-Bayesians.

Example : Let X be a random variable with range {1, 2, 3, 4} and probability distribution

X :	1	2	3	4
Prob :	$(1 - \theta)/6$	$(1 + \theta)/6$	$(2 - \theta)/6$	$(2 + \theta)/6$

where $0 < \theta < 1$. We have n independent observations on X. The cell frequencies $x = (n_1, n_2, n_3, n_4)$ constitute the minimum sufficient statistic. The likelihood function is

$$L(\theta) = (1 - \theta)^{n_1}(1 + \theta)^{n_2}(2 - \theta)^{n_3}(2 + \theta)^{n_4} .$$

Let us write Bin (n, p) for the Binomial distribution with parameters n and p. Observe that $Y = n_1 + n_2$ is an ancillary statistic with probability distribution Bin $(n, \frac{1}{3})$ and that $U = n_1 + n_4$ is another ancillary with distribution Bin $(n, \frac{1}{2})$. If we condition x by Y then we can look upon the data as a pair of independent random variables n_1 and n_3 that are distributed as

$$\text{Bin}\ (Y, (1 - \theta)/2) \quad \text{and} \quad \text{Bin}\ (n - Y, (2 - \theta)/4),$$

respectively. However, if we choose to condition x by the other ancillary U, then we simplify the data to two independent variables n_1 and n_3 distributed respectively as

$$\text{Bin}\ (U, (1 - \theta)/3) \text{ and } \text{Bin}\ (n - U, (2 - \theta)/3).$$

In either case, the sample-space analysis of the conditioned data will be fairly easy and straightforward. But can anyone give a convincing argument for the choice of either Y or U as the conditioning ancillary?

We can easily introduce a nuisance parameter into the foregoing example by incorporating into the data, say, the result z of an independent coin-tossing experiment with an unknown bias ϕ in the coin. (George Barnard once remarked that when he retires he will go into business manufacturing biased coins and selling them to people like Basu!) In this case both (Y, z) and (U, z) will be S-ancillaries and we shall be back in the choice dilemma.

In contrast to the conditionality argument, the marginalization argument, in terms of the sufficiency or the generalized sufficiency principles, does not suffer from the above kind of a choice dilemma. With the kind of models that we work with in statistics, the existence of an essentially unique minimum sufficient statistic is always assured, and if the class of statistics that are p-sufficient for θ is not vacuous, then there will exist an essentially unique minimum such statistic.

6. A Conflict

The two elimination methods, namely, the one that marginalizes to a statistic T that is p-sufficient for θ and the one that conditions with respect to a statistic Y that is S-ancillary for θ, owe their origin to the same statistical intuition that guided us through (i) to (v) in Section 4. However, this does not mean that the two methods can co-exist in logical harmony. The possibility of a natural conflict between the methods was pointed out to me by Philip Dawid (1975). We give below a simple example along the lines of the dilemma example of the previous section to highlight this conflict.

Example : Let T and Y be two random variables with the same range {1, 2, 3, 4} and a joint distribution as described in the following table.

To simplify the argument let us suppose that we have only one observation x = (t, y) on the pair (T, Y) — the general case where we have n observations on (T, Y) is very similar. Observe that the statistic T, defined as T(x) = t, is p-sufficient for θ and that the statistic Y, defined as Y(x) = y, is S-ancillary for θ . The trouble is that T and Y are not stochastically independent in this example. The marginal distribution of T is very different from its conditional distribution for any given value of Y.

Joint Distribution of T and Y

	T				
Y	1	2	3	4	Total
1	$(1-\theta)(1-\phi)/12$	$(1+\theta)(1-\phi)/12$	0	0	$(1-\phi)/6$
2	$(1-\theta)(1+\phi)/12$	$(1+\theta)(1+\phi)/12$	0	0	$(1+\phi)/6$
3	0	0	$(2-\theta)(2-\phi)/24$	$(2+\theta)(2-\phi)/24$	$(2-\phi)/6$
4	0	0	$(2-\theta)(2+\phi)/24$	$(2+\theta)(2+\phi)/24$	$(2+\phi)/6$
Total	$(1-\theta)/6$	$(1+\theta)/6$	$(2-\theta)/6$	$(2+\theta)/6$	1

It is thus clear that we can have a pair of stochastically dependent statistics T and Y such that (T, Y) is sufficient for (θ,ϕ), T is p-sufficient for θ , and Y is S-ancillary for θ . The nuisance parameter ϕ can be eliminated from the argument either by marginalizing to T or by conditioning (T, Y) — that is, T — by the S-ancillary Y. The two elimination methods cannot be reconciled in such cases.

What went wrong? Should we blame the statistical intuition that guided us through (i) to (v) in Section 4? The above conflict is only a manifestation of the difficulties that we have to face when we try to interpret data in some sample-space terms.

7. Rao-Blackwell Type Theorems

In Section 4, our case for the Sufficiency Principle $\mathcal{S}$, the Conditionality Principle $\mathcal{C}$ and their generalizations $\mathcal{S}^*$ and $\mathcal{C}^*$ rested on the highly nonmathematical phrase, "It makes good statistical sense." The author does not know how else to argue in non-Bayesian terms for these essentially Bayesian principles of data analysis. A large majority of the statisticians belonging to the Fisher-Neyman school of thought seem to agree wholeheartedly with $\mathcal{S}$ although most of them are quite wary of $\mathcal{C}$. This almost universal faith in $\mathcal{S}$ is there, partly because it makes good statistical sense, but mainly because of the widespread belief that principle $\mathcal{S}$ has been mathematically proved in the Rao-Blackwell theorem. On p. 17 of Basu (1973) we briefly examined this mathematical proof of a statistical principle. Now, let us turn the spotlight on a similar proof of $\mathcal{S}^*$ given by Fraser (1956).

Let $a(\theta)$ be a real valued function of θ. We are looking for a reasonable point estimate of $a(\theta)$ on the basis of the data $(\mathcal{E}, x)$. Let us suppose that the loss $W(t, \theta)$, when $a(\theta)$ is estimated by t, is convex in t for each θ. Let $\mathcal{U}$ be the class of all estimators U of $a(\theta)$ such that the risk function

$$r_U(\theta) = r_U(\theta, \phi) = E[W(U, \theta) \mid \theta, \phi]$$

is well-defined and θ-oriented, that is, depends on $\omega = (\theta, \phi)$ only through θ.

Theorem (Fraser) **:** If T is p-sufficient for θ then, for each $U \varepsilon \mathcal{U}$, there exists an estimator $U_0 = U_0(T)$ such that $r_{U_0}(\theta) \leqslant r_U(\theta)$ for all $\theta \varepsilon \Theta$.

Proof : The statistic T is θ-oriented by definition, so the risk function generated by any function of T, if well-defined, must be θ-oriented. Now, choose and fix $\phi_0 \varepsilon \Phi$ and define

$$U_0 = E(U \mid T, \theta, \phi_0).$$

Since T, by definition, is sufficient for θ when ϕ is fixed, it follows that U_0 is well-defined as an estimator; that is, the unknown θ does not enter into the definition of U_0. From Jensen's inequality it follows that, for all $\theta \varepsilon \Theta$,

$$r_{U_0}(\theta) = r_{U_0}(\theta, \phi_0) \leqslant r_U(\theta, \phi_0) = r_U(\theta),$$

and thus the theorem is proved. The Rao-Blackwell theorem clearly corresponds to the particular case where ϕ is a known constant ϕ_0.

The above theorem may be generalized further along the following lines suggested by Hájek (1965). Let $\mathcal{U}'$ be the class of all estimators U such that the risk function $r_U(\theta, \phi)$ is well-defined (but not necessarily θ-oriented). Using the so-called minimax principle (see paragraph 7 of Section 1) let us define

$$R_U(\theta) = \sup_\phi r_U(\theta, \phi)$$

as the eliminated risk function associated with U. If $U \varepsilon \mathcal{U}$ then $r_U(\theta) = r_U(\theta, \phi)$ is θ-oriented and thus $R_U(\theta) = r_U(\theta)$. Now, if we define U_0 as in the Fraser theorem, then it follows (in view of the fact that U_0 is θ-oriented) that $R_{U_0}(\theta) = r_{U_0}(\theta, \phi_0) \leqslant r_U(\theta, \phi_0) \leqslant R_U(\theta)$. This generalizes the Fraser theorem to the following result :

Theorem (Hájek) **:** If T is p-sufficient for θ, then for each $U \in \mathcal{U}'$ there exists an $U_o = U_o(T)$ such that $R_{U_o}(\theta) \leqslant R_U(\theta)$ for all $\theta \in \Theta$.

The proofs of the preceding two theorems do not make full use of the supposition that T is p-sufficient for θ. They rest heavily on the supposition that T is θ-oriented but require T to be sufficient for θ for just one specific value ϕ_o of ϕ. This suggests the following generalization of the notion of partial sufficiency. For each $\theta \in \Theta$, let us define $\bar{\mathcal{P}}_\theta$ to be the convex hull of the family $\mathcal{P}_\theta = \{P_{\theta,\phi} : \theta$ fixed, $\phi \in \Phi\}$ of measures on $(\mathcal{X},\mathcal{A})$. In other words, $\bar{\mathcal{P}}_\theta$ is the family of all measures Q of the form :

$$Q(A) = \int_\Phi P_{\theta,\phi}(A)\, d\,\xi(\phi) \quad \text{for all } A \in \mathcal{A}, \tag{7.1}$$

where ξ is an arbitrary probability measure on Φ . The following definition is due to Hájek (1965).

Definition (H-Sufficiency) **:** The statistic T is H-sufficient (partially sufficient in the sense of Hájek) for θ if, for each $\theta \in \Theta$, there exists a choice of a measure $Q_\theta \in \bar{\mathcal{P}}_\theta$ such that, with $\mathcal{Q} = \{ Q_\theta : \theta \in \Theta \}$, T is sufficient in the model $(\mathcal{X}, \mathcal{A}, \mathcal{Q})$, and T is θ-oriented in the model $(\mathcal{X},\mathcal{A},\mathcal{P})$.

It should be noted that for the definition of H-sufficiency it is not necessary for θ and ϕ to be variation independent. Clearly, p-sufficiency implies H-sufficiency. We have only to choose and fix $\phi_o \in \Phi$ and then define $Q_\theta = P_{\theta,\phi_o}$. Let us check now that the Fraser-Hájek theorems remain true even if we replace the requirement of p-sufficiency for the statistic T by the less stringent requirement of H-sufficiency. For any U, we define U_o as $E(U \mid T, Q_\theta)$. Observe that U_o is an estimator in view of the definition of H-sufficiency. We then invoke Jensen's inequality to prove that, for all $\theta \in \Theta$,

$$\int_{\mathcal{X}} W(U_o, \theta)\, dQ_\theta \leqslant \int_{\mathcal{X}} W(U, \theta) dQ_\theta \,.$$

Now, if we look back on the supposition that Q_θ is in the form (7.1) above, then, from the fact that U_o — being a function of T — is θ-oriented, it follows at once that the left side of the above inequality is equal to

$$r_{U_o}(\theta) = R_{U_o}(\theta)$$

for all θ. Similarly, the right side is equal to $r_U(\theta)$ if $U \in \mathcal{U}$ and is clearly not greater than $R_U(\theta)$ if $U \in \mathcal{U}'$. Thus the two preceding theorems may be finally restated as :

Theorem (Fraser-Hájek) **:** If T is H-sufficient for θ, then for any $U \in \mathcal{U}$ there exists a $U_o = U_o(T)$ such that $r_{U_o}(\theta) \leqslant r_U(\theta)$ for all θ. Furthermore, for any $U \in \mathcal{U}'$ it is true that

$R_{U_0}(\theta) \leqslant R_U(\theta)$ for all θ .

How much comfort can an advocate of the generalized sufficiency principle $\mathcal{S}^*$ derive from the Fraser-Hájek theorem? Before answering this question, let us take a brief look at the question of how and where the notion of H-sufficiency fits into the ten-fold factorization scheme of the likelihood that we laid out in Section 4.

In order for T to be H-sufficient for θ it is necessary that T is θ-oriented; that is, we have a factorization of the form

$$p(x|\ \theta,\ \phi\ ) = g(T|\ \theta)\ f(x|\ T,\ \theta,\ \phi). \tag{7.2}$$

It is also necessary (in view of the sufficiency condition for T) that there exists a family $\{\xi_\theta : \theta \in \Theta\}$ of probability measures on Φ such that the "mixed" frequency function

$$q(x|\ \theta) = \int_\Phi p(x|\ \theta,\ \phi)\ d\ \xi_\theta(\phi)$$

factors as

$$q(x|\ \theta) = G(T,\ \theta)\ F(x)\ . \tag{7.3}$$

Let us look back at the classical problem where the sample $x = (x_1, x_2, \ldots, x_n)$ consists of n independent observations on an $N(\mu,\sigma)$. Clearly $\bar{x}$ is not H-sufficient for μ—indeed, no T can be H-sufficient for μ . But is $s^2 = \Sigma(x_i - \bar{x})^2$ H-sufficient for σ? Can we find a family $\{\xi_\sigma : 0 < \sigma < \infty\}$ of "mixing measures" on R_1 that will lead to a factorization of the type (7.3) above with $T = s^2$? Observe that

$$p(x|\ \mu,\ \sigma) = A(\sigma)\ \exp\ (\ -\frac{s^2}{2\sigma^2})\ \exp\ [\ -\frac{n(\bar{x}-\mu)^2}{2\sigma^2}\ ],$$

where $A(\sigma) = ((2\pi)^{1/2}\ \sigma)^{-n}$.

We, therefore, need a family of mixing measures ξ_σ such that

$$\int_{-\infty}^{\infty} \exp\ [\ -\frac{n(\bar{x}-\mu)^2}{2\sigma^2}\ ]\ d\xi_\sigma(\mu) = B(\bar{x})C(\sigma). \tag{7.4}$$

The above factorization clearly holds if we choose for ξ_σ the uniform distribution (the Lebesgue measure) over the whole real line. But, with such improper mixings, it is easily seen that the Fraser-Hájek theorem will fall to pieces. If the range of σ is the whole of the positive half line, then there cannot exist a family of proper mixing measures ξ_σ for which the factorization (7.4) will hold.

So how are we going to prove that we ought to marginalize to s when the parameter of interest is σ? Hájek (1965) came up with the following ingenious mathematical argument. In any particular situation, we should always be able to limit (on a priori considerations) the parameter σ to some finite interval (0, k). With σ

restricted to such a finite interval, the statistic s becomes H-sufficient for σ . Just check that the factorization (7.4) holds if we choose for ξ_σ the Normal measure with mean zero and variance $(k^2 - \sigma^2)/n$.

Hájek's definition of partial sufficincy is intriguing and full of mathematical possibilities. But, what are the statistical contents of Hájek's definition of partial sufficiency and his generalized Rao-Blackwell theorem? Hájek's 'proof,' that we should marginalize to s when we do not know μ , certainly does not scandalize our statistical intuition. In the language of R.A. Fisher, if we throw away $\bar{x}$ and marginalize to s, then our loss of information on σ has the measure of only one degree of freedom in the worst possible case (when μ is fully known). Of the total information available on σ, the fraction of information summarized in s is at least (n-1)/n. Let us now look at the following celebrated example due to Neyman and Scott (1948) :

Example (Neyman & Scott) : The data x consists of 2n observations x_1, x'_1, x_2, x'_2, . . . , x_n, x'_n. The statistical model here corresponds to 2n independent normal variables with equal variance σ^2 and with x_i, x'_i having common mean μ_i(i = 1,2, . . . , n). The parameter of interest is σ , the nuisance parameter is the vector $\mu = (\mu_1, \mu_2, \ldots, \mu_n)$.

With $S^2 = \Sigma (x_i - x'_i)^2$, $\bar{x}_i = (x_i + x'_i)/2$ and $A(\sigma) = ((2\pi)^{1/2}\sigma)^{-2n}$, we then have

$$p(x|\mu, \sigma) = A(\sigma) \exp\left(-\frac{S^2}{4\sigma^2}\right) \exp\left[-\frac{\Sigma(\bar{x}_i - \mu_i)^2}{\sigma^2}\right].$$

The statistic S^2 is clearly σ-oriented. Is it H-sufficient for σ ? Again the answer is no if σ is unrestricted, but it is yes if we restrict σ to a finite interval (0,k). For the mixing measure ξ_σ on R_n, we now choose the one for which $\mu_1, \mu_2, \ldots, \mu_n$ are iid normal variables with means zero and variances $(k^2 - \sigma^2)/2$.

Of course, we are prepared to assume that $0 < \sigma < k$ for some k. The Hájek proof notwithstanding, how secure do we really feel about marginalizing to S without taking a hard look at $\bar{x}_1$, $\bar{x}_2$, . . . , $\bar{x}_n$? If μ were known, then the sample would contain 2n units (degrees of freedom) of information on σ , out of which S summarizes in itself only n units. Are we really prepared to sacrifice n degrees of freedom at the altar of ignorance on μ ? The issues raised in this example of Neyman and Scott are all very complex.

We close this section with one more jab at the notion of H-sufficiency and the Rao-Blackwell type proof of our generalized sufficiency principle in sample-space terms.

Example : Let $x = (x_1, x_2, \ldots, x_m; y_1, y_2, \ldots, y_n)$ be $m + n$ independent normal variables all with unit variances. It is known that $Ex_i = \theta, (i = 1, 2, \ldots, m)$, and $Ey_j = \theta\phi, (j = 1, 2, \ldots, n)$, where $-\infty < \theta < \infty$ is the parameter of interest and $\phi (= 0 \text{ or } 1)$ is the nuisance parameter.

The likelihood function neatly factors as

$$p(x|\theta, \phi) = A(x) \exp[-m(\bar{x} - \theta)^2/2]. \exp[-n(\bar{y} - \theta\phi)^2/2].$$

Clearly, the pair $(\bar{x}, \bar{y})$ constitutes the minimal sufficient statistic. The statistic $\bar{x}$ is θ-oriented. It is also sufficient (for θ) when ϕ is fixed at the value zero. Therefore, $\bar{x}$ is H-sufficient for θ and so the Fraser-Hájek theorems proved earlier recommend marginalization to $\bar{x}$. However, the reduction of the data from $(\bar{x}, \bar{y})$ to $\bar{x}$ will mean a substantial loss of information on θ in the event $\phi = 1$. From the full data we should be able to tell (with a reasonable amount of certainty if m and n are large) whether $\phi = 0$ or 1. (If E stands for the event $m(\bar{x} - \bar{y})^2 > (m + n)\bar{y}^2$, then it is easy to check that the maximum likelihood (ML) estimator $\hat{\phi}$ of ϕ is the indicator of E and that the ML estimator of θ is $\hat{\theta} = (1 - \hat{\phi})\bar{x} + \hat{\phi}\cdot(m\bar{x} + n\bar{y})/(m + n)$.)

This example does not contradict the good statistical sense that led us to the generalized (or partial) sufficiency principle $\mathcal{S}^*$, but only tells us not to be unduly impressed with Fraser's mathematical proof of the principle. The statistical literature is full of this kind of proof (see for instance Lehmann (1959)) where we start on the wrong foot either by delimiting the discussion to a conveniently small (and nice) class of decision procedures or by simplifying the hypothetical risk function by an ad hoc maximization process. I am very sceptical about the relevance of this kind of statistical mathematics in theoretical statistics.

8. The Bayesian Way

After a long journey through a whole forest of confusing ideas and examples, we seem to have lost our way. Let us now see if our Bayesian guide can find a way out of this wilderness.

According to a Bayesian, the role of the data $(\mathcal{E}, x)$ is to act as an operator on the experimenter's prior opinion q (a probability measure on Ω) and to transform it into a posterior opinion q_x^*.

With $\omega = (\theta, \phi)$, where θ is the parameter of interest and ϕ is the nuisance parameter, the Bayesian analysis of data is always firmly anchored to the posterior marginal distribution $q_x^\dagger$ on Θ defined as

$$q_x^{\dagger}(\theta) = \sum_{\phi} q_x^* (\theta, \phi),$$

where $q_x^*(\omega) = L(\omega)\, q(\omega)/\Sigma_\omega L(\omega)q(\omega)$. As we said in paragraph (10) of Section 1, the Bayesian way of eliminating the nuisance parameter from the argument is to integrate it out from the posterior distribution of (θ, ϕ).

In 1942, A. N. Kolmogorov defined the notion of a sufficient statistic in the following Bayesian terms :

Definition : The statistic T is sufficient if, for every prior q on Ω, the posterior q_x^* depends on x only through T; that is, $T(x) = T(x')$ implies that $q_x^* = q_{x'}$.

In the discrete setup, there is no difficulty in proving the equivalence of the above definition and the classical Fisher definition of sufficiency. In the same 1942 paper, we find Kolmogorov suggesting the following definition of partial sufficiency.

Definition (K-Sufficiency) : The statistic T is partially sufficient for θ if, for all prior q on Ω, the posterior marginal distribution $q_x^{\dagger}$ on Θ depends on x only through T. (Let us call such a statistic K-sufficient for θ.)

At last we seem to have something for which we have been looking for so long. However, it was demonstratred by Hájek (1965) that the definition of K-sufficiency is vacuous in the following sense:

Theorem (Hájek) : If the parameter θ is not a constant in ω, then every T that is K-sufficient for θ is sufficient (in the usual sense).

Proof : Pretending as always that we are dealing with a discrete model, we first recall that if T is not sufficient then there must exist x, x' such that $T(x) = T(x')$, but the likelihood ratio $p(x \mid \omega)/p(x' \mid \omega)$ is not a constant in ω. Therefore, if T is not sufficient, then we must have x, x' and ω_1, ω_2 such that $T(x) = T(x')$ and

$$p(x \mid \omega_1)/p(x' \mid \omega_1) \neq p(x \mid \omega_2)/p(x' \mid \omega_2). \qquad (8.1)$$

Let $\omega_1 = (\theta_1, \phi_1)$ and $\omega_2 = (\theta_2, \phi_2)$. There is no loss of generality in supposing that $\theta_1 \neq \theta_2$. (Otherwise, we choose $\omega_3 = (\theta_3, \phi_3)$, with $\theta_3 \neq \theta_1 = \theta_2$, and consider the ratio $p(x \mid \omega_3)/p(x' \mid \omega_3)$ along with any one of the two ratios in (8.1) that differs from it.) Now, consider the prior q whose entire mass is equally distributed over the two points ω_1 and ω_2. Observe that

$$q_x^{\dagger}(\theta_1) = q_x^*(\omega_1) = p(x \mid \omega_1)/\sum_{i=1}^{2} p(x \mid \omega_i),$$

and that a similar expression holds true for $q_{x'}^{\dagger}(\theta_1)$. In view of (8.1) it follows that

$$q_x^\dagger(\theta_1) \neq q_{x'}^\dagger(\theta_1) \text{ even though } T(x) = T(x') .$$

Thus, T not-sufficient implies T not K-sufficient. This proves the theorem. (Observe that we do not require θ and ϕ to be variation independent either in the definition of K-sufficiency or in the proof of the above theorem.)

The fault in Kolmogorov's definition of partial sufficiency is easily detected. We may try to correct this by restricting the discussion to a relatively small class Q of prior measures q on Ω. We find the following definition in Raiffa and Schlaifer (1961) :

Definition (Q-Sufficiency : The statistic T is Q-sufficient for θ if, for all $q \in Q$, the posterior marginal distribution $q_x^\dagger$ on Θ depends on x only through T. (In the language of Raiffa and Schlaifer, such a T is called marginally sufficient with respect to Q.)

From the beginning, we have been concerned with the problem of eliminating a parameter ϕ that is "unrelated" to the parameter of interest θ . However, we have not as yet clearly stated what we mean by two unrelated parameters. Is it enough to say that θ and ϕ are unrelated if they are variation independent and if the loss depends only on the terminal decision and the parameter θ? Clearly not, but what else can a non-Bayesian say? Just ask a non-Bayesian what he means when he agrees that the unknown true height ϕ of Mount Everest is unrelated to the unknown number θ of civilians who lost their lives in the Vietnam war! A Bayesian has no problem in defining the term. He calls θ and ϕ unrelated parameters if, apart from the condition on the loss function, his prior q for $\omega = (\theta, \phi)$ is of the form

$$q(\theta, \phi) = q_1(\theta)\, q_2(\phi) .$$

Let Q_0 be the class of all (independent) priors q of the form $q(\theta, \phi) = q_1(\theta)q_2(\phi)$. When is a statistic T going to be Q_0-sufficient for θ in the sense of our modified Kolmogorov definition of partial sufficiency? We find the following result in Raiffa and Schlaifer (1961) :

Theorem (Raiffa and Schlaifer) : If, for all $x \in \mathcal{X}$, the likelihood function factors as

$$p(x \mid \theta, \phi) = G(T, \theta)\, F(x, \phi) ,$$

then T is Q_0-sufficient.

Proof : If the prior distribution is $q(\theta, \phi) = q_1(\theta)\, q_2(\phi)$, then

$$q_x^{\dagger}(\theta) = \sum_{\phi} q_x^{*}(\theta, \phi)$$

$$= G(T, \theta) q_1(\theta) / \sum_{\theta} G(T, \theta)\, q_1(\theta)$$

depends on x only through T.

The above theorem suggests the following definition.

Definition (L-Sufficiency) : The statistic T is L-sufficient for θ if, for all $x \in \mathcal{X}$, the likelihood factors as in the statement of the previous theorem.

We just proved that L-sufficiency implies Q_0-sufficiency. Is the converse true? The answer is, of course, no. If T sufficient, in the sense of Fisher or Kolmogorov, then it is Q-sufficient for every Q and in particular for Q_0. Raiffa and Schlaifer's definition of Q_0-sufficiency for θ suffers from a defect very similar to that of Kolmogorov's definition of partial sufficiency (K-sufficiency). The definition is too wide and fails to pinpoint the exact notion of partial sufficiency we are after. Perhaps an example will make clear the point we are driving at.

Example : Let $x_1, x_2, \ldots, x_n$ be iid $N(\theta\phi, 1)$, where $0 < \theta < \infty$ is the parameter of interest and ϕ (= - 1 or 1) is the nuisance parameter. Just imagine μ, $(-\infty < \mu < \infty)$, to be the common mean and then let $\theta = |\mu|$ and $\phi = \text{Sgn}\ \mu$.

The statistic $T = |\bar{x}|$ is θ-oriented. It is a reasonable estimator of θ, but is it, in some sense, partially sufficient for θ? Check that T is not Q_0-sufficient for θ. Indeed, the notion of Q_0-sufficiency leads us to $\bar{x}$ which is sufficient. If, however, we agree to restrict our discussion to the smaller class $Q_0' \subset Q_0$ of (independent) priors q of the form $q(\theta, \phi) = q_1(\theta) q_2(\phi)$ such that q_2 is the uniform prior on $\Phi = \{-1, 1\}$, then it is easy to check that $T = |\bar{x}|$ is Q_0'-sufficient for $\theta = |\mu|$.

If we look back on the proof of the one-way implication theorem above, then it will be clear that L-sufficiency takes us far beyond Q_0-sufficiency. If T is L-sufficient for θ then the posterior marginal $q_x^{\dagger}$ on Θ depends on the sample x only through T and on the prior $q = q_1 q_2$ only through q_1. In Bayesian terms, we may redefine the notion of L-sufficiency as follows :

Definition (B-Sufficiency) : The statistic T is B-sufficient (partially sufficient in a restricted Bayes sense) for θ if, for $q = q_1 q_2 \in Q_0$ and $x \in \mathcal{X}$, the posterior marginal $q_x^{\dagger}$ on Θ depends on x only through T and on q only through q_1. (Indeed, one may try to further generalize the above notion of partial sufficiency by restricting q to an arbitrary but fixed class Q of priors on $\Omega = \Theta \times \Phi$ and calling q_1 the prior marginal on Θ. In the present context we have, however, no use for such a generalization.) In the next section we develop the theme of B-sufficiency to its natural conclusion.

9. Unrelated Parameters

Let us consider a rather loosely formulated question : Under what circumstances can we recognize the nuisance parameter ϕ to be so "unrelated" to the parameter of interest θ that we can meaningfully isolate the whole of the relevant information about the parameter θ contained in the data $(\mathcal{E}, x)$?

This is a good test question with which we can try to classify a statistician into one or another of the numerous feuding groups (or mutual admiration societies) that divide the current community of statisticians. For instance, a pucca (fully baked) Bayesian will probably dismiss the question out of hand as naive, incompetent and unnecessarily argumentative. This is because a pucca-Bayesian has no use for the notion of "information in the data." According to him natural dwelling place for information is the head of a homo sapien, and he recognizes only two kinds of statistical information — prior and posterior. Being a pucca-Bayesian, he always knows his prior q as a well-defined probability measure on $\Omega = \Theta \times \Phi$. Given the data he can, therefore, compute the posterior information q_x^* and then isolate the information $q_x^\dagger$ on θ by integration.

In the pucca-Bayesian statistical theory of Bruno de Finetti and L. J. Savage, there is no room for a family Q of prior distributions. However, having examined the question from various angles, the author has come to recognize the merit of Kolmogorov's half-baked Bayesian approach to the problem at hand. In the spirit of Kolmogorov, Raiffa and Schlaifer, let us put down the following definition for unrelated parameters. Let $\theta \varepsilon \Theta$ and $\phi \varepsilon \Phi$ be two parameters that enter into the statistical structure or model of an experiment $\mathcal{E}$, and let Q_0 be the class of all product probability distributions $q = q_1 q_2$ on $\Omega = \Theta \times \Phi$.

Definition (Unrelatedness) : The parameters θ , ϕ are unrelated relative to a model of the experiment $\mathcal{E}$ if, for all prior $q \varepsilon Q_0$ and all sample outcomes x of $\mathcal{E}$, the posterior distributions q_x^* also belong to the class Q_0.

If the likelihood function $L(\theta, \phi | x) = p(x | \theta, \phi)$ factors as

$$p(x| \theta, \phi) = A(\theta, x)\, B(\phi, x) , \tag{9.1}$$

then, for any prior $q(\theta, \phi) = q_1(\theta) q_2(\phi)$, it is easily seen that the posterior factors as

$$q_x^* (\theta, \phi) = q_x^\dagger(\theta)\; q_x'(\phi),$$

where

$$q_x^\dagger(\theta) = A(\theta, x) q_1(\theta) / \sum_\theta A(\theta, x) q_1(\theta),$$

with a similar expression holding true for $q_x'(\phi)$. Conversely, if

$$q_x^* (\theta, \phi) = p(x| \theta, \phi)\, q_1(\theta) q_2(\phi) / \sum_{\theta,\phi} p q_1 q_2$$

belongs to Q_0 then it is equally clear that $p(x|\theta, \phi)$ must factor in the manner of (9.1) above. We thus have the

Theorem : The parameters θ, ϕ are unrelated relative to a model of the experiment $\mathcal{E}$ if and only if the likelihood function factors in the manner (9.1).

It is then easy to recognize whether the parameter of interest is unrelated (in the preceding sense) to the nuisance parameter or not. With such a recognition of unrelatedness, (and, of course, with the further condition that the nuisance parameter has nothing to do with the hazards of incorrect decisions) the Bayesian will not waste his time in figuring out his prior q_2 for ϕ as long as he is satisfied that his prior q for (θ, ϕ) must be in the class Q_0. He will carefully figure out his prior q_1 for θ and then work out his posterior for θ as

$$q_x^{\dagger}(\theta) = A(\theta, x)q_1(\theta)/ \sum_{\theta} A(\theta, x)q_1(\theta).$$

In Basu (1973) we examined in depth the question of information in the data. Our conclusion was that, relative to a particular statistical model for the experiment $\mathcal{E}$ in question, Fisher's notion of the "whole of the relevant information" about $\omega = (\theta, \phi)$ that is contained in the data $(\mathcal{E}, x)$ may be identified with the likelihood function

$$L(\theta, \phi | x) = p(x | \theta, \phi) .$$

What we are saying now is that when the likelihood comes factored as in (9.1), when, on a priori considerations, we are willing to regard θ and ϕ as independent entities, and when information gained on ϕ is of no direct relevance to the decision problem on hand (i.e., ϕ does not enter into the loss function), then we may regard the function

$$L^{\dagger}(\theta | x) = A(\theta, x)$$

as the "whole of the relevant information" on θ that is supplied by the data $(\mathcal{E}, x)$. This may be regarded as a generalized likelihood principle.

The generalized sufficiency principle $\mathcal{S}^*$ and the generalized conditionality principle $\mathcal{C}^*$ are in conformity with the above principle. The existence of a statistic T that is p-sufficient for θ or of a statistic Y that is S-ancillary for θ presupposes a factorization of the likelihood as in (9.1). The principles $\mathcal{S}^*$ and $\mathcal{C}^*$ are indirectly advising us to concern ourselves with the factor of $L(\theta, \phi | x)$ that involves only θ. This is precisely why the p-sufficiency and the S-ancillarity arguments do not lead us astray.

Also observe that we can have a statistic T that is L-sufficient (B-sufficient) for θ if and only if θ and ϕ are unrelated in the sense of the likelihood factoring as in (9.1). If and when the likelihood factors in the above manner, we can always fashion a statistic T that is minimal L-sufficient for θ and a statistic Y that is minimal L-sufficient for ϕ. For example, T will be defined in terms of the equivalence relation : $x' \sim x''$ if $A(\theta, x') = C(x', x'')A(\theta, x'')$ for all $\theta \varepsilon \Theta$. In general, such a T will fail to be θ-oriented; that is, T will not be p-sufficient for θ. Similarly, Y will, in general, fail to be S-ancillary for θ. Indeed, we shall give an example where T and Y are the same. In such an example the same statistic T is in some sense isolating all the relevant information about θ and also all the information about the unrelated parameter ϕ.

A major source of our confusion on the important question of when and how we can isolate the information on the parameter of interest, is the fact of our arguing (in the manner of Sir Ronald) in terms of statistics. The notion of a statistic as a measurable map has hardly any relevance at the data analysis stage. It was Sir Ronald who distorted the question "what is information?" to the question "what (statistic) has all the information?". He taught us that a statistic is sufficient if and only if it summarizes in itself all the relevant information in the data. In the same spirit, we have been looking for a statistic T that is partially sufficient for θ—a statistic that summarizes in itself all the relevant and usable information about θ in the event of ignorance about the nuisance parameter ϕ.

We end this marathon discussion with three examples of statistical models where the parameters come naturally separated in the manner of (9.1), and yet we cannot take advantage of the fact (and isolate the information on the parameter of interest) in terms of either the generalized sufficiency or the conditionality principle.

Example 1 : We have a multinomial distribution with three categories and with probabilities

$$\theta\phi, (1 - \theta)(1 + \phi)/2 \text{ and } (1 + \theta)(1 - \phi)/2,$$

where $0 < \theta < 1$ and $0 < \phi < 1$. With n observations, the three frequencies (n_1, n_2, n_3) constitute the minimal sufficient statistic, and the likelihood factors as

$$2^{-(n_2+n_3)} [\theta^{n_1}(1 - \theta)^{n_2}(1 + \theta)^{n_3}] [\phi^{n_1}(1 + \phi)^{n_2}(1 - \phi)^{n_3}].$$

We do not have any statistic that is p-sufficient, H-sufficient or S-ancillary for θ. The statistic $T = (n_1, n_2, n_3)$ is minimal L-sufficient (B-sufficient) for θ and also for ϕ. The (likelihood) information in the data on the parameter of interest θ is crying to be isolated as

$$L^{+}(\theta) = \theta^{n_1}(1 - \theta)^{n_2}(1 + \theta)^{n_3}.$$

If θ and ϕ are independent a priori and if ϕ does not enter into the loss function, then a Bayesian will analyze the data in the same manner as he would have done in the hypothetical case when ϕ were known to be equal to $\frac{1}{2}$, say. Can anyone suggest a reasonable sample-space analysis of the data?

Example 2 : Let $0 < \theta < \infty$ and $0 < \phi < \infty$. Let X and Y be two random variables with probability density functions

$$\theta e^{-\theta(x-\phi)} I(x - \phi) \quad \text{and} \quad \phi e^{-\phi(y+\theta)} I(y+\theta),$$

respectively, where I(.) stands for the indicator of the positive half of the real line. The sample consists of n independent observations $x_1, x_2, \ldots, x_n$ on X together with an independent set $y_1, y_2, \ldots, y_n$ of n independent observations on Y. Observe that the likelihood neatly factors as

$$[\theta^n \exp(- n \theta \bar{x}) I(y_{(1)} + \theta)] \cdot [\phi^n \exp(- n \phi \bar{y}) I(x_{(1)} - \phi)],$$

where $x_{(1)} = \min x_i$ and $y_{(1)} = \min y_i$. Clearly, the two parameters θ, ϕ are unrelated relative to the model. The statistic $(\bar{x}, y_{(1)})$ is B-sufficient (L-sufficient) for θ. The Bayesian analysis of the data is very simple as ϕ gets eliminated almost by itself. Can anyone suggest how to deal with the nuisance parameter in non-Bayesian terms?

Anyone who would sneer at the last two examples, on the grounds that they are not apparently related to any real life problem, is advised to take a hard look at the next example.

Example 3 : The experiment consists of the observation, for each of n week days in a large metropolitan area, of the number of accidents involving one or more automobiles and also the corresponding number of such accidents involving one or more fatalities. The parameter of interest is the proportion θ of automobile accidents that result in death. The mean number ϕ of auto accidents per working day is the nuisance parameter.The statistical model for our record,

$$x = \{(x_1, y_1), \ldots, (x_n, y_n)\},$$

of the number of accidents x_i and the corresponding number of fatal accidents y_i on the ith day $(i = 1, 2, \ldots, n)$ is that we have a set of n independent observations on a pair of random variables (X, Y) such that X is a Poisson variable with mean ϕ and Y, given X, is a Binomial variable Bin (X, θ). Now with $N = \Sigma x_i$ and $T = \Sigma y_i$, the likelihood neatly factors as

$$p(x| \theta, \phi) = A(x) \{\phi^N \exp(- n\phi)\} \{\theta^T (1 - \theta)^{N-T}\}. \tag{9.2}$$

If n were a preselected constant, then the statistic N, distributed as a Poisson variable with mean $n\phi$, would qualify as an S-ancillary for θ. In this case the generalized conditionality principle will eliminate ϕ and will permit us to argue in some sample-space terms. Sir Ronald would have advised us to reduce the data to the minimal sufficient statistic (N, T), hold the ancillary N as fixed (at its observed value), and then look upon T as an observation on a Binomial variable with parameters N (known) and θ (unknown).

What happens if we do not preselect n but let it be determined by the very system that was under observations? Suppose we continue our observations until $T = \Sigma y_i$ exceeds a preselected number, say 10. How should we analyze the data then? Observe that our stopping rule has no effect on the likelihood function which comes factored in the same form as (9.2) above. Now the triple (n, N, T) constitutes the minimal sufficient statistic — the statistic T is nearly a constant but not quite. The statistics (N, T) and (n, N) are B-sufficient (L-sufficient) for θ and ϕ , respectively, but the notions of p-sufficiency and S-ancillarity are vacuous in this instance.

CHAPTER VIII

SUFFICIENCY AND INVARIANCE

0. Notes

Editor's Note : This chapter is based on Basu, D. (1969a). It is shown that if a σ-field is invariant under a group of transformations preserving each probability measure in a given family, then it is sufficient. This fact was known to ergodic theorists at the time Basu rediscovered it, but the language of sufficiency was new. This language has turned out to be quite convenient in representing invariant measures. See, e.g., Maitra (1977, Tran. Am. Math. Soc., 229, 209-225).

A second interesting fact is that in a few cases the minimal sufficient σ-field is identified as the maximal invariant σ-field under a suitably chosen group of measure preserving transformations. In these examples, invariance alone leads to the maximum reduction possible by sufficiency.

SUFFICIENCY AND INVARIANCE

1. Summary

Let $(\mathcal{X}, \mathcal{A}, \mathcal{P})$ be a given statistical model and let $\mathcal{G}$ be the class of all one-to-one, bimeasurable maps g of $(\mathcal{X}, \mathcal{A})$ onto itself such that g is measure-preserving for each $P \varepsilon \mathcal{P}$, i.e. $Pg^{-1} = P$ for all P. Let us suppose that there exists a least (minimal) sufficient sub-field $\mathcal{L}$. Then, for each $L \varepsilon \mathcal{L}$, it is true that $g^{-1}L$ is $\mathcal{P}$-equivalent to L for each $g \varepsilon \mathcal{G}$, i.e., the least sufficient sub-field is almost $\mathcal{G}$-invariant. It is demonstrated that, in many familiar statistical models, the least sufficient sub-field and the sub-field of all almost $\mathcal{G}$-invariant sets are indeed $\mathcal{P}$-equivalent. The problem of data reduction in the presence of nuisance parameters has been discussed very briefly. It is shown that in many situations the principle of invariance is strong enough to lead us to the standard reductions. For instance, given n independent observations on a normal variable with unknown mean (the nuisance parameter) and unknown variance, it is shown how the principle of invariance alone can reduce the data to the sample variance.

2. Definitions and Preliminaries

(a) The basic probability model is denoted by $(\mathcal{X}, \mathcal{A}, \mathcal{P})$, where $\mathcal{X} = \{x\}$ is the sample space, $\mathcal{A} = \{A\}$ the σ-field of events and $\mathcal{P} = \{P\}$ the family of probability measures.

(b) By set we mean a typical member of $\mathcal{A}$. By function (usually denoted by f) we mean a measurable mapping of $(\mathcal{A}, \mathcal{A})$ into the real line.

(c) A set A is $\mathcal{P}$-null if P(A) = 0 for all $P \varepsilon \mathcal{P}$. Two sets A_1 and A_2 are $\mathcal{P}$-equivalent if their symmetric difference is $\mathcal{P}$-null. Two functions f_1 and f_2 are $\mathcal{P}$-equivalent if the set of points where they differ is $\mathcal{P}$-null. The relation symbol $\sim$ stands for $\mathcal{P}$-equivalence.

(d) By sub-field we mean a sub-σ-field of $\mathcal{A}$. A statistic is a measurable mapping of $(\mathcal{X}, \mathcal{A})$ into any measurable space. We identify a statistic with the sub-field it induces (see pp. 36-39 of Lehmann (1959)).

(e) By the $\mathcal{P}$-completion $\bar{\mathcal{A}}_0$ of a sub-field $\mathcal{A}_0$ we mean the least sub-field that contains $\mathcal{A}_0$ and all $\mathcal{P}$-null sets. Observe that $\bar{\mathcal{A}}_0$ may also be characterized as the class of all sets that are $\mathcal{P}$-equivalent to some member of $\mathcal{A}_0$. Two sub-fields are $\mathcal{P}$-equivalent if they have identical $\mathcal{P}$-completions.

(f) By transformation (usually denoted by g) we mean a one-to-one, bimeasurable mapping g of $(\mathcal{X},\mathcal{A})$ onto itself such that the family

$$\mathcal{P} g^{-1} = \{Pg^{-1}|P\varepsilon\mathcal{P}\}$$

of induced probability measures is the same as the family $\mathcal{P}$. A transformation g is called model-preserving if

$$Pg^{-1} \equiv P, \quad \text{for all} \quad P\varepsilon\,\mathcal{P}.$$

Observe that, if g is any transformation, then so also is g^n for each integral (positive or negative) n and that the identity map is always model-preserving. Also observe that any transformation carries $\mathcal{P}$-null sets into $\mathcal{P}$-null sets.

(g) Given a transformation g, the sub-field $\mathcal{A}(g)$ of g-invariant sets is defined as

$$\mathcal{A}(g) = \{A|g^{-1} A = A\}.$$

The $\mathcal{P}$-completion $\bar{\mathcal{A}}(g)$ of $\mathcal{A}(g)$ is then the class of all essentially g-invariant sets, i.e., sets that are $\mathcal{P}$-equivalent to some g-invariant set.

(h) The set A is almost g-invariant if $g^{-1}A \sim A$. It is easy to demonstrate that every almost g-invariant set is also essentially g-invariant and vice versa (see Lemma 1 for a sharper result).

Thus, $\bar{\mathcal{A}}(g)$ is also the class of all almost g-invariant sets.

(i) Given a class $\mathcal{G}$ of transformations g, the three sub-fields of α) $\mathcal{G}$-invariant, β) essentially $\mathcal{G}$-invariant and γ) almost $\mathcal{G}$-invariant sets are defined as follows :

α) $\mathcal{A}(\mathcal{G}) = \bigcap \mathcal{A}(g)$, ($\mathcal{G}$-invariant)

β) $\bar{\mathcal{A}}(\mathcal{G}) = \mathcal{P}$-completion of $\mathcal{A}(\mathcal{G})$, (essentially $\mathcal{G}$-invariant)

and γ) $\tilde{\mathcal{A}}(\mathcal{G}) = \bigcap \bar{\mathcal{A}}(g)$, (almost $\mathcal{G}$-invariant).

Observe that $\mathcal{A}(\mathcal{G}) \subset \bar{\mathcal{A}}(\mathcal{G}) \subset \tilde{\mathcal{A}}(\mathcal{G})$. With some assumptions on $\mathcal{G}$, one can prove (see Theorem 4 on p. 225 in Lehmann (1959)) the equality of $\bar{\mathcal{A}}(\mathcal{G})$ and $\tilde{\mathcal{A}}(\mathcal{G})$. That $\bar{\mathcal{A}}(\mathcal{G})$ can be a very small sub-field compared to $\tilde{\mathcal{A}}(\mathcal{G})$ is shown in example 1.

(j) A function f is α) $\mathcal{G}$-invariant, β) essentially $\mathcal{G}$-invariant, or γ) almost $\mathcal{G}$-invariant, according as

α) f = f(g), for all g ε $\mathcal{G}$,

β) f ~ some $\mathcal{G}$-invariant function,

or

γ) f ~ f(g), for all g ε $\mathcal{G}$.

Observe that f satisfies the definitions α), β) or γ) above if and only if f is measurable with respect to the corresponding sub-field defined in (i).

(k) When the class $\mathcal{G}$ of transformations g happens to be a group (with respect to the operation of composition of transformations), the sub-field $\mathcal{A}(\mathcal{G})$ of $\mathcal{G}$-invariant sets is easily recognized as follows. For each $x \varepsilon \mathcal{X}$, define the orbit 0_x as

$$0_x = \{x' | x' = gx \quad \text{for some} \quad g \varepsilon \mathcal{G}\}.$$

The orbits define a partition of $\mathcal{X}$, and the sub-field $\mathcal{A}(\mathcal{G})$ is the class of all (measurable) sets that are the unions of orbits. The sub-field of essentially $\mathcal{G}$-invariant sets is then the $\mathcal{P}$-completion of $\mathcal{A}(\mathcal{G})$. Our main concern, in this paper, is the sub-field $\tilde{\mathcal{A}}(\mathcal{G})$ and this is not so easily understood in terms of the orbits – unless $\mathcal{G}$ has a structure simple enough to ensure the equality of $\bar{\mathcal{A}}(\mathcal{G})$ and $\tilde{\mathcal{A}}(\mathcal{G})$ (see lemma 1 and example 1).

3. A Mathematical Introduction

Let $(\mathcal{X}, \mathcal{A}, \mathcal{P})$ be a given probability model and let g be a fixed model-preserving transformation [definition 2(f)]. We remark that the identity map is trivially model-preserving. In many instances of statistical interest there exist fairly wide classes of such transformations. However, it is not difficult to construct examples where no non-trivial transformation is model-preserving. See for instance example 5.

For each bounded function f [definition 2(b)], define the associated sequence $\{f_n\}$ of (uniformly bounded) functions as follows :

$$f_n(x) = [f(x) + f(gx) + \dots + f(g^n x)]/(n+1), \qquad n = 1, 2, 3, \dots .$$

Since g is measure-preserving for each $P \varepsilon \mathcal{P}$, the pointwise ergodic theorem tells us that the set N_f where $\{f_n(x)\}$ fails to converge is $\mathcal{P}$-null [definition 2(c)].

If we define f* as

$$f^*(x) = \begin{cases} \lim f_n(x) & \text{when } x \notin N_f, \\ 0 & \text{otherwise} \end{cases}$$

then, it is easily seen that

i) f* is $\mathcal{A}(g)$-measurable [definition 2(g)], i.e., it is g-invariant [definition 2(i)], and that

ii) for all $B \varepsilon \mathcal{A}(g)$ and $P \varepsilon \mathcal{P}$

$$\int_B f dP = \int_B f^* dP.$$

This is another way of saying that f* is the conditional expectation of f, given the sub-field $\mathscr{A}(g)$ of g-invariant sets. Since the definition of f* does not involve P, we have the following theorem (lemma 2 in Farrell (1962)).

Theorem 1. If the transformation g preserves the model ($\mathscr{X}$, $\mathscr{A}$, $\mathscr{P}$), then the sub-field $\mathscr{A}(g)$ is sufficient.

[Remark : Note that in the proof of theorem 1 we have not used the assumptions that g is a one-to-one map and that it is bimeasurable. The proof remains valid for any measurable mapping of ($\mathscr{X}$, $\mathscr{A}$) into itself that is measure-preserving for each $P \varepsilon \mathscr{P}$. A similar remark will hold true for a number of other results to be stated later. However, in a study of the statistical theory of invariance (see Lehmann (1959)) it seems appropriate to restrict our attention to one-to-one, bimeasurable maps of ($\mathscr{X}, \mathscr{A}$) onto itself that preserve the model either wholly or partially.]

Now, given a class $\mathscr{G}$ of model-preserving transformations g, what can we say about the sufficiency of the sub-field

$$\mathscr{A}(\mathscr{G}) = \bigcap \mathscr{A}(g)$$

of $\mathscr{G}$-invariant [definition 2(i)] sets? The intersection of two sufficient sub-fields is not necessarily sufficient. However, it is known (see Theorem 4 and Corollary 2 of Burkholder (1961)) that the intersection of the $\mathscr{P}$-completions [definition 2(e)] of a countable number of sufficient sub-fields is sufficient. Using this result, we have the following theorem (theorem 2 in Farrell (1962)).

Theorem 2. If the class $\mathscr{G}$ of model-preserving transformations g is countable, then the sub-field

$$\tilde{\mathscr{A}}(\mathscr{G}) = \bigcap \bar{\mathscr{A}}(g)$$

of almost $\mathscr{G}$-invariant sets [definition 2(i)(γ)] is sufficient.

Given a countable class $\mathscr{G}$ of transformations, consider the larger class $\mathscr{G}^*$ of transformations of the type $\alpha_1 \alpha_2 \ldots \alpha_n$, where each α_i is such that either α_i or α_i^{-1} belongs to $\mathscr{G}$, and n is an arbitrary positive integer. The following properties of $\mathscr{G}^*$ are easy to check :

a) $\mathscr{G}^*$ is a group (the group operation being composition of transformations),
b) $\mathscr{G}^*$ is countable,
c) $\mathscr{A}(\mathscr{G}^*) = \mathscr{A}(\mathscr{G})$ and $\tilde{\mathscr{A}}(\mathscr{G}^*) = \tilde{\mathscr{A}}(\mathscr{G})$.

Now, let A be an arbitrary almost $\mathscr{G}^*$-invariant set, i.e., $g^{-1}A \sim A$ (equivalently, $gA \sim A$) for all $g \varepsilon \mathscr{G}^*$. Consider the set

$$B = \bigcap gA$$

where the intersection is taken over all $g \varepsilon \mathcal{G}^*$. Since $\mathcal{G}^*$ is a group, the set B must be $\mathcal{G}^*$-invariant. Again, since $\mathcal{G}^*$ is countable, and each gA is $\mathcal{P}$-equivalent to A, we have $B \sim A$. We have thus established the following lemma.

Lemma 1. For any countable class $\mathcal{G}$ of transformations [definition 2(f)], every almost $\mathcal{G}$-invariant set is essentially $\mathcal{G}$-invariant, i.e., $\tilde{\mathcal{A}}(\mathcal{G}) = \bar{\mathcal{A}}(\mathcal{G})$.

Theorem 2, together with lemma 1 and the observation that a sub-field that is $\mathcal{P}$-equivalent to a sufficient sub-field is itself sufficient, leads to the following theorem.

Theorem 3. If $\mathcal{G}$ is a countable class of model-preserving transformations, then the sub-field $\mathcal{A}(\mathcal{G})$ of $\mathcal{G}$-invariant sets is sufficient.

[Remark : Note that in Theorem 3 we have used the one-to-oneness and bi-measurability of our transformations.]

Before proceeding further, let us consider an example which shows that Theorem 3 is no longer true if we drop the condition that $\mathcal{G}$ is countable.

Example 1. Let $\mathcal{P}$ be a family of continuous distributions on the real line $\mathcal{X}$ where $\mathcal{A}$ is the σ -field of Borel-sets. Let $\mathcal{G} = \{g\}$ be the class of all one-to-one maps of $\mathcal{X}$ onto $\mathcal{X}$ such that $\{x \mid gx \neq x\}$ is finite. Clearly, $\mathcal{G}$ is a group and every member of it is model-preserving. It is easy to check that the sub-field $\mathcal{A}(\mathcal{G})$ of $\mathcal{G}$-invariant sets consists of ϕ and $\mathcal{X}$ only, and hence is not sufficient. The sub-field $\tilde{\mathcal{A}}(\mathcal{G})$ of almost $\mathcal{G}$-invariant sets is the same as $\mathcal{A}$ and is sufficient.

Consider another example.

Example 2. Let $\mathcal{X}$ be the n-dimensional Euclidean space and $\mathcal{P}$ the class of all probability measures (on the σ-field $\mathcal{A}$ of all Borel-sets) that are symmetric (in the co-ordinates). If $\mathcal{G} = \{g\}$ is the group of all permutations (of the co-ordinates) then $\mathcal{A}(\mathcal{G})$ is the sub-field of all sets that are symmetric (in the co-ordinates) and is sufficient. Since the empty set is the only $\mathcal{P}$-null set, the two sub-fields $\mathcal{A}(\mathcal{G})$ and $\tilde{\mathcal{A}}(\mathcal{G})$ are the same here.

Given a probability model $(\mathcal{X}, \mathcal{A}, \mathcal{P})$ we ask ourselves the following questions.

1. How wide is the group $\mathcal{G}$ of all model-preserving transformations?
2. Is the sub-field $\tilde{\mathcal{A}}(\mathcal{G})$ of almost $\mathcal{G}$-invariant sets sufficient?
3. If a least (minimal) sufficient sub-field $\mathcal{L}$ exists, then is it true that $\mathcal{L} \sim \tilde{\mathcal{A}}(\mathcal{G})$?
4. What is $\tilde{\mathcal{A}}(\mathcal{G})$, when $\mathcal{G}$ is the class of all transformations that partially preserve the model in a given manner?

4. Statistical Motivation

Suppose we have two different systems of measurement co-ordinates for the outcome of a statistical experiment, so that, if a typical outcome is recorded as x under the first system, the same outcome is recorded as gx under the second system. Let us suppose further that the two statistical variables x and gx have the same domain (sample space) $\mathcal{X}$, the same family $\mathcal{A}$ of events, and the same class of probability measures $\mathcal{P}$. Furthermore, if $P \varepsilon \mathcal{P}$ holds for x, then the same probability measure P also holds for gx. The second system of measurement co-ordinates may then be represented mathematically as a model-preserving transformation g of the statistical model $(\mathcal{X}, \mathcal{A}, \mathcal{P})$. The principle of invariance then stipulates that the decision rule should be invariant with respect to the transformation g – and this irrespective of the actual decision problem. If the choice of a new system of measurement co-ordinates leaves the problem (whatever it is) entirely unaffected, then so also should be every reasonable inference procedure.

Thus, if g is model-preserving, the principle of invariance leads to the invariance reduction of the model $(\mathcal{X}, \mathcal{A}, \mathcal{P})$ to the simpler model $(\mathcal{X}, \mathcal{A}(g), \mathcal{P})$, where $\mathcal{A}(g)$ is the sub-field of g-invariant sets. To put it differently, the principle of invariance requires every decision function to be g-invariant or $\mathcal{A}(g)$-measurable. Consider now the class $\mathcal{G}$ of all model-preserving transformations g. Must we, following the principle of invariance, insist that every decision rule be $\mathcal{G}$-invariant (i.e. g-invariant for every $g \varepsilon \mathcal{G}$)? We have already noted in example 1 that, when $\mathcal{P}$ consists of a family of non-atomic measures, the class $\mathcal{G}$ is large enough to reduce the sub-field $\mathcal{A}(\mathcal{G})$ of $\mathcal{G}$-invariant sets to the trivial one (consisting of only the empty set and the whole space). Obviously, we cannot (must not) reduce $\mathcal{A}$ all the way down to $\mathcal{A}(\mathcal{G})$ or even to $\bar{\mathcal{A}}(\mathcal{G})$– the sub-field of essentially $\mathcal{G}$-invariant sets. A logical compromise (with the principle) would be to reduce $\mathcal{A}$ to the sub-field $\tilde{\mathcal{A}}(\mathcal{G})$ of all almost $\mathcal{G}$-invariant sets – that is to insist upon the decision function to be almost $\mathcal{G}$-invariant.

The principle of sufficiency is another reduction principle of the omnibus type. If $\mathcal{S}$ be a sufficient sub-field, then this principle tells us not to use a decision function that is not $\mathcal{S}$-measurable. Suppose there exists a least (minimal) sufficient sub-field $\mathcal{L}$. Following the principle of sufficiency, we reduce the model $(\mathcal{X}, \mathcal{A}, \mathcal{P})$ to the model $(\mathcal{X}, \mathcal{L}, \mathcal{P})$.

Which of the two reductions (invariance or sufficiency) is more extensive? In other words, what is the relation between $\tilde{\mathcal{A}}(\mathcal{G})$ and $\mathcal{L}$?

Theorem 1 tells us that $\mathcal{A}(g)$ is sufficient (for each g in $\mathcal{G}$). Since $\mathcal{L}$ is the least sufficient sub-field we have (from the definition of $\mathcal{L}$)

$$\bar{\mathcal{L}} \subset \bar{\mathcal{A}}(g) \text{ for all } g \in \mathcal{G}.$$

Theorem 4 follows at once.

Theorem 4. $$\bar{\mathcal{L}} \subset \bigcap \bar{\mathcal{A}}(g) = \tilde{\mathcal{A}}(\mathcal{G}).$$

[Remark : Theorem 4 does not establish the sufficiency of $\tilde{\mathcal{A}}(\mathcal{G})$. (See example 1 in Burkholder (1961).)]

We thus observe that the invariance reduction (in terms of the group $\mathcal{G}$ of all model-preserving transformations) of a model can never be more extensive than its maximal sufficiency reduction (if one such reduction is available). The principal question raised in this paper is,"When is $\tilde{\mathcal{A}}(\mathcal{G})$ equal to $\bar{\mathcal{L}}$?" We shall show later that in many familiar situations the sub-fields $\tilde{\mathcal{A}}(\mathcal{G})$ and $\mathcal{L}$ are essentially equal. This raises the question about the nature of $\tilde{\mathcal{A}}(\mathcal{G})$, where $\mathcal{G}$ is the class of all transformations that preserve the model partially in some well-defined manner. This question is discussed in a later section. In the next two sections we give two alternative approaches to Theorem 4.

5. When a boundedly complete sufficient sub-field exists

Let us suppose that the sub-field $\mathcal{L}$ is sufficient and boundedly complete. We need the following lemma.

Lemma 2. If z is a bounded $\mathcal{A}$-measurable function such that

$$E(z\,|P) \equiv 0 \quad \text{for all} \quad P \varepsilon \mathcal{P},$$

then, for all bounded $\mathcal{L}$-measurable functions f, it is true that

$$E(zf|P) \equiv 0 \quad \text{for all} \quad P \varepsilon \mathcal{P}.$$

The proof of this well-known result is omitted. Now, let S be an arbitrary member of $\mathcal{L}$, and let g be an arbitrary model preserving transformation. Let $S_0 = g^{-1}S$. Since g is model preserving we have

$$P(S) \equiv Pg^{-1}S \equiv P(S_0) \quad \text{for all} \quad P \varepsilon \mathcal{P}.$$

Writing I_S for the indicator of S, and noting that $I_S - I_{S_0}$ and I_S satisfy the conditions for z and f in Lemma 2, we at once have

$$E[(I_S - I_{S_0})\, I_S | P] \equiv 0, \quad \text{for all } P \varepsilon \mathcal{P},$$

or

$$P(S) \equiv P(SS_0), \quad \text{for all } P \varepsilon \mathcal{P}.$$

Hence,

$$\begin{aligned} P(S \Delta S_0) &\equiv P(S) + P(S_0) - 2P(SS_0) \\ &\equiv 2[P(S) - P(SS_0)] \\ &\equiv 0, \quad \text{for all } P \varepsilon \mathcal{P}. \end{aligned}$$

That is, for all $S \varepsilon \mathcal{L}$, the two sets S and g^{-1} S are $\mathcal{P}$-equivalent. In other words,

$$\mathcal{L} \subset \bar{\mathcal{A}}(g),$$

and since g is an arbitrary element of $\mathcal{G}$ (the class of all model-preserving transformations) we have the following theorem.

Theorem 4 (a). If $\mathcal{L}$ is a boundedly complete sufficient sub-field then $\mathcal{L} \subset \tilde{\mathcal{A}}(\mathcal{G})$.

[Remark : Since bounded completeness of $\mathcal{L}$ implies that $\mathcal{L}$ is the least sufficient sub-field, Theorem 4(a) is nothing but a special case of Theorem 4. The proof of Theorem 4(a) is simple and amenable to a generalization to be discussed later.]

6. The dominated case

We make a slight digression to state a useful lemma. Let T be a measurable map of $(\mathcal{X}, \mathcal{A})$ into $(\mathcal{Y}, \mathcal{B})$. Let P and Q be two probability measures on $\mathcal{A}$ and let PT^{-1} and QT^{-1} be the corresponding measures on $\mathcal{B}$. Suppose that Q dominates P and let $f = dP/dQ$ be the Randon-Nikodym derivative defined on $\mathcal{X}$. It is then clear that QT^{-1} dominates PT^{-1}. Let $h = (dPT^{-1})/(dQT^{-1})$. The function hT on $\mathcal{X}$ (defined as $hT(x) = h(Tx)$) is $T^{-1}(\mathcal{B})$-measurable and satisfies the following relation.

Lemma 3. $$hT = E(f|T^{-1}(\mathcal{B}), Q),$$

i.e., hT is the conditional expectation of f, given $T^{-1}(\mathcal{B})$ and Q.

The proof of this well-known lemma consists of checking the identity

$$\int_B f dQ \equiv \int_B hT dQ, \text{for all } B \varepsilon T^{-1}(\mathcal{B}).$$

Corollary. If $T^{-1}(\mathcal{B})$ is Q-equivalent to $\mathcal{A}$, then f and hT are Q-equivalent.

Now, returning to our problem, let $\mathcal{G}$ be the class of all transformations g that preserves the model $(\mathcal{X}, \mathcal{A}, \mathcal{P})$. Let us suppose that $\mathcal{P}$ is dominated by some σ-finite measure. It follows that there exists a countable collection $P_1, P_2, \ldots$, of elements in $\mathcal{P}$ such that the convex combination

$$Q = \Sigma c_i P_i \,, \quad c_i > 0, \;\; \Sigma c_i = 1,$$

dominates the family $\mathcal{P}$. Let $f_p = dP/dQ$ be a fixed version of the Radon-Nikodym derivative of P with respect to Q. The factorization theorem for sufficient statistics asserts that a sub-field $\mathcal{A}_0$ is sufficient if and only if f_p is $\bar{\mathcal{A}}_0$ -measurable for every $P \in \mathcal{P}$, where $\bar{\mathcal{A}}_0$ is the $\mathcal{P}$-completion of $\mathcal{A}_0$. We now prove that f_p is $\tilde{\mathcal{A}}(\mathcal{G})$-measurable for every P. (The $\mathcal{P}$-completion of $\tilde{\mathcal{A}}(\mathcal{G})$ is itself).

Since $Pg^{-1} = P$ for all $P \in \mathcal{P}$, it follows that $Qg^{-1} = Q$. From Lemma 3 we have

$$f_p g = E(f_p | g^{-1}(\mathcal{A}), Q).$$

The assumption that g is one-to-one and bimeasurable implies that $g^{-1}(\mathcal{A}) = \mathcal{A}$. Hence $f_p g = f_p$ a.e.w. [Q]. Since Q dominates $\mathcal{P}$, it follows that f_p is almost g-invariant, i.e., is $\tilde{\mathcal{A}}(g)$ measurable. Since, in the above argument, g is an arbitrary element of $\mathcal{G}$, we have the following theorem. (See problem 19 on p.253 in Lehmann (1959)).

Theorem 4(b). If $\mathcal{P}$ is dominated, then $\tilde{\mathcal{A}}(\mathcal{G})$ is sufficient.

[Remark : Since, in the dominated set-up, the least sufficient sub-field $\mathcal{L}$ always exists and since, in this set-up, any sub-field containing $\mathcal{L}$ is necessarily sufficient, it is clear that Theorem 4(b) is nothing but an immediate corollary to Theorem 4. Also note that in the present proof (of Theorem 4(b)) we had to draw upon our supposition that g is one-to-one and bimeasurable. This section has been written only with the object of drawing attention to some aspects of the problem.]

7. Examples

Example 3 : Let y be a real random variable having a uniform distribution over the unit interval. For each c in [0, 1) define the transformation g_c as

$$g_c y = y + c \pmod 1$$

In this example, $\mathcal{P}$ consists of a single measure and each g_c is model-preserving (measure-preserving). If $\mathcal{G}_0 = \{ g_c | 0 < c < 1 \}$, then the only $\mathcal{G}_0$-invariant sets are the empty set and the whole of the unit interval. Here, $\mathcal{A}(\mathcal{G}_0)$ is sufficient.

[Remark : In this case, there are a very large number of measure-preserving transformations that are not one-to-one maps of [0, 1] onto itself. For example, let $a_n(y)$ be the nth digit in the decimal representation of y and let

$$gy = \sum_{k=1}^{\infty} \frac{a_{n_k}(y)}{10^k}$$

where $\{n_k\}$ is a fixed increasing sequence of natural numbers.]

Again, if x has a fixed continuous distribution on the real line with cumulative distribution function F, then the class $\mathcal{G}_0$ of transformations g_c defined as

$$g_c x = F^{-1}[F(x) + c \ (\mathrm{mod}\ 1)], \quad 0 < c < 1$$

are all model-preserving for x. [In case F is not a strictly increasing function of x, we define $F^{-1}(y) = \inf\{x \mid F(x) = y\}$.]

Thus, for any fixed continuous distribution on the real line, there always exists a large class of measure-preserving transformations.

Example 4. Let $\mathcal{X} = [0, \infty)$ and let x have a uniform distribution over the interval $[0,\theta)$, where θ is an unknown positive integer. Here $\mathcal{P}$ consists of a countable infinity of probability measures. For each c in [0, 1), define the transformation g_c as

$$g_c x = [x] + \{x - [x] + c \ (\mathrm{mod}\ 1)\},$$

where [x] is the integer-part of x.

Here, each g_c is model-preserving. The minimal (least) sufficient statistic [x] is also the maximal-invariant with respect to the group $\mathcal{G}_0 = \{g_c\}$ of model-preserving transformations defined above. Thus, if $\mathcal{L}$ is the sub-field (least sufficient) generated by [x] and $\mathcal{A}(\mathcal{G}_0)$ is the sub-field of $\mathcal{G}_0$-invariant sets, we have

$$\mathcal{L} = \mathcal{A}(\mathcal{G}_0). \tag{a}$$

Now, if $\mathcal{G}$ is the class of all model-preserving transformations, then (as we have seen in Example 1) the sub-field $\mathcal{A}(\mathcal{G})$ will reduce to the level of triviality. (It will consist of only the empty set and the whole set $\mathcal{X}$.) However, from Theorem 4 we have

$$\bar{\mathcal{L}} \subset \tilde{\mathcal{A}}(\mathcal{G}). \tag{b}$$

Since the group $\mathcal{G}_0$ has a decent structure, we can apply Stein's theorem (theorem 4 on p. 225 in Lehmann (1959)) to prove that

$$\mathcal{A}(\mathcal{G}_0) \sim \tilde{\mathcal{A}}(\mathcal{G}_0). \tag{c}$$

Since $\mathcal{G}_0 \subset \mathcal{G}$, we at once have

$$\tilde{\mathcal{A}}(\mathcal{G}) \subset \tilde{\mathcal{A}}(\mathcal{G}_0). \tag{d}$$

Putting the relations (a), (c) and (d) together we have

$$\tilde{\mathscr{A}}(\mathscr{G}) \subset \bar{\mathscr{L}} \tag{e}$$

Putting (b) and (e) together we finally have

$$\bar{\mathscr{L}} = \tilde{\mathscr{A}}(\mathscr{G}),$$

i.e., the least-sufficient sub-field (rather, the $\mathscr{P}$-completion of any version of the least sufficient sub-field) and the sub-field of almost $\mathscr{G}$-invariant sets are identical.

The chain of arguments, detailed as above, is of a general nature and will be used repeatedly in the sequel.

Example 5. Let x be a normal variable with unit variance and mean equal to either μ_1 or μ_2. Does there exist a non-trivial transformation of $\mathscr{X}$(the real line) into itself that preserves each of the two measures? That the answer is "no" is seen as follows. Let $\mathscr{G}$ be the class of all model-preserving transformations. In view of theorem 4, $\tilde{\mathscr{A}}(\mathscr{G})$ contains the least sufficient sub-field $\mathscr{L}$. But, in this example, the likelihood ratio (which is the least sufficient statistic) is

$$\exp\,[(\mu_1 - \mu_2)\,x - \tfrac{1}{2}(\mu_1^2 - \mu_2^2)],$$

and this is a one-to-one measurable function of x. Thus, every set is almost $\mathscr{G}$-invariant. And this implies that every g in $\mathscr{G}$ must be equivalent to the identity map.

Example 6. Let $x_1, x_2, \ldots, x_n$ be n independent observations on a normal variable with known mean and unknown standard deviation σ. Without loss of generality we may assume the mean to be zero. Let $\mathscr{G}_0$ be the group of all linear orthogonal transformations of the n-dimensional Euclidean space onto itself. Clearly, every member of $\mathscr{G}_0$ is model-preserving. That the class $\mathscr{G}$ of all model-preserving transformations is much wider than $\mathscr{G}_0$ is seen as follows. Let

$$y_i = \phi_i(x_i)|x_i|, \quad i = 1, 2, \ldots, n,$$

where $\phi_1, \phi_2, \ldots, \phi_n$ are arbitrary skew-symmetric (i.e., $\phi(x) = -\phi(-x)$ for all x) functions on the real line that take only the two values -1 and +1. It is easily checked that, whatever the value of σ, the two vectors $(x_1, x_2, \ldots, x_n)$ and $(y_1, y_2, \ldots, y_n)$ are identically distributed, i.e., the above transformation (though non-linear) is model-preserving. However, the sub-group $\mathscr{G}_0$ is large enough to lead us to Σx_i^2 – which is the least sufficient statistic - as the maximal invariant. In view of the decent structure of the sub-group $\mathscr{G}_0$, the arguments given for example 4 are again available to prove the equality of $\bar{\mathscr{L}}$ and $\tilde{\mathscr{A}}(\mathscr{G})$.

8. Transformations of a Set of Normal Variables

This section is devoted to a study of the special case (model) of n independent normal variables $x_1, x_2, \ldots, x_n$ with equal unknown means μ and equal unknown standard deviations σ. Hence Σx_i and Σx_i^2 jointly constitute the least sufficient statistic.

If $\bar{\mathcal{L}}$ is the $\mathcal{P}$-completion of the sub-field $\mathcal{L}$ induced by $(\Sigma x_i, \Sigma x_i^2)$, then we know from Theorem 4, that

$$\bar{\mathcal{L}} \subset \tilde{\mathcal{A}}(\mathcal{G})$$

where $\mathcal{G}$ is the class of all the model-preserving transformations of $\mathbf{x} = (x_1, x_2, \ldots, x_n)$ to $\mathbf{y} = (y_1, y_2, \ldots, y_n)$. For any model-preserving transformation from $\mathbf{x}$ to $\mathbf{y}$, we, therefore, have

$$\Sigma x_i \sim \Sigma y_i$$

and

$$\Sigma x_i^2 \sim \Sigma y_i^2$$

i.e, the statistic $(\Sigma x_i, \Sigma x_i^2)$ is almost $\mathcal{G}$-invariant. If we can demonstrate the existence of a 'decent' sub-group $\mathcal{G}_0$ of $\mathcal{G}$ for which the statistic $(\Sigma x_i, \Sigma x_i^2)$ is the maximal invariant, then (following the method of proof indicated in examples 4 and 6) we can show that $\bar{\mathcal{L}}$ is indeed equal to $\tilde{\mathcal{A}}(\mathcal{G})$.

Let $\mathcal{G}_0$ be the sub-group of all linear model-preserving transformations. Do there exist non-trivial linear model-preserving transformations (i.e., linear transformations that are not a permutation of the co-ordinates)? That the answer is "yes" is seen as follows. Let $\mathcal{M} = \{\mathbf{M}\}$ be the family of all orthogonal $n \times n$ matrices with the initial row as

$$\left(\frac{1}{\sqrt{n}}, \frac{1}{\sqrt{n}}, \ldots, \frac{1}{\sqrt{n}}\right).$$

Now, if $\mathbf{x} = (x_1, x_2, \ldots, x_n)$ are independent $N(\mu, \sigma)$'s and we define

$$\mathbf{y'} = \mathbf{Mx'}$$

(where $\mathbf{x'}$ is the corresponding column vector), then the y_i's are independent normal variables with equal standard deviation σ and with means as follows :

$$E(y_1) = \sqrt{n}\,\mu, \quad E(y_i) = 0, \qquad i = 2, 3, \ldots, n.$$

Thus, for an arbitrary pair of members $\mathbf{M}_1$ and $\mathbf{M}_2$ in $\mathcal{M}$, we note that $\mathbf{M}_1\mathbf{x'}$ and $\mathbf{M}_2\mathbf{x'}$ are identically distributed (whatever the values of μ and σ). Therefore, the two vectors $\mathbf{x'}$ and $\mathbf{M}_2^{-1}\mathbf{M}_1\mathbf{x'}$ are identically distributed for each μ and σ. In other words, the linear transformation defined by the matrix

$$M^{-1}_2 M_1 \quad (= M'_2 M_1,$$

since M_2 is orthogonal) is model-preserving for each pair M_1, M_2 of members from $\mathcal{M}$.

[Remark : Later on we shall have some use for this way of generating members of $\mathcal{G}_0$.]

For example, the 4×4 matrix

$$\begin{bmatrix} 1/2 & 1/2 & 1/2 & -1/2 \\ 1/2 & 1/2 & -1/2 & 1/2 \\ 1/2 & -1/2 & 1/2 & 1/2 \\ -1/2 & 1/2 & 1/2 & 1/2 \end{bmatrix}$$

defines a member of $\mathcal{G}_0$ for $n = 4$.

Let H be a typical $n \times n$ matrix that defines a model-preserving linear transformation. We are going to make a brief digression about the nature of H. From the requirement that each co-ordinate of

$$\mathbf{y}' = H\mathbf{x}'$$

has mean μ and standard deviation σ, we at once have that the elements in each row of H must add up to unity and that their squares also must add up to unity. From the mutual independence of the y_i's, it follows that the row vectors of H must be mutually orthogonal. Thus, H must be an orthogonal matrix with unit row sums. Now, for each model-preserving H, its inverse

$$H^{-1}(= H', \text{ since H is orthogonal})$$

is necessarily also model-preserving and, thus, the columns of H must also add up to unity, etc.

[It is easily checked that the sub-group $\mathcal{G}_0 = \{H\}$ of linear model-preserving transformations of the n-space onto itself may also be characterized as the class of all linear transformations that preserve both the sum and the sum of squares of the co-ordinates. The author came to learn that G.W. Haggastrom of the University of Chicago had come upon these matrices from this point of view and had a brief discussion on them in an unpublished work of his. We call such matrices by the name Haggastrom-matrix.]

Going back to our problem, we have to demonstrate that the statistic $T = (\Sigma x_i, \Sigma x_i^2)$ is a maximal invariant with respect to the sub-group $\mathcal{G}_0 = \{H\}$ of model-preserving linear transformations. For this we have only to prove that if $\mathbf{a} = (a_1, a_2, \ldots, a_n)$ and $\mathbf{b} = (b_1, b_2, \ldots, b_n)$ are any two points in n-space such that

$T(\mathbf{a}) = T(\mathbf{b})$ then there exists a Haggastrom-matrix H such that

$$\mathbf{b}' = H\mathbf{a}'.$$

Let M_1 and M_2 be two arbitrary members of $\mathcal{M}$ — the class of all orthogonal $n \times n$ matrices with the leading row as $(1/\sqrt{n}, \ldots, 1/\sqrt{n})$. If $\alpha = M_1\mathbf{a}'$ and $\beta = M_2\mathbf{b}'$, then we have, from $T(\mathbf{a}) = T(\mathbf{b})$, that $\alpha_1 = \beta_1$ and that

$$\sum_2^n \alpha_i^2 = \sum_2^n \beta_i^2 .$$

It follows that there exists an $(n-1) \times (n-1)$ orthogonal matrix that will transform $(\alpha_2, \alpha_3, \ldots, \alpha_n)$ to $(\beta_2, \beta_3, \ldots, \beta_n)$. And this, in turn, implies that there exists an $n \times n$ orthogonal matrix K, with the first row as $(1, 0, 0, \ldots, 0)$, such that

$$\boldsymbol{\beta}' = K\boldsymbol{\alpha}' .$$

Thus, the transformation $M_2^{-1}KM_1$ takes **a** into **b**. Now, note that, since K is an orthogonal matrix with the initial row as $(1, 0, 0, \ldots, 0)$, the matrix KM_1 is orthogonal with its initial row as $(1/\sqrt{n}, 1/\sqrt{n}, \ldots, 1/\sqrt{n})$, i.e., $KM_1 \in \mathcal{M}$. In other words, there exists a matrix of the form $M_2^{-1}M$ (with M_2 and M belonging to $\mathcal{M}$) which transforms **a** into **b**. We have already noted that all such matrices belong to $\mathcal{G}_0$, and this completes our proof that $(\Sigma x_i, \Sigma x_i^2)$ is a maximal invariant with respect to $\mathcal{G}_0$.

In example 6 of the previous section, we considered the particular case of the foregoing problem where μ is known. Consider now the other particular case where σ is known and μ is the only unknown parameter. In this case Σx_i is the least sufficient statistic. Now, it is no longer possible to produce a sub-group $\mathcal{G}_0$ of linear model-preserving transformations such that Σx_i is the maximal invariant with respect to $\mathcal{G}_0$. This is because every linear model-preserving transformation must necessarily be orthogonal and would thus preserve Σx_i^2 also. We proceed as follows.

Suppose (without loss of generality) that $\sigma = 1$. Let $\mathbf{y}' = M\mathbf{x}'$ where M is a fixed member of the class $\mathcal{M}$ of orthogonal $n \times n$ matrices with the initial row as $(1/\sqrt{n}, \ldots, 1/\sqrt{n})$. Observe that the y_i's are mutually independent and that $y_2, y_3, \ldots, y_n$ are standard normal variables. Let F stand for the cumulative distribution function of a standard normal variable. Let $c_2, c_3, \ldots, c_n$ be arbitrary constants in $[0, 1)$. If we define $z_1 = y_1$ and

$$z_i = F^{-1}\,[F(y_i) + c_i \,(\text{mod } 1)], \qquad i = 2, 3, \ldots, n,$$

then $\mathbf{z} = (z_1, z_2, \ldots, z_n)$ has the same distribution as that of $\mathbf{y}$ and so it follows that $\mathbf{x}'$ and $M^{-1}\mathbf{z}'$ are identically distributed. Observe that we have described above a model-preserving transformation corresponding to each $(n-1)$-vector $(c_2, c_3,$

. . . , c_n) with the c_i's in [0, 1). It is easily checked that Σx_i is a maximal invariant with respect to the above class of model-preserving transformations.

9. Parameter-preserving transformations

Let $\gamma = \gamma(P)$ be the parameter of interest. That is, the experimenter is interested only in the characteristic $\gamma(P)$ of the measure P (that actually holds) and considers all other details about P to be irrelevent (nuisance parameters). We define a γ-preserving transformation as follows.

Definition. The transformation [see Definition 2(f)] g is γ-preserving if $\gamma(Pg^{-1}) \equiv \gamma(P)$, for all $P \in \mathcal{P}$.

If g is γ-preserving then so also is g^{-1}. The composition of any two γ-preserving transformations is also γ-preserving. Let $\mathcal{G}_\gamma$ be the group of all γ-preserving transformations.

In the particular case where $\gamma(P) = P$, the γ-preserving transformations are what we have so far been calling model-preserving. If $\mathcal{G}$ is the class of all model-preserving transformations, then note that $\mathcal{G} \subset \mathcal{G}_\gamma$ for every γ.

If γ is the parameter of interest, then the principle of invariance leads to the reduction of $\mathcal{A}$ to the sub-field $\tilde{\mathcal{A}}(\mathcal{G}_\gamma)$. This section is devoted to a study of the sub-field $\tilde{\mathcal{A}}(\mathcal{G}_\gamma)$

Since $\mathcal{G} \subset \mathcal{G}_\gamma$ we have

$$\tilde{\mathcal{A}}(\mathcal{G}_\gamma) \subset \tilde{\mathcal{A}}(\mathcal{G}).$$

Let us suppose that the least sufficient sub-field $\bar{\mathcal{L}}$ exists. We have discussed a number of examples where $\tilde{\mathcal{A}}(\mathcal{G})$ and $\bar{\mathcal{L}}$ are identical. [The author believes that the above identification is true under very general conditions.] In all such situations we therefore have

$$\tilde{\mathcal{A}}(\mathcal{G}_\gamma) \subset \bar{\mathcal{L}} .$$

That is, when the interest of the experimenter is concentrated on some particular characteristic $\gamma(P)$ or P, the principle of invariance will usually reduce the data more extensively than the principle of sufficiency.

Theorems 4, 4(a) and 4(b) tell us that

$$\bar{\mathcal{L}} \subset \tilde{\mathcal{A}}(\mathcal{G}).$$

The following theorem gives us a similar lower bound for $\tilde{\mathcal{A}}(\mathcal{G}_\gamma)$.

A set A is called γ-oriented if, for every pair $P_1, P_2 \in \mathcal{P}$ such that $\gamma(P_1) = \gamma(P_2)$, it is true that $P_1(A) = P_2(A)$. In other words, a γ-oriented set is one whose probability depends on P through $\gamma(P)$. A sub-field is γ-oriented if every member of the sub-field is. The following theorem* is a direct generalization of Theorem 4(a).

Theorem 5. Let $\mathcal{G}_\gamma$ be the class of all γ-preserving transformations and let $\mathcal{B}$ be a sub-field that is

i) γ-oriented and

ii) contained in the boundedly complete sufficient sub-field $\mathcal{L}$ (which exists).

Then,

$$\mathcal{B} \subset \tilde{\mathcal{A}}(\mathcal{G}_\gamma).$$

Proof : Let $g \varepsilon \mathcal{G}_\gamma$ and $B \varepsilon \mathcal{B}$ and let $B_0 = g^{-1} B$.

Then

$$P(B_0) = P(g^{-1}B) \qquad (\because B_0 = g^{-1}B)$$
$$= Pg^{-1}(B).$$

Since g is γ-preserving, we have $\gamma(Pg^{-1}) = \gamma(P)$. And since B is γ-oriented we have

$$Pg^{-1}(B) = P(B).$$

Therefore,

$$P(B_0) = P(B) \quad \text{or} \quad E(I_B - I_{B_0} \mid P) \equiv 0 .$$

The rest of the proof is the same as in Theorem 4(a).

In the next section we repeatedly use the above theorem.

10. Some Typical Invariance Reductions

Let $x_1, x_2, \ldots, x_n$ be n independent and identical normal variables with unknown μ and σ. Suppose the parameter of interest is σ. Let $\mathcal{G}_\sigma$ be the class of all σ-preserving transformations. If $\bar{x}$ is the mean and s, the standard deviation of the n observations, then the pair $(\bar{x}, s)$ is a complete sufficient statistic. Also s is σ-oriented. From Theorem 5, we then have that s is almost $\mathcal{G}_\sigma$-invariant. Thus, the principle of invariance cannot reduce the data beyond s. That s is indeed the exact (upto $\mathcal{P}$-equivalence) attainable limit of invariance reduction is shown as follows.

Let {H} be the class of all n × n Haggastrom matrices (see section 8), i.e., each H is an orthogonal matrix with unit row and column sums. Let c be an arbitrary real

*The author wishes to thank Professor W.J. Hall of the University of North Carolina for certain comments that eventually led to this theorem.

number and let $\mathbf{i}$ stand for the n-vector (1, 1, . . . , 1). Consider all linear transformations of the type

$$\mathbf{y}' = H\mathbf{x}' + (c\mathbf{i})'.$$

If $\mathcal{G}_\sigma^*$ is this class of transformations, then it is easily verified that $\mathcal{G}_\sigma^*$ is a sub-group of $\mathcal{G}_\sigma$. That $\Sigma(x_i - \bar{x})^2$ is a maximal invariant with respect to the sub-group $\mathcal{G}_\sigma^*$ of σ-preserving transformations is seen as follows.

Let $\mathbf{a} = (a_1, a_2, \ldots, a_n)$ and $\mathbf{b} = (b_1, b_2, \ldots, b_n)$ be two arbitrary n-vectors such that

$$\Sigma(a_i - \bar{a})^2 = \Sigma(b_i - \bar{b})^2 .$$

Let $c = \bar{b} - \bar{a}$. Then the two vectors $\mathbf{a} + c\mathbf{i}$ and $\mathbf{b}$ have equal sums and sums of squares (of co-ordinates). Hence there exists a Haggstrom matrix H that transforms $\mathbf{a}+c\mathbf{i}$ into $\mathbf{b}$. In other words, the transformation

$$\mathbf{y}' = H\mathbf{x}' + (c\mathbf{i})'$$

maps $\mathbf{a}$ into $\mathbf{b}$.

If $\mathcal{B}(s)$ is the sub-field generated by s, then we have just proved that $\mathcal{B}(s) = \mathcal{A}(\mathcal{G}_\sigma^*)$. The proof of the $\mathcal{P}$-equivalence of

$$\mathcal{B}(s) \quad \text{and} \quad \tilde{\mathcal{A}}(\mathcal{G}_\sigma)$$

will follow the familiar pattern set up earlier for example 4 in section 7.

Now, suppose that our parameter of interest is $\gamma = \mu/\sigma$. The statistic $\bar{x}/s$ is γ-oriented and hence (Theorem 5) is almost $\mathcal{G}_\gamma$-invariant, where $\mathcal{G}_\gamma$ is the group of all γ-preserving transformations.

Consider now the sub-group $\mathcal{G}_\gamma^*$ of all linear transformations of the type

$$\mathbf{y}' = cH\mathbf{x}'$$

where $c > 0$ and H is a Haggastrom-matrix.

It is easily checked that the maximal invariant with respect to the sub-group $\mathcal{G}_\gamma^*$ is the statistic $\bar{x}/s$ and hence, the sub-field $\mathcal{B}(\bar{x}/s)$ generated by $\bar{x}/s$ is $\mathcal{P}$-equivalent to the sub-field $\tilde{\mathcal{A}}(\mathcal{G}_\gamma)$ of all almost $\mathcal{G}_\gamma$-invariant sets.

The case where our parameter of interest is $|\mu|/\sigma$ is very similar. Once again we observe that the statistic $|\bar{x}|/s$ is oriented towards $|\mu|/\sigma$. Hence, making use of Theorem 5 and observing that $|\bar{x}|/s$ is the maximal invariant with respect to the sub-group of linear parameter-preserving transfromations of the form

$$\mathbf{y}' = cH\mathbf{x}' \ (c \neq 0, \text{ H a Haggastrom matrix}),$$

we are able to show that the invariance reduction of the data is to the statistic $|\bar{x}|/s$.

11. Some Final Remarks

a)In statistical literature,the principle of invariance has been used in a rather half-hearted manner (see, e.g., Hall et al (1965) and Lehmann (1959)). We do not find any consideration given to the present project of reducing the data with the help of the whole class $\mathcal{G}_\gamma$ of γ-preserving transformations. In fact, the question of how extensive the class $\mathcal{G}_\gamma$ can be has escaped general attention. One usually works in the framework of a relatively small and simple sub-group of $\mathcal{G}_\gamma$ and invokes simultaneously the two different principles of sufficiency and invariance for the purpose of arriving at a satisfactory data reduction. The main object of the present article is to investigate how far the principle of invariance by itself can take us.

b) The main limitation of the principle of sufficiency is that it does not recognize nuisance parameters. Several attempts have been made to generalize the idea of sufficiency so that one gets an effective data reduction in the presence of nuisance parameters. Not much success has, however, been achieved in this direction.

c) On the other hand, the invariance principle usually falls to pieces when faced with a discrete model. The Bernoulli experimental set-up is one of the rare discrete models that the invariance principle can tackle. If $x_1, x_2, \ldots, x_n$ are n independent zero-one variables with probabilities θ and $1-\theta$, then all permutations (of co-ordinates) are model-preserving. And they reduce the data directly to the least sufficient statistic $r = x_1 + \ldots + x_n$. Take, however, the following simple example. Let x and y be independent zero-one variables, where

$$P(x = 0) = \theta, \quad 0 < \theta < 1,$$

and

$$P(y = 0) = 1/3.$$

Now, the identity-map is the only available model-preserving transformation. The principle of sufficiency reduces the data immediately to x.

d) The object of this article was not to make a critical evaluation of the twin principles of data reduction. Yet, the author finds it hard to refrain from observing that both the principles of sufficiency and invariance are extremely sensitive to changes in the model. For example, the spectacular data reductions we have achieved in the many examples considered here become totally unavailable if the basic normality assumption is changed ever so slightly.

CHAPTER IX

ANCILLARY STATISTICS, PIVOTAL QUANTITIES AND CONFIDENCE STATEMENTS

1. Introduction

The most commonly used expression in Statistics is information; yet, we have no agreement on the definition or usage of this concept. However, in the particular situation where the problem is to predict a future value of a random variable X with a known probability distribution $p(.)$, we all seem to agree that the information on the yet unobserved future value of X may be characterized by the function $p(.)$ itself. And if we have another variable Y such that the conditional distribution $p(.|Y)$ of X, given Y, is also known then, having observed Y, we can claim that the information on X has shifted from $p(.)$ to $p(.|Y)$. [To avoid a multiplicity of notations, we do not distinguish between a random variable X, an observed value of X and a typical point in the sample space of X.] If $p(.|Y)$ is the same for all values of Y, then X is stochastically independent of Y. In this case Y is said to have no information on X. And we know how to prove then that X has no information on Y.

A problem of statistical inference is somewhat similar. We have a parameter of interest θ and an observable random variable X. The argument begins with a choice of a model $\{p(.|\theta)\}$ for X, with the interpretation that $p(.|\theta)$ is the conditional probability distribution of X given θ. [In this article we do not consider the case where a nuisance parameter exists.] Typically, the distributions $p(.|\theta)$ are different for different values of the parameter θ. So θ has information on X. The converse proposition that X has information on θ is a reasonable one. However this cannot be proved in probabilistic terms unless we take the Bayesian route and regard the parameter θ as a random variable. For five decades R. A. Fisher tried to set up a non-Bayesian theory of information in the data. This gave rise to a set of novel ideas like sufficiency, likelihood, information function, ancillarity, reference sets, conditionality argument, pivotal quantities and fiducial probability. Many learned discussions by contemporary statisticians and philosophers on Fisher's theory have illuminated as well as clouded the statistical literature. In the hope of dispelling some of the still lingering clouds, I propose to take yet another look at the twin concepts of ancillary statistics and pivotal quantities and the related issue of confidence statements.

2. Ancillary Statistics

A statistic T is ancillary if the probability distribution of T, given θ, is the same for all θ. If the problem is to predict a future value of T, then the parameter θ has no information on T. Is it reasonable then to say that T has no information on θ? Can we generate any information on an unknown State of Nature by, say, rolling a fair coin a number of times? When we are arguing within the framework of a particular model for an observable X, and T = T(X) is ancillary, then the act of recording only the T-value of the sample X seems to be quite as useless a statistical exercise as that of rolling the fair coin. R. A. Fisher must have been guided by a reasoning of the above kind to arrive at the conclusion that an ancillary statistic by itself cannot possibly have any information on θ . To a Bayesian the proposition is almost self-evident, because, in the context of a prior opinion q(.) on θ , whatever q might be, the posterior distribution of θ , given T = t, will work out as

$$\begin{aligned} q(\theta | t) &= q(\theta)\, p_T(t|\theta) / \sum_\theta q(\theta)\, p_T(t|\theta) \\ &= q(\theta) / \sum_\theta q(\theta), \text{ since } p_T(t|\theta) \text{ does not involve } \theta, \\ &= q(\theta) \end{aligned}$$

for all θ and t. Since an observation on T in isolation cannot change any opinion on θ , the Bayesian must regard T as uninformative in itself. Strange as it may seem, this hardcore statistical intuition on ancillarity has been challenged by V. P. Godambe (1979a,1980) with his repeated assertions that there are situations where an ancillary statistic by itself can yield a quantum of information on the parameter of interest. This Godambe intuition on ancillary information will be examined in the final section of the essay.

The notion of sufficiency is closely related to the above notion of ancillarity. A statistic S = S(X) is sufficient if the sample X becomes ancillary when it is conditioned by S. In other words, S is sufficient if the conditional distribution $p(\cdot \mid S,\theta)$ of X, given S and θ , depends on (S, θ) only through S. Thus, once we know the S-value of X, the parameter θ has no further information on the sample X. The Fisher intuition that "a sufficient statistic exhausts (summarises) all the relevant information on θ in X" was perhaps based on an unconscious reversal of the roles of θ and X in the previous sentence. Kolmogorov (1942) presented us with the correct Bayesian perspective on sufficiency with the following

Definition : The statistic S is sufficient if, for every prior q(.) for θ , the posterior distribution $q(\cdot \mid X)$ of θ , given X, depends on X only through S = S(X).

In other words, S is sufficient if, for every prior q, the variables X and θ are conditionally independent given S. Likewise T is ancillary if, for every prior q, the variables T and θ are independent of each other.

Following Neyman-Pearson (1936), we call an X-event A (a measurable subset of the sample space) **similar** (a similar region) if $p(A \mid \theta)$ is a constant, say α , in θ. Despite the fact that the sample space is endowed (by the model) with a multiplicity of probability distributions, a similar event A with $P(A \mid \theta) \equiv \alpha$, like the impossible and the sure event, seems to be endowed with an absolute (unconditional, that is) probability α . As before, a Bayesian characterization of a similar event is that A is similar if it is independent of θ for every prior q. [It should be understood that we are always arguing in the context of a fixed model.]

The class $\mathscr{A}$ of similar regions is endowed with the following closure properties. It (i) contains the whole space $\mathscr{X}$, (ii) is closed for differences (if $A \subset B$ and both belong to $\mathscr{A}$ then so also does (B - A) and (iii) is a monotone class (closed for monotone limits). It is what Dynkin (1965) calls a **λ-system.** A λ-system that is closed for intersection is a Borel field. The class $\mathscr{A}$ is typically not a Borel field. A statistic T is ancillary if and only if every T-event (a subset of X defined in terms of T(X)) belongs to $\mathscr{A}$. Whenever $\mathscr{A}$ is not a Borel field, we can find two ancillary statistics T_1 and T_2 such that the pair (T_1, T_2) is not ancillary. In such a situation we cannot have an ancillary statistic T_0 that is a maximum ancillary in the sense that every other ancillary statistic is a function of T_0.

Example 1: Let $\underline{X} = (X_1\ X_2)$ be a pair of i.i.d. random variables whose common distribution on the real line is known excepting for a location parameter θ . Clearly, $T = X_1 - X_2$ is ancillary and, therefore, so also is every function h(T) of T. Contrary to popular impression T is not the maximum ancillary. Indeed, a maximum ancillary can never exist in a situation like this. That the family $\mathscr{A}$ of similar regions is not a Borel field is seen as follows. For any $\underline{X}$-event A let A^c denote its complement and A^* the event obtained from A by interchanging X_1, X_2 in its definition. We call A symmetric in the co-ordinates if $A^* = A$. Clearly $(A^*)^* = A$ and $(AB)^* = A^*B^*$ for all A,B. Since the distribution of $\underline{X} = (X_1, X_2)$ is symmetric in the co-ordinates for each θ , we have $p(A^* \mid \theta) = p(A \mid \theta)$ for all θ and A. Now, let A be an arbitrary T-event that is not symmetric, e.g., $A = \{(x_1, x_2) : x_1 - x_2 > 1\}$ and let B be an arbitrary symmetric event that is not a T-event, e.g., $B = \{(x_1, x_2): x_1^2 + x_2^2 < 1\}$. By definition A and $A^* = \{(x_1, x_2) : x_2 - x_1 > 1\}$ are similar regions. Consider $E = AB \cup A^*B^c$. Since $(A^*B^c)^* = A^{**}B^c = AB^c$, we have $p(A^*B^c|\theta) \equiv p(AB^c|\theta)$ and so

$$\begin{aligned} p(E|\theta) &= p(AB|\theta) + p(A^*B^c|\theta) \\ &= p(AB|\theta) + p(AB^c|\theta) \\ &= p(A|\theta) \end{aligned}$$

for all θ. That is, E is a similar region even though it is not a T-event.

Observe that in this particular instance
$AE = AB = \{(x_1, x_2) : x_1^2 + x_2^2 < 1, x_1 - x_2 > 1\}$ and that this, being a bounded set,

cannot be a similar region for a location parameter family. This is, $\mathscr{A}$ is not closed for intersection. The family of ancillary statistics is very large.

Example 2 : Let X be uniformly distributed over the interval $(\theta, \theta + 1)$. With a single observation on X, can we have a nontrivial ancillary statistic? Let [X] be the integer part of X and $\Psi(X) = X - [X]$, the fractional part. It is then easy to verify that $\Psi(X)$ is uniformly distributed over the interval (0, 1) for all θ and so it is an ancillary statistic. In this case $\Psi(X)$ is the maximum ancillary. If $X_1, X_2, \ldots, X_n$ are n independent observations on X, then the statistic $T = (\Psi(X_1), \Psi(X_2), \ldots, \Psi(X_n))$ is ancillary and so also is the difference statistic $D = (X_2 - X_1, X_3 - X_1, \ldots, X_n - X_1)$.

3. Ancillary Information

An ancillary statistic T by itself carries no information on the parameter θ, but in conjunction with another statistic Y (which, as we shall see later, may even be ancillary itself) may become fully informative (sufficient). Fisher's controversial theory of recovery of ancillary information (Basu, 1964) is based on a recognition of the fact that an ancillary statistic can carry a lot of potent information. A simple example is given to elucidate the Fisher method.

Example 3 : Consider a threefold multinomial model with cell probabilities $\theta/2$, $(1 - \theta)/2$ and 1/2. Let the observed cell frequencies be n_1, n_2 and $n - n_1 - n_2$. Now the statistic $S = (n_1, n_2)$ is sufficient and $T = n_1 + n_2$ being distributed as Bin(n,1/2), is ancillary.The statistic T by itself does not tell us anything about θ that we did not know already, but it tells us how informative the sample is. For instance, if T is zero then we know that the likelihood function generated by the data is the constant $1/2^n$ and so the sample is void of any information on θ. In a sense it seems clear that larger the observed value of T the more informative the sample is. The maximum likelihood (ML) estimate $\hat{\theta}$ of θ is n_1/T if $T \neq 0$ and is undefined if $T = 0$. The ML estimator $\hat{\theta}$, assuming that we have suitably defined it for the case $T = 0$ is not a sufficient statistic and so, according to Fisher, there will be a loss of information if we evaluate $\hat{\theta}$ in terms of its (marginal) distribution $p_{\hat{\theta}}(\cdot|\theta)$. The statistic T is an ancillary complement to $\hat{\theta}$ in the sense that the pair $(\hat{\theta}, T)$, being equivalent to $S = (n_1, n_2)$, is fully informative. This is the kind of situation where Fisher (1935c) wants us to evaluate the ML estimator $\hat{\theta}$ by conditioning it by the ancillary statistic T. The conditional distribution of n_1, given T, θ, is Bin (T, θ), when $T \neq 0$. Thus $E(\hat{\theta}|T, \theta) = \theta$, $V(\hat{\theta}|T, \theta) = \theta(1 - \theta)/T$ and the Fisher information content of $\hat{\theta}$, conditionally on T, is $T/\theta(1 - \theta)$. Fisher would regard $\hat{\theta}$ to be an unbiased estimate of θ with small variance and large information if T is large. The smaller the observed value of T, the less informative $\hat{\theta}$ will be as an estimate of θ. In the extreme case when T = 0, there is no information at all.

This example brings out the Fisher dilemma. He recognized that as a rule not all samples are equally informative and so a sample space analysis of the data may not be quite appropriate. He realized that the information in the data obtained is fully summarized in the corresponding likelihood function but he did not quite know how to analyze the likelihood function in a non-Bayesian fashion. [In Fisher (1956) we do find some half-hearted advice on how to analyze the likelihood function. This has been fully scrutinized (and rejected) by the author in Basu (1973).] So he had to go for a compromise solution : Analyze the data in sample space terms, but suitably restrict the sample space by clustering together all sample points that are, in some qualitative sense, as informative as the sample actually observed.

In the present case the Fisher solution seems pretty attractive. With $n = 15$, the two samples (1, 2, 12) and (12, 2, 1) are qualitatively quite different. It seems reasonable enough to say that the data (1, 2, 12) should be interpreted as 1 success in a succession of 3 Bernoulli trials with θ as the probability of success. Similarly the data (12, 2, 1) should be looked upon as 12 successes in 14 Bernoulli trials. In the first case 1/3 is an unbiased estimate of θ with variance $\theta(1 - \theta)/3$, whereas in the second case 12/14 is an unbiased estimate of θ with variance $\theta(1 - \theta)/14$.

Admittedly, it is not easy to make an unconditional sample space analysis of the data $X = (n_1, n_2, n_3)$. Our heart reaches out for the ML estimator $n_1/(n_1+n_2)$ but we worry about it being undefined when $T = n_1 + n_2 = 0$. We do not relish the idea of figuring out the bias, variance and the information content of the estimator. We recognize many unbiased estimators like $2n_1/n$, $1 - 2n_2/n$ and $1/2 + (n_1 - n_2)/n$ but we like none of them as we recognize that they can take values outside the parameter space (0, 1). The minimum sufficient statistic $S = (n_1, n_2)$ is not complete, there does not exist a minimum variance unbiased estimator of θ, no unbiased estimator can be admissible and so on. All the difficulties seem to be stemming from the fact that T is ancillary. So why not cut the Gordian knot by holding fixed the observed value of $T = n_1 + n_2$ just as we do for the sample size $n = n_1 + n_2 + n_3$?

The sample size analogy for an ancillary statistic comes from Fisher (1935c) who wrote : "It is shown that some, or sometimes all of the lost information may be recovered by calculating what I call ancillary statistics, which themselves tell us nothing about the value of the parameter, but instead, tell us how good an estimate we have made of it. Their function is, in fact, analogous to the part which the size of the sample is always expected to play, in telling us what reliance to place on the result".

Example 4 : An experiment consists of n Bernoulli trials where n = 5 or 100 depending on the flip of a fair coin. The sample is $X = (Y, n)$, where Y is the number of successes in n trials. The ML estimate of θ is $\hat{\theta} = Y/n$. The sample size n is an ancillary statistic now. How to analyze the data X = (3, 5)? Clearly, the data is qualitatively very different from the possible data (60,100), although the ML estimate is the same in either case. As Fisher would have put it, the ML estimate is insufficient (in this case) to identify the full information content of the data, namely, the likelihood function. In terms of the conditional model $p(\cdot|n,\theta)$ for the data, the ML estimator $\hat{\theta}$ is fully sufficient.

Suppose we slightly alter the experiment in Example 4 by determining the sample size n, not by the flip of a fair coin, but by the outcome of the first trial – the sample size is 5 or 100 according as the first trial yields a success or a failure. Now, n is no longer an ancillary statistic even though the quality of the sample depends very much on n. The likelihood function generated by an observed sequence of n successes and failures is $\theta^y(1-\theta)^{n-y}$, where y is the number of successes in the sequence. Therefore, (y, n) is the minimum sufficient statistic and $\hat{\theta} = y/n$ is still the ML estimate of θ. To evaluate $\hat{\theta}$ in terms of its full (unconditional) model $p_{\hat{\theta}}(\cdot|\theta)$ will clearly entail a substantial loss of information. The recovery of information argument does not work here in view of the fact that we do not have an ancillary complement to $\hat{\theta}$.

In Basu (1964) there are other illustrations of how the argument can fail. For instance, there can be a multiplicity of ancillary statistics with no clear cut choice for one to condition the data with. There can also be situations where conditioning X with an ancillary statistic T reduces it to a degenerate random variable with the point of degeneracy depending on T and θ. In Example 2 we have such a situation because the conditional distribution of X, given $\psi(X)$ and θ, is degenerate. In Section 6 we discuss a similar situation. A traditional sample space analysis of data with a degenerate sample space is unheard of.

With his recovery of ancillary information argument, Fisher was seeking for a via media between the Bayesian and the Neyman-Pearson way. As a compromise solution it failed as most compromises do. From the point of view of history, it is more important to recognize what Fisher was attempting to do than the fact of his failure to do so. Clearly, he was trying to cut down the sample space to size. Why? Because he realized that not all sample points in the same sample space are equally informative. With a data of poor quality in hand, e.g. X = (3, 5) in Example 4, it makes little sense to derive some comfort from the thought of a might have been excellent data, e.g., X = (60, 100) or any other sample with n = 100. In this situation, why not evaluate the data (3, 5) in the context of the reduced (conditional) sample space (reference set) consisting of only the six points (0, 5), (1, 5),... ,(5,.5)?

Fisher (1936) wrote : "The function which this ancillary information is required to perform is to distinguish among samples of the same size those from which more or less accurate estimates can be made, or, in general, to distinguish among samples having different likelihood functions, even though they may be maximized at the same value". Clearly, Fisher though that two samples, even though they may be of the same size, may be different in their information content in that a more (or less) accurate estimate of the parameter can be made from the one than from the other. It is not at all clear why he thought that his ancillary statistic can "distinguish among samples with different likelihood functions". The minimum sufficient statistic S is the one that distinguishes between samples with different likelihood functions. [Two likelihood functions are equivalent if they differ only by a multiplicative constant. The minimum sufficient statistic S partitions the sample space into sets of points with equivalent likelihood functions.] The M.L estimator $\hat{\theta}$ is a function of S. In Example 4, the ancillary statistic T complements $\hat{\theta}$, that is, $(\hat{\theta}, T)$ is a one-one function of S. Thus, $(\hat{\theta}, T)$ can distinguish among samples with different likelihood functions. If $\hat{\theta}(x_1) = \hat{\theta}(x_2)$, then x_1 and x_2 generate different likelihood functions if and only if $T(x_1) \neq T(x_2)$. Of course, T by itself cannot distinguish among samples with different likelihood functions, for, if it could, it would have been the minimum sufficient statistic.

In the thirties, when Fisher came up with the recovery of information argument, he was still trying to justify the method of maximum likelihood in some sample space terms. In the mid-fifties, however, we find Fisher (1956, pp 66-73) recognize particular situations where he thought that the data ought to be interpreted only in terms of the particular likelihood function generated by it.

Apart from recovering ancillary information, Fisher found two other uses for the conditionality argument. In the next section we briefly discuss how he extended the scope of his fiducial argument by conditioning pivotal quantities that were not based on what he called "sufficient estimates". Fisher's conditioning method for elimination of nuisance parameters has been discussed by me at some length in Basu (1977).

4. Pivotal Quantities

The notion of a pivotal quantity, perhaps the most innovative idea that came from Sir Ronald, is an extension of the concept of an ancillary statistic. A quantity $Q = Q(\theta, X)$ is a measurable function of the parameter $\theta \in \Theta$ and the sample $X \in \mathcal{X}$, that is, Q is a map of the product space $\Omega = \Theta \times \mathcal{X}$ into a range space R that is usually taken to be the real line. A statistic is a quantity that is constant in θ.

Definition : $Q : \Omega \to R$ is a **pivotal quantity** in short, a **p-quantity,** if the conditional distribution of $Q(\theta, X)$, given θ, is the same for all θ.

Thus, an ancillary statistic is the extreme case of a p-quantity that is constant in θ. If $Q : \Omega \to R$ is a p-quantity then so also is $h(Q)$ for every measurable function h on R. In the location parameter models of Examples 1 and 2, typical examples of p-quantities are $X_1 - \theta$, $\bar{X} - \theta$, $\tilde{X} - \theta$, etc., where $\tilde{X}$ is an equivariant statistic like the mean, median or the mid-range. A more complex example of a p-quantity involving a location parameter is as follows :

Example 5 : Let $X_1, X_2, \ldots, X_n$ be i.i.d. random variables with common c.d.f. $F(x - \mu)$ where about F we only know that it is continuous at the origin and that $F(0) = 1/2$. With μ as the parameter of interest and $X = (X_1, X_2, \ldots, X_n)$, define the quantity $Q(\mu, X)$ as the number of i's for which $X_i - \mu > 0$, $i = 1,2, \ldots, n$. Then Q is a p-quantity since the conditional distribution of Q, given $\theta = (\mu, F)$, is Bin(n, 1/2).

A Bayesian definition of a p-quantity would run along the following lines:

Definition : $Q(\theta, X)$ is a pivotal quantity with respect to a model $\{p(\cdot|\theta) : \theta \in \Theta\}$ for the sample X, if, for every prior distribution q of θ, the quantity Q and the parameter θ are stochastically independent.

Equivalently, Q is a p-quantity if its predictive distribution depends on the prior q and the model $\{p(\cdot|\theta)$ only through the later.

With n i.i.d. observations $X_1, X_2, \ldots, X_n$ on $N(\theta, 1)$, Fisher derived the fiducial distribution of θ as $N(\bar{X}, 1/n)$ by using the p-quantity $\bar{X} - \theta$ as a pivot. It is not the purpose of this article to examine the fiducial argument once again. During the past five decades the argument has been thoroughly examined many times and has been declared, with a few exceptions, as logically invalid. We propose to examine here the confidence statement argument that Neyman-Pearson in the early thirties, synthesized from the fiducial argument. But before we get into that it will be useful to examine why Fisher chose the particular p-quantity $\bar{X} - \theta$ as the pivot.

Fisher regarded the p-quantity $\bar{X} - \theta$ as the correct pivot for the fiducial argument because i) $\bar{X}$ is a natural estimate of θ, ii) $\bar{X}$ is a (minimum) sufficient statistic and iii) the range of variation of $\bar{X}$ is the same as that of the parameter θ, namely, the whole real line. Fisher's fiducial logic would not permit us to use $X_1 - \theta$ as a pivot because X_1 is not sufficient.

The weak p-quantity $X_1 - \theta$ can, however, be made strong by proper conditioning! The difference statistic $D = (X_2 - X_1, X_3 - X_1, \ldots, X_n - X_1)$ is an ancillary complement of X_1. Consider the conditional distribution of $X_1 - \theta$ for fixed D and θ. Since $X_1 - \theta = \bar{X} - \theta + (X_1 - \bar{X})$, $X_1 - \bar{X}$ is a function of D, and $\bar{X}$ is independent of D for fixed θ, it follows that $X_1 - \theta$, given D and θ, is distributed as $N(X_1 - \bar{X}, 1/n)$. Thus, $X_1 - \theta$ remains a p-quantity even when it is conditioned by D. If the fiducial argument is based on the conditioned p-quantity $X_1 - \theta | D$ then we arrive at the correct fiducial distribution $N(\bar{X}, 1/n)$ for θ.

Fisher generalized the above conditional pivotal quantity argument to the case of a location parameter model as follows. Let $X = (X_1, X_2, \ldots, X_n)$ be the sample with density function $f(x_1 - \theta, x_2 - \theta, \ldots, x_n - \theta)$. There are many p-quantities like $X_1 - \theta$, $\bar{X} - \theta$, $\tilde{X} - \theta$, etc., where $\tilde{X}$ is any equivariant estimator of θ. On which one of these shall we base the fiducial argument? It does not matter as long as we condition the chosen p-quantity by the difference statistic! Suppose we choose the sickly p-quantity $X_1 - \theta$, condition it by D and θ, thus arriving at a density function, say, $p(t)$. Invoking the fiducial argument we then arrive at the fiducial density function $p(X_1 - t)$ for the parameter θ. But had we started off with a stronger looking p-quantity $\tilde{X} - \theta$, where $\tilde{X}$ is an equivariant estimate of θ like the mean or the median, then, in view of the fact that $\tilde{X} - X_1$ is a function of D, the conditional density function for $\tilde{X} - \theta$, given D and θ, would have been $p(t - \tilde{X} + X_1)$. Therefore, the fiducial distribution of θ, with $\tilde{X} - \theta$ held as the pivot, would also have worked out as $p(X_1 - t)$.

Many of our contemporary statisticians — Pitman, Barnard, Kempthorne, Fraser, to name only a few — have been greatly (and very diversely) influenced by the above conditional pivot argument of Fisher. For one thing, the extensive study of invariance as a data reduction principle originated in the fiducial distribution $p(X_1 - t)$ for the location parameter θ. Note that the mean

$$\begin{aligned}
\hat{\theta} &= \int t p(X_1 - t)\, dt \\
&= \int (X_1 - t)\, p(t)\, dt \\
&= X_1 - \int t p(t)\, dt \\
&= X_1 - E(X_1 - \theta | D, \theta) \\
&= X_1 - E(X_1 | D, \theta = 0)
\end{aligned}$$

of the fiducial distribution is the Pitman estimator (the best equivariant estimator with squared error loss function) of θ. Similarly, the median of the fiducial distribution is the best equivariant estimator when the absolute error is taken as the loss function. If the proof of the pudding is in the eating, then one may argue that Fisher's fiducial argument is plausible at least in the case of a location parameter.

I regard the following proposition as an empirically established statistical metatheorem:

No inferential argument in statistics has anything going for it unless a sensible Bayesian interpretation can be found for it.

It so happens that the fiducial distribution $p(X_1 - t)$ for the location parameter θ may be interpreted as the Bayes posterior with the (improper) uniform prior over the real line. So it is possible to condition the sample X all the way down to its observed value and derive a distribution for θ directly from the likelihood function. However, the uniform prior over the unbounded real line can hardly be regarded as a sensible representation of anyone's ignorance about the parameter θ.

Sir Ronald is no longer with us, but his pivotal qunatities are very much alive. As we shall see in the next section, all confidence statements are based on p-quantities. Fisher tried hard but failed to establish a coherent theory for the choice of an appropriate p-quantity. Currently there appears to be no law and order regarding the choice of the pivot. Often a sizeable part of the data is ignored to arrive at a particular pivot. Stein's (1945) classical work on fixed width confidence interval is an example of this kind. Sometimes post-randomization variables are deliberately introduced in the data so as to create a p-quantity in terms of the extended data. For instance, if the data generated by a sequence of n Bernoulli trials be enhanced by a randomization variable uniformly distributed over (0, 1), then we can construct a 95% confidence interval for the probability parameter p. Finally, there is the curious (Fisher inspired) method of holding a substantial part of the data as fixed and then recognizing that a quantity $Q(\theta, X)$ becomes a p-quantity in terms of the restricted reference set. Godambe and Thompson (1971) and Seheult (1980) are only two of many such licentious use of the fiducial argument that Fisher himself thought to be of rather limited coverage.

5. Confidence Statements

A quantity Q was defined earlier as a measurable function of (θ, X). A measurable subset of $\Omega = \Theta \times \mathcal{X}$ will be called a **quantal.** Let $Q: \Omega \to R$ be an arbitrary pivotal quantity(p-quantity) and let B be a measurable subset of R. Consider the quantal $E = \{(\theta, X) : Q(\theta, X) \varepsilon B\}$. For each $\theta \varepsilon \Theta$, let $E^\theta = \{X: (\theta, X) \varepsilon E\}$ denote the θ-section of E. For a given θ, the statements $Q(\theta, X) \varepsilon B$ and $X \varepsilon E^\theta$ are identical. Since Q is a p-quantity, the conditional probability of the former statement, given θ, is a constant in θ, and therefore, so also is Prob($X \varepsilon E^{\theta} | \theta$). This motivates the following

Definition : In the context of a model $\{p(.|\theta)\}$ for the sample X, a subset E of $\Omega = \Theta \times \mathcal{X}$ is called a **p-quantal** if $p(E^{\theta} | \theta)$ is a constant in θ. The constant value of $p(E^{\theta} | \theta)$ is called the size of the p-quantal E.

The empty set and the whole space Ω are trivial examples of p-quantals of size 0 and 1 respectively. If A is a similar region of size α then $\Theta \times A$ is a p-quantal of size α. Like the family $\mathcal{A}$ of similar regions, the family $\mathcal{E}$ of p-quantals is a λ-system of sets. The indicator function of a p-quantal is a p-quantity. A function $Q(\theta, X)$ is a p-quantity if and only if every subset of $\Theta \times \mathcal{X}$ defined in terms of Q is a p-quantal. Analogous to our Bayesian definition of a p-quantity, the notion of a p-quantal may be redefined as

Definition : The set $E \subset \Omega$ is a p-quantal if, irrespective of the scientist's prior distribution (or belief) on Θ, the event E is independent of the random variable θ.

Equivalently, E is a p-quantal if its probability is well defined in terms of the model $\{p(.|\theta)\}$ alone.

R.A. Fisher's fiducial argument was severely restrictive in that only a few simplistic statistical models could cope with his stringent requirements for the right pivotal quantity. Jerzy Neyman and E.S. Pearson rejected the fiducial logic but they nevertheless accepted a part of it and generalized it to the limit. Neyman-Pearson's confidence statement argument is based not on the limited stock of the Fisherian p-quantities but on the plentiful supply of p-quantals. Any p-quantal E can be the basis of a confidence statement as described below.

Let $E_X = \{\theta : (\theta, X) \varepsilon E\}$ denote the X-section of a p-quantal E. Clearly, the three statements $(\theta, X) \varepsilon E$, $X \varepsilon E^{\theta}$ and $\theta \varepsilon E_X$ are logically equivalent. If we regard E_X as a random set determined by the random variable X, then, prior to the observation of X, the probability of $\theta \varepsilon E_X$ is well defined and is independent of any prior opinion (or lack of it) that the scientist may have on the parameter θ. Suppose the p-quantal E is of size 0.95. Then, prior to the observation of X, the scientist is 95% sure (ordinary probability) that the true θ will belong to the set E_X that is going to be determined by the observance of X. Once X is observed and the particular E_X determined, then is it still reasonable for the scientist to assert that he or she is still 95% sure that $\theta \varepsilon E_X$? Fisher would have answered the question with a firm negative if he found that the p-quantal E has not been defined in the right manner in terms of the right pivotal quantity. Neyman-Pearson's theory of confidence statements suffers from no such inhibition. Any p-quantal of size 0.95 is a generator of a 95% confidence set estimator $\{E_X : X \in \mathcal{X}\}$ of θ. The converse proposition that every confidence set estimator corresponds to a p-quantal is easily seen to be also true.

If we consider the hypothetical population of a sequence of observations $X_1, X_2, \ldots$ on X with θ fixed at its true value, then is it not correct to say that 95% of the sets $E_{X_1}, E_{X_2}, \ldots$ will cover θ and only 5% will not?

The frequency interpretation of the confidence statement $\theta \in E_X$ that is implicit in the above hypothetical question is the cornerstone of the Neyman-Pearson argument. Fisher always maintained that the argument was a logical error. I vividly recall an occasion (Winter of 1955, Indian Statistical Institute, Calcutta) when Professor Fisher bluntly asserted that a confidence interval, unless it coincides with a fiducial interval, cannot be interpreted in frequency probability terms. At the end of the Fisher seminar, Professor Mahalanobis, who was chairing the session, invited me to comment on the proposition. So, shaking in my shoes, I rose to defend Professor Neyman and gave the standard frequency interpretation of confidence statements. Professor Fisher summarily dismissed my explanation as yet another example of the "acceptance test" type argument.

It is important for us to understand the Fisher point of view. According to Fisher (1956, p. 77) the population (an hypothetical sequence of repeated trials with certain elements held fixed) to which a certain inferential statement is referred to for probabilistic interpretation is a figment of the mind. Only in some "acceptance test" type situations, Fisher would concede that the population (reference set) is well-defined. In problems of scientific inference it should be recognized that the data can be differently interpreted in terms of different reference sets. Of course, Fisher alone knew how to artfully choose the reference set to suit an individual scientific problem!

Let us go back to a situation that we have already discussed in Section 3. Suppose with a sample $X = (X_1, X_2, \ldots, X_n)$ of n i.i.d. $N(\theta, 1)$'s, a misguided scientist chooses to make a confidence statement based on the first observation, X_1 alone. Since the p-quantity $X_1 - \theta \sim N(0,1)$, the 95% confidence interval is $I = (X_1 - 1.96, X_1 + 1.96)$. Following Neyman the scientist will claim that the random interval I covers the true θ with 95% probability. But following Fisher the scientist may recognize quite a different probability for I covering θ. Since X_1 is not a sufficient statistic, the scientist will have to look for an ancillary complement to X_1. Suppose $d = \bar{X} - X_1$ is chosen as the ancillary complement to X_1. [Note that $\bar{X} - X_1$ is an ancillary statistic and that $(X_1, \bar{X} - X_1)$ is a sufficient statistic.] The conditional probability

$$\begin{aligned} &\Pr(|X_1 - \theta| < 1.96 | d, \theta) \\ &= \Pr(|\bar{X} - \theta - d| < 1.96 | d, \theta) \\ &= \Pr[|N(d, 1/n)| < 1.96] \end{aligned}$$

and this is the same as Fisher's fiducial probability that θ is in I. That the choice of the reference set (or the conditioning statistic) can dramatically affect the confidence co-efficient is evident from this example.

Most of my statistical colleagues can never cease to admire the sheer elegance and simplicity of Neyman-Pearson's confidence statement argument. And then it is so wonderfully easy to construct a p-quantal E. Choose and fix a number α in (0, 1); for each $\theta \varepsilon \Theta$, choose and fix a subset E^θ of $\mathcal{X}$ such that $p(E^\theta | \theta) = \alpha$; and then define E as

$$E = \{(\theta, X) : \theta \varepsilon \Theta, X \varepsilon E^\theta\}.$$

By construction E is a p-quantal of size α and so the family $\{E_X : X \varepsilon \mathcal{X}\}$ of X-sections of E is a 100α % confidence set estimator of θ. With E_X as the confidence set corresponding to the observed sample X, can any **evidential meaning** be attached to the assertion $\theta \varepsilon E_X$? Suppose on the basis of sample X one can construct a 95% confidence interval estimator for the parameter θ, then does it mean that (the random variable) X has **information** on θ in some sense?

Anyone who fails to answer both the questions quickly and firmly in the negative is invited to take a look at the following simple example.

Example 6 : The parameter θ lies somewhere in the unit interval (0, 1) and the sample X is a pure randomization variable having the θ-free uniform distribution over (0, 1). Surely X has no information on θ. No evidential meaning can be attributed to any inference about θ based on X. That there is a plentiful supply of 95% confidence intervals for θ is seen as follows. Choose and fix any subset B of (0,1) and then define the quantal E as the union of the two sets $B \times (0, .95)$ and $B^c \times (.05, 1)$. For each θ the section E^θ is either (0, .95) or (.05, 1), and so E is a p-quantal of size 0.95. The 95% confidence intervals $\{E_X\}$ based on the p-quantal E are

$$E_X = \begin{cases} B & \text{if } 0 < X \leq .05 \\ (0, 1) & \text{if } .05 < X < .95 \\ B^c & \text{if } .95 \leq X < 1. \end{cases}$$

Of course it does not make any sense to say that 95% confidence can be placed on the statement $\theta \in E_X$ irrespective of what X turns out to be.

As I said in the last paragraph of the previous section, the statistical literature is full of many singular kinds of confidence statements. But has anyone ever dared to base confidence statements on no data, that is, on an uninformative part of the data? Surprisingly, the answer is, yes. In the next section we briefly consider the Godambe (1979a,1980) contention that the label part of the survey data, despite being an ancillary statistic, is informative by itself.

6. Ancillarity in Survey Sampling

In current survey literature a lot of confusion and controversy surround the notion of the **label set,** that is, the set of names or label identities of the surveyed units. Let us denote the population to be surveyed as $\{1, 2, \ldots, N\}$ of unit labels. The universal parameter is $\theta = (Y_1, Y_2, \ldots, Y_N)$, where Y_j is an unknown characteristic of unit j. The survey design (sampling plan) selects a subset **s** of the population lebels $\{1, 2, \ldots, N\}$ with a known selection probability p(**s**). The survey fieldwork determines the set $\mathbf{y} = \{ Y_i : i \in \mathbf{s} \}$ of Y-values of the units in the label set **s**. We write $\mathbf{x} = (\mathbf{s}, \mathbf{y})$ to denote the sample generated by the survey.

For a typical survey design the probability p(**s**) of the label set **s** does not depend on the parameter θ , and so **s** is an ancillary statistic. Can we then discard **s** and marginalize the data to the set of observed Y-values? Since **s** is ancillary, should we not condition the data **x** by holding **s** fixed? If we factor the likelihood function $L(\theta)$ as

$$L(\theta) = \Pr(\mathbf{s}, \mathbf{y}|\theta) = p(\mathbf{s})\Pr(\mathbf{y}|\ \mathbf{s}, \theta)$$

and discard the θ -free factor p(**s**) from the likelihood, then it becomes clear [see Basu (1969) for more on this] that the sampling plan is not a determinant of the likelihood. Invoking the **Likelihood Principle** should we not declare then that at the data analysis stage we need not concern ourselves with the details of the particular survey sampling plan? These are some of the hotly debated issues in survey theory.

It is useful to note the close similarity between the survey setup and Example 2, where the sample X is uniformity distributed over the interval $(\theta, \theta + 1)$ and $-\infty<\theta<\infty$ is the parameter.The label set **s** of the survey sample **x** corresponds to the fractional part $\varphi(X)$ of the sample X in Example 2, both being "very large" ancillary statistics in the following sense. The conditional distribution of the sample X, given the ancillary statistic $\varphi(X)$ and the parameter θ is degenerate in Example 2, that is, X is a function of $\varphi(X)$ and θ. In precisely the same sense, the sample **x** is a function of the label set **s** and the universal parameter θ of our survey setup.

The fact that the survey sample **x** has a degenerate conditional distribution, given **s** and θ, has the simple consequence that no unbiasedly estimable parametric function (unless it is a constant) can have a uniformly minimum variance unbiased estimator (UMVUE). This is seen as follows.

Let T be an unbiased estimator of $g(\theta)$. Choose and fix a particular parameter value θ_0, and consider the estimator

$$T_0 = T - E(T|\, \mathbf{s}, \theta_0) + g(\theta_0).$$

The second term on the right hand side, being a function of the ancillary **s**, has a constant (θ-free) mean. Considering the particular case $\theta = \theta_0$, it is then clear that the constant mean is $g(\theta_0)$. Hence T_0 is an unbiased estimator of $g(\theta)$. Since the distribution of T, given **s** and θ, is degenerate, the first two terms on the right hand side are the same when $\theta = \theta_0$. We have thus established that for each unbiasedly estimable parametric function $g(\theta)$ and for any prefixed parameter value θ_0, there exists an unbiased estimator T_0 for $g(\theta)$ with zero variance at $\theta = \theta_0$. As in Example 2, we cannot talk of UMVUE's in survey theory.

In Example 2, the likelihood function is flat over the interval (X - 1, X) and is zero outside. [With n observations on X, the likelihood is flat over the interval (M - 1, m), where m and M are, respectively, the minimum and the maximum sample.] Exactly the same thing is true for the survey setup. It is rather curious that the flat likelihood for the survey parameter is sometimes characterised as "uninformative". I have not met anyone yet who would regard the equally flat likelihood in Example 2 as uninformative.

Survey theory is full of all kinds of confusing ideas. We end this essay with a good look at a mystifying Godambe proposition that was stated at the end of the previous section. It will be useful to consider a simplified version of the Godambe example first.

Example 7 : The population consists of 100 individuals labeled as 1,2, ..., 100. It is known that only one of the 100 individuals is black, the rest being all white. It is also known that the black individual is either 1 or 2. Writing 1 for black and 0 for white, the universal parameter then takes one of the two values $\theta_1 = (1, 0, 0, 0, \ldots, 0)$ and $\theta_2 = (0, 1, 0, 0, \ldots, 0)$. An individual is selected at random (equal probabilities) and the sample is **x** = (**s**, **y**), where **s** and **y** are, respectively, the label index and the colour value of the chosen individual. In this case both **s** and **y** are ancillary statistics, an example of how two ancillaries can jointly be sufficient. As we noted earlier, the statistic **y** gets fully determined in terms of the parameter θ and the label **s**. In other words, the ancillary statistic **y** may be represented as a p-quantity $Q(\theta, \mathbf{s})$. Since $\Pr(\mathbf{y} = 0 | \theta) = 0.99$ for both θ, the set $E=\{(\theta, \mathbf{s}) : Q(\theta, \mathbf{s}) = 0\}$ is a p-quantal of size 0.99. Defining $E_{\mathbf{s}}$ as the **s**-section of E,we then have in $\{E_{\mathbf{s}}\}$ a 99% confidence set estimator of θ in the sense of Neyman-Pearson. Note that $E_1 = \{\theta_2\}$, $E_2 = \{\theta_1\}$ and $E_{\mathbf{s}} = \{\theta_1, \theta_2\}$ for all other **s**. Does it make any sense to say that when $\mathbf{s} = 1$, we should be 99% confident that $\theta = \theta_2$? How confident should we be in the proposition $\theta \varepsilon E_{\mathbf{s}}$ when $\mathbf{s} = 3$?!

Example 8 (V. P. Godambe) : The population consists of four individuals 1, 2, 3, 4, the universal parameter θ is either

$$\theta_1 = (-1, 1, -1, 1) \text{ or } \theta_2 = (-1, 1, 1, -1).$$

With a simple random sample of size 2, let $\mathbf{s} = (s_1, s_2)$ be the label set and let $\mathbf{y} = (y_1, y_2)$ be the sample Y-values. It is then easy to see that $T = | y_1 + y_2 |$ is an ancillary statistic taking the two values 0 and 2 with probabilities 4/6 and 2/6 respectively. As in the previous example, we represent the ancillary statistic T as a pivotal quantity

$$Q(\theta, \mathbf{s}) = | \sum_{i \varepsilon \mathbf{s}} Y_i | .$$

The set $E = \{ (\theta, \mathbf{s}) : Q(\theta, \mathbf{s}) = 0\}$ is then a p-quantal of size 4/6 and the family $\{E_{\mathbf{s}}\}$ of **s**-sections of E constitute a confidence set estimator of θ with a confidence level of 4/6. Note that $E_{(1,3)} = \{\theta_2\}$. According to Godambe, the partial data $\mathbf{s}=$ (1, 3), despite being ancillary in the sense of Fisher, is informative (in itself) about the parameter in the sense that it makes the parameter value θ_2 "more plausible" than the value θ_1.

Example 6 of the previous section and the two examples of this section are really the same. They all demonstrate how it is possible to construct sizeable confidence statements with meaningless data. We must recognize that the confidence statement argument when presented in its classical p-quantal form is a mistake. Sir Ronald was certainly more right on this controversial issue than Professor Neyman.

Like a blind man in a dark room groping for a black cat that is perhaps not there, we statisticians are still seeking for the true meaning of statistical information. This article was written in the hope of sharing with my colleagues my imperfect intuition on (and limited understanding of) the subject. If in doing so I have hurt anyone's feelings, I am sorry.

PART II
SURVEY SAMPLING AND RANDOMIZATION

CHAPTER X

SUFFICIENCY IN SURVEY SAMPLING

1. Introduction and Summary

In the present context the term 'statistical structure' denotes a triplet $(X, \mathcal{A}, \mathcal{P})$ where

(i) $X = \{x\}$ is the sample space — the set of all the 'possible outcomes' of the statistical (sampling) experiment,

(ii) $\mathcal{A} = \{ A \}$ is a Borel field of subsets of X — the class of 'events' that are recognized as 'measurable', and

(iii) $\mathcal{P} = \{P_\theta\}$ is a family of probability measures (on $\mathcal{A}$) indexed by the 'parameter' θ.

The parameter θ is an unknown physical state or entity. The statistician performs a statistical experiment (and is thus led on to the consideration of a statistical structure) with a view to eliciting some 'information' about the parameter θ. The information is a 'sample' — a point x in the sample space X. The statistician 'knows' his statistical structure, i.e., he knows the set of 'possible' values (the parameter space) for the unknown parameter θ and, for each such value of θ and for each $A \varepsilon \mathcal{A}$, he knows the probability $P_\theta(A)$ of the sample lying in A. The statistician uses his knowledge about the statistical structure and the actual observation (the sample) x in making a 'reasonable' inference about the parameter θ. One of the questions that intrigued R.A. Fisher quite early in his research career is 'what part of the observation x is really relevant to inference-making about θ?' For instance, if the observation x incorporates in itself the outcome of a fair die rolling experiment, then it stands to reason to claim that this part of the observation ought to be thrown out as irrelevant. What then is the relevant 'information core' of the information (sample) x? The idea of 'sufficient statistic' is perhaps the most important discovery of Fisher in the field of statistical inference. The so-called 'principle of sufficiency', implicitly contained in the writings of Fisher, may be stated as follows:

> 'If $T = T(x)$ is a sufficient statistic and if $T(x) = T(x')$, then the inference arrived at on the basis of an observation x ought to be the same as that based on an observation x'.'

In other words, if T be sufficient then the statistician should throw out all other characteristics of the observation x — excepting its T-value $T(x)$ — as useless and irrelevant to the problem of inference making.

Now, if the family $\{P_\theta\}$ of probability measures be dominated by a σ-finite measure, then, for each x, we have a likelihood function $L(\theta) = L_x(\theta)$. We can look upon the likelihood function as a statistic — a transformation that maps the sample x into a function on the parameter space. The sufficiency of $L(\theta)$ was intuitively recognized by Fisher. It is now recognized that if we 'sieve out' the constant (θ-free) part from the likelihood functions $L(\theta)$, we arrive at the minimal (smallest) sufficient statistic. The minimal[1] sufficient statistic (if it exists) is then the information core of the sample. Unfortunately such a statistic need not always exist. In section 3, we describe two pathological examples constructed by Pitcher (1957) and Burkholder (1961). In each of the two examples we are concerned with an undominated family of discrete distributions on the real line. In each case the Borel sets are considered to be measurable. In the first example no minimal sufficient statistic exists. In the second example we have the surprising phenomenon that there exists a sufficient statistic T and an insufficient statistic T_1 that is 'more informative' than T in the sense that given the T_1-value of x we always know its T-value.

Our main concern in this paper are statistical structures that resemble the Pitcher and Burkholder pathologies in the following respects :

(a) X and $\mathcal{P}$ are uncountable;

(b) Each $P \in \mathcal{P}$ is a discrete measure;

(c) The empty set ϕ is the only set that has zero P-measure for each $P \in \mathcal{P}$.
We shall see in Section 4 that these conditions are typically fulfilled in sample survey situations. We impose the following additioal condition on our statistical structures:

(d) All sets are measurable.

Remark : Since each $P \in \mathcal{P}$ is discrete it is entirely natural to take all sets as measurable. The pre-condition that ϕ is the only $\mathcal{P}$-null set is, in reality, no restriction at all.

In Section 4, we characterize sufficient statistics and sub-fields for statistical structures of the above type. We prove, among other things, that there always exists a minimum sufficient statistic (sub-field), that any statistic that is 'more informative' than a sufficient statistic must necessarily be sufficient and that every sufficient sub-field induces and is induced by a sufficient statistic.

2. Sufficient Statistics and Sub-Fields

Let $(X, \mathcal{A}, \mathcal{P})$ be a fixed statistical structure. By 'statistic' we mean a 'partition' of the sample space, i.e. a decomposition $\Pi = \{\pi\}$ of X into a number (possibly

[1]Burkholder (1961) has pointed out that if a sufficient statistic (sub-field) is minimal then it is also minimum.

uncountable) of mutually exclusive and collectively exhaustive 'parts' or sub-sets π. In general, the parts π need not be $\mathscr{A}$-measurable. For each statistic (partition) Π, we define $\mathscr{A}(\Pi)$ to be the class of all $\mathscr{A}$-measurable sets that may be expressed as a union of a number (possibly uncountable) of parts of Π. It is easily checked that $\mathscr{A}(\Pi)$ is a sub-field (sub- σ-field) of $\mathscr{A}$. We shall refer to $\mathscr{A}(\Pi)$ as the sub-field induced by the statistic Π.

It is well known that, in general, not every sub-field is inducible by some partition. For example, let $\mathscr{A}$ be the σ-field of all Borel sets on the real line X and let $\mathscr{C}$ be the sub-field of all countable sets and their complements. Since $\mathscr{C}$ contains all single point sets, it is obvious that any partition Π that induces $\mathscr{C}$ must have all its parts as single point sets. The sub-field $\mathscr{A}(\Pi)$ induced by Π must then be the same as $\mathscr{A}$.

On the other hand, each sub-field $\mathscr{B}$ induces a partition as follows. For each $x \in X$, define π_x to be the intersection of all sets B such that $x \in B \in \mathscr{B}$. It is easy to check that the class of distinct π_x's forms a partition. We shall denote this partition by $\Pi(\mathscr{B})$. Again it is easy to see that not every partition is inducible by some sub-field. Also note that in general two different sub-fields may induce the same partition and vice versa.

If every member of the sub-field $\mathscr{B}_1$ is also a member of the sub-field $\mathscr{B}_2$ then we say that $\mathscr{B}_1$ is 'smaller' than $\mathscr{B}_2$ and write $\mathscr{B}_1 \subset \mathscr{B}_2$. If every part of the partition Π_1 is a union of parts of Π_2 then we say that Π_1 is 'thicker' than Π_2 and write $\Pi_1 < \Pi_2$. If $\mathscr{B}_1 \subset \mathscr{B}_2$ then $\Pi(\mathscr{B}_1) < \Pi(\mathscr{B}_2)$ and conversely, $\Pi_1 < \Pi_2$ implies $\mathscr{A}(\Pi_1) \subset \mathscr{A}(\Pi_2)$. Let us also observe that Π is 'thinner' then the partition induced by $\mathscr{A}(\Pi)$. Also $\mathscr{B}$ is smaller than the sub-field induced by $\Pi(\mathscr{B})$.

The sub-field $\mathscr{B}$ is said to be 'sufficient' if, for every $A \in \mathscr{A}$, there exists a $\mathscr{B}$-measurable function $f(A|\,.)$ such that, for all $B \in \mathscr{B}$ and $P \in \mathscr{P}$,

$$P(A \cap B) = \int_B f(A|\,.)\, dP(.)$$

The partition Π is sufficient if the induced sub-field $\mathscr{A}(\Pi)$ is sufficient.

3. Pitcher and Burkholder Pathologies

In this section we describe two pathological examples due to Pitcher (1957) and Burkholder (1961). In both the examples we take X to be the real line and $\mathscr{A}$ the σ-field of all Borel sub-sets of X. We also choose and fix a non-empty, non-Borel set B that excludes the origin but is symmetric about the origin, i.e. $B = -B = \{x : -x \in B\}$.

Example 1 (Pitcher) : Define $\mathcal{P} = \{ P_\theta \mid \theta \varepsilon X \}$, the family of probability measures, as follows. If $\theta \varepsilon B$ then P_θ is the discrete measure allotting probabilities 1/2 and 1/2 to the two points $-\theta$ and θ. If $\theta \not\varepsilon B$ then P_θ is degenerate at θ, i.e. it allots probability 1 to the single point θ. Observe that the empty set is the only $\mathcal{P}$-null set. It is now easy to check that in the present set-up there cannot exist a (minimal) sufficient sub-field or a thickest sufficient partition. The existence of such a sub-field or statistic would contradict the initial supposition that B is non-measurable.

Example 2 (Burkholder). For all $\theta \varepsilon X$, define P_θ as the symmetric discrete measure that allots probabilities 1/2 and 1/2 to the two points $-\theta$ and θ: if $\theta = 0$ then P_θ is degenerate at 0. As in the previous example, the empty set is the only $\mathcal{P}$-null set. Since each member of $\mathcal{P} = \{ P_\theta \}$ is symmetric (about the origin) it follows that the sub-field $\mathcal{C}$ of symmetric Borel sets is sufficient and so also is the corresponding partition Π consisting of $\{0\}$ and all two point sets $\{-x, x\}$ with $x \neq 0$. Indeed, $\mathcal{C}$ is the smallest sufficient sub-field and Π the thickest sufficient partition. Now let Π^* be the partition consisting of all single point sets x with $x \not\varepsilon B$ and all two point sets $\{-x, x\}$ with $x \varepsilon B$. And let $\mathcal{C}^*$ be the sub-field induced by Π^*— we may also characterize $\mathcal{C}^*$ as the class of all Borel sets that intersect B in a symmetric set.Follow ing Burkholder (1961) we can now show that Π^*,in spite of being thinner than the sufficient partition Π, is not itself sufficient. A similar remark is true of $\mathcal{C}^*$.

4. Sufficiency in Typical Sampling Situations

Let us describe an elementary sampling situation. We have a population of N units (individuals) and suppose they are listed in our sampling frame as 1, 2,. . ., N. Associated with unit i we have an unknown character θ_i(i = 1, 2, . . . , N). The unknown state of nature (the parameter) is then the N-tuple,

$$\theta = (\theta_1, \theta_2, \ldots, \theta_N).$$

Consider now the simplest possible sampling scheme. A simple random sample of size one is drawn and we observe only the θ-character x of the sample unit obtained. What is the smaple space? It is clear that x takes the N values $\theta_1, \theta_2, \ldots, \theta_N$ with equal probabilities. But it would mess up everything if we say that the sample space consists of the N points $\theta_1, \theta_2, \ldots \theta_N$. That way, the parameter would determine the sample space and not the probability distribution and the statistical models that we have been considering here would be entirely inapplicable. The difficulty can be easily overcome if we look at the problem from a slightly different angle. If, for instance, the θ-characters of the population units are real valued, then we say that the sample space X is the real line and that the parameter $\theta = (\theta_1, \theta_2, \ldots, \theta_N)$ defines a discrete probability measure P_θ that allots equal probabilities to the N points $\theta_1, \theta_2, \ldots, \theta_N$ and zero probabilities to all other points. We then have a fixed sample space and a family of probability measures P_θ indexed by the parameter θ. Now, we have to face the difficulty that the set of possible values of θ(the parameter space) may be uncountable; this is so, at least in theory, in typical sampling situations with continuous θ-characters. In such cases, the family $\mathcal{P} = \{P_\theta\}$ of probability measures would be undominated.

Is it then possible that the Pitcher and the Burkholder anomalies would persist in realistic sampling situations? A careful re-examination of the two pathological examples would reveal the fact that, though we are dealing with discrete measures, we are regarding only the Borel sets as 'measurable'. The natural domain of definition of a discrete measure is the family of all sub-sets of the sample space X. In this section we are trying to underline the fact that if we take all sets as measurable then we need not worry about such anomalies.

It is realistic to impose the following three conditions on statistical structures $(X, \mathcal{A}, \mathcal{P})$ that are suitable as sample survey models.

(C_1) Every member of $\mathcal{P}$ is a discrete probability measure;

(C_2) $\mathcal{A}$ Consists of all sub-sets of X;

(C_3) The empty set is the only $\mathcal{P}$-null set.

Observe that there is no loss of generality because of (C_3). We can always remove the $\mathcal{P}$-null points (if any) from X and work with the reduced sample space. Also observe that if either X or $\mathcal{P}$ is countable, then the family $\mathcal{P}$ would be a dominated family and in this case we have no problem. We tacitly assume that both X and $\mathcal{P}$ are uncountable.

Now, let $\Pi = \{\pi\}$ be a partition (statistic) and let $\mathcal{A}(\Pi)$ be the sub-field induced by Π. For each $x \in X$, let π_x be that part of Π that includes x. Let us write P(x) for the P-measure of the single point set $\{x\}$. We prove the following factorization theorem.

Theorem 1 : A necessary and sufficient condition in order that the statistic Π is sufficient is that there exists a function $g = g(x)$ on X such that

$$P(x) \equiv g(x)\, P(\pi_x) \qquad (1)$$

for all $x \in X$ and $P \in \mathcal{P}$.

The 'necessary' part is proved as follows. Let Π be sufficient. Corresponding to each $A \in \mathcal{P}$ there exists an $\mathcal{A}(\Pi)$-measurable function $f(A|\,.)$ such that

$$P(A \cap B) \equiv \int_B f(A|\,.)\, dP(.) \qquad (2)$$

for all $B \in \mathcal{A}(\Pi)$ and $P \in \mathcal{P}$.

Since all sets are $\mathcal{A}$-measurable, it follows that each π_x is an atom of $\mathcal{A}(\Pi)$. Hence, the function $f(A|\,.)$, being $\mathcal{A}(\Pi)$-measurable, is constant on each π_x. Putting $A = \{x\}$ and $B = \pi_x$ in (2) and defining $g(x)$ to be the constant value of $f(\{x\}|\,.)$ on π_x we at once have the identity (1).

The 'sufficient' part of Theorem 1 is proved as follows. Suppose there exists a function $g = g(x)$ for which identity (1) holds. From condition (C_3) it then follows that $g(x) > 0$ for all $x \in X$. Now, let π be a typical part of Π. For each $x \in \pi$ we have, from (1)

$$P(x) = g(x)\, P(\pi)$$

and hence summing both sides over all $x \in \pi$ and noting that $P(\pi) > 0$ for at least one $P \in \mathcal{P}$ we have

$$\sum_{x \in \pi} g(x) \equiv 1 \qquad \text{for all } \pi.$$

This proves incidentally that each part π of the partition Π must be countable.

Now, for each $A \in \mathcal{A}$, define the function $f(A|\,.)$ in the following manner:

$$f(A|\,x) = \sum_{y \in A \cap \pi_x} g(y)$$

where π_x is that part of Π that contains x.

Clearly, the function $f(A|\,.)$ is constant on each part π of Π. That is, $f(A|\,.)$ is $\mathcal{A}(\Pi)$-measurable. The constant value of $f(A|\,.)$ on the part π is denoted by $f(A|\,\pi)$. Note that

$$f(A|\,\pi) = \sum_{y \in A \cap \pi} g(y).$$

That the function $f(A|\,.)$ satisfies the integral identity (2) is checked as follows.

$$\int_B f(A|\,.)\, dP(.) = \sum_{x \in B} f(A|\,x)\, P(x) \qquad \text{(since P is discrete)}$$

$$= \sum_{\pi \subset B} f(A|\,\pi)\, P(\pi) \qquad \text{(B, being } \mathcal{A}(\Pi)\text{-measurable, is a union of parts of } \Pi)$$

$$= \sum_{\pi \subset B} \sum_{y \in A \cap \pi} g(y)\, P(\pi)$$

$$= \sum_{\pi \subset B} \sum_{y \in A \cap \pi} P(y) \qquad \text{(from our supposition that (1) holds)}$$

$$= \sum_{y \in A \cap B} P(y)$$

$$= P(A \cap B)$$

And this completes the proof of Theorem 1.

For each $x \in X$, let

$$\mathcal{P}_x = \{P \mid P \in \mathcal{P}, P(x) > 0\}.$$

From condition (C_3), no $\mathcal{P}_x$ is vacuous. Consider now the binary relation on x :

$x \sim y$ if $\mathcal{P}_x = \mathcal{P}_y$ and $P(x) / P(y)$ is a constant in P for all $P \in \mathcal{P}_x$.

It is easy to check that the relation $\sim$ is an equivalence relation.

Let $P \in \mathcal{P}_x$ and let $\pi_x^* = \{y \mid y \sim x\}$. Since, for each $y \in \pi_x^*$, $P(y)/P(x)$ is a constant in P, it follows that so also is $P(\pi_x^*)/P(x)$.

Hence, defining

$$g(x) = P(x)/P(\pi_x^*)$$

we have the identity

$$P(x) = g(x)P(\pi_x^*) \text{ for all } x \in X \text{ and } P \in \mathcal{P}.$$

From the sufficient part of Theorem 1, we now know that the partition $\Pi^* = \{\pi_x^*\}$ is a sufficient partition.

Conversely, if $\Pi = \{\pi\}$ be a sufficient partition then the necessary part of Theorem 1 tells us that $x \sim y$ whenever x and y belong to the same part π of Π. Thus, we see that the partition Π^* is thicker than any other sufficient partition and we have the following

Theorem 2 : There always exists a thickest sufficient partition (minimum sufficient statistic).

Let us now turn our attention to sufficient sub-fields. A partition (statistic) is sufficient (by definition) if the induced sub-field is sufficient. In general it is not true that every sufficient sub-field is induced by a partition. However, in the present set-up we prove that every sufficient sub-field is so induced.

Let $\mathcal{B}$ be a sufficient sub-field and let $\Pi = \{\pi\}$ be the partition induced by $\mathcal{B}$ in the manner described in Section 2. Let π be a typical part of Π and let $\phi(.)$

be the conditional probability (function) of π given $\mathcal{B}$. Since $\mathcal{B}$ is sufficient it follows that the function ϕ does not involve P. Since ϕ is $\mathcal{B}$-measurable and Π is induced by $\mathcal{B}$, it follows that $\phi(x)$ is constant, say c, on π. Let $C = \phi^{-1}(\{c\})$. Noting that $C \in \mathcal{B}$ and $\pi \subset C$, we have

$$P(\pi) \equiv P(\pi \cap C) \equiv \int_C \phi(.)\, dP(.)$$

$$\equiv c\, P(C) \text{ for all } P \in \mathcal{P}.$$

Since $P(\pi) > 0$ for some $P \in \mathcal{P}$, it follows that $c > 0$.

Now let π_1 be another part of Π. We claim that the constant value that the function ϕ takes on π_1 must be different from c. Suppose, on the contrary, that $\phi(x) \equiv c$ for all $x \in \pi_1$, i.e. $\pi_1 \subseteq C = \phi^{-1}(c)$. Since π and π_1 are different parts of the partition induced by $\mathcal{B}$ it follows that there exists $B \in \mathcal{B}$ that separates π and π_1, that is

$$\pi_1 \subset B \text{ and } \pi \cap B = \phi, \text{ the empty set.}$$

Then, for all $P \in \mathcal{P}$,

$$0 = P(\pi \cap B \cap C) = \int_{B \cap C} \phi(.)\, dP(.)$$

$$= c\, P(B \cap C)$$

$$\geqslant cP(\pi_1), \text{ (since } \pi_1 \subset B \cap C).$$

Since $c > 0$, it follows that $P(\pi_1) \equiv 0$ for all $P \in \mathcal{P}$, which contradicts condition (C_3).

Thus, we have established that

$$\pi = C = \phi^{-1}(\{c\}) \in \mathcal{B}.$$

Now, let B be an arbitrary union of parts of Π. We prove that $B \in \mathcal{B}$. Let λ be the conditional probability (function) of B given $\mathcal{B}$. Since $\mathcal{B}$ is sufficient, λ does not involve P. Let π be a typical part of Π and let e be the constant value of λ on π (λ is $\mathcal{B}$-measurable). Since $\pi \in \mathcal{B}$ we have, for all $P \in \mathcal{P}$,

$$eP(\pi) = \int_\pi \lambda(.)\, dP(.)$$

$$= P(B \cap \pi) = \begin{cases} P(\pi) & \text{if } \pi \subset B \\ 0 & \text{otherwise.} \end{cases}$$

Since $P(\pi) > 0$ for some P, we have $e = 1$ or 0 according as $\pi \supset B$ or otherwise. And this proves that λ is the indicator function of the set B and hence $B \in \mathcal{B}$.

That every member of $\mathcal{B}$ is a union of parts of Π is obvious. We thus see that $\mathcal{B}$ is precisely the family of all unions of parts of Π, that is, it is induced by Π.

This proves the following :

Theorem 3 : Every sufficient sub-field is induced by a sufficient partition.

In the present set-up the converse of Theorem 3 is trivially true.

We end this section with a few remarks.

Remark 1 : An immediate consequence of Theorem 1 is the fact that if Π be a sufficient partition then so also is any thinner partition Π^* .

Let π_x be that part of Π that includes x. Similarly define π_x^*. Since Π^* is thinner than Π we have $\pi_x^* \subset \pi_x$ for all $x \in X$. Since Π is sufficient, it follows from Theorem 1 that $P(x)/P(\pi_x)$ is a constant in P for all $P \in \mathcal{P}_x$. It follows then that $P(x)/P(\pi_x^*)$ is also a constant in P for all $P \in \mathcal{P}_x$ and this implies the sufficiency of Π^*.

Remark 2 : Let us observe that a similar remark does not hold true for sub-fields. Under the present set-up also it may happen that $\mathcal{C} \subset \mathcal{B}$ and that the smaller sub-field is sufficient without the larger one being so.

Remark 3 : In the present set-up it is very easy to prove that the intersection of two sufficient sub-fields is always sufficient. Let $\mathcal{B}_1$, $\mathcal{B}_2$ be sufficient sub-fields. From Theorem 3 we know that $\mathcal{B}_1$ and $\mathcal{B}_2$ are induced by sufficient partitions Π_1 and Π_2 respectively. Let $\Pi_1 \wedge \Pi_2$ stand for the thinnest partition that is thicker than both Π_1 and Π_2. (Such a partition always exists.) It is not difficult to see that $\mathcal{B}_1 \cap \mathcal{B}_2$ is induced by $\Pi_1 \wedge \Pi_2$. From Theorem 2 we know that there exists a thickest sufficient partition Π_0. It follows then that Π_0 is thicker than $\Pi_1 \wedge \Pi_2$ and hence (Remark 1) $\Pi_1 \wedge \Pi_2$ is sufficient.

Remark 4 : The existence of a minimal sufficient sub-field in our case could also be deduced from results of LeCam (1964) and Pitcher (1965). In a subsequent note the relation of this work with their work and some extensions and applications will be examined.

CHAPTER XI

LIKELIHOOD PRINCIPLE AND SURVEY SAMPLING

0. Notes

Editor's Note : The material covered in this chapter is based on an article that was presented in part at a Statistics Session of the Indian Science Congress in January, 1968 and then published in Sankhyā, 31, 1969, 441-454. This is the first article of Basu in which he took an unequivocal stand in favour of the Likelihood Principle and pointed out its revolutionary implications in Survey Theory.

Author's Note : As I said in my note on Chapter II, when George Barnard tried to sell the Likelihood Principle to me in the mid-fifties, I tried to discredit the principle by pointing at the flat likelihood in finite population sampling. I wrote this article only when I recognized my error. A flat likelihood does not necessarily mean an uninformative likelihood, though this has been repeatedly asserted by some of my esteemed colleagues. A likelihood is uninformative only if it is flat over the whole parameter space.

LIKELIHOOD PRINCIPLE AND SURVEY SAMPLING

1. Introduction

This article was written with the object of emphasizing the following four points.

(a) The first point is only of pedagogical interest. Recently, a series of interesting papers have appeared [Pathak (1964), Godambe (1966), Hanurav (1968), Joshi (1968) to mention only a few] in which the statistical model for sample surveys has been so formulated as to confuse conventional statistical mathematicians [ordinarily incapable of speculating about anything excepting the trinity of $(X, \alpha, \mathcal{P})$!] into the belief that the analysis of survey type data falls outside the mainstream of the theory of statistical analysis. In these formulations, one sees on the surface a 'sample space' S (of possible samples s) with just one probability measure p on S. [How can there be any inference with just one measure?!] The pair (S, p) is called the sampling design. A typical sample $s \in S$ is a subset of (or a finite sequence with its members drawn from) a fixed population Π of individuals 1, 2, 3, . . . , N. The parameter is an unknown vector $\theta = (Y_1, Y_2, \ldots, Y_N)$. A statistic is a very special kind of a function of the sample s and the parameter θ. [How can a statistic be anything but a function defined on the sample space?!] And so on and on it (the new formulation) goes, apparently blazing a new trail in the wilderness of statistical thought. In this article we point out that it is not really necessary to formulate the survey model in the above 'unfamiliar' manner. We need not abandon the trinity $(X, \alpha, \mathcal{P})$!

(b) The second point emphasized here is also of a purely academic nature. If we assume that the set of 'possible' values for the parameter θ is uncountable, then the family $\mathcal{P}$ in the sample survey model $(X, \alpha, \mathcal{P})$ would be typically undominated. This raises the possibility that there may not exist a maximal sufficiency reduction of the survey data (and other hair raising possibilities!). But the saving grace for the survey model is that each member of $\mathcal{P}$ is always a discrete measure. The existence of the maximal sufficiency reduction of the data (the minimal sufficient statistic) is always assured if we take every set as measurable. Also it is very easy to characterize and use the minimal sufficient statistic.

(c) In this article we examine the role of the twin principles of sufficiency and likelihood in the analysis of survey data and arrive at the revolutionary but entirely reasonable conclusion that at the analysis stage the statistician should not pay any attention to the nature of the sampling design. Indeed, the analyst need not even know the sampling design that produced the data.

(d) It goes without saying that there is a great need for designing the survey very carefully. How else can we expect to get a good (representative) sample? Currently, survey statisticians make extensive use of the random number tables. In this article, the author very briefly examines the randomization principle and comes to the conclusion that there is very little (if any) use for it in survey designs.

2. Statistical Models and Sufficiency

The notion of a sampling (or statistical) experiment is idealized as a statistical model $(X, \alpha, \mathcal{P})$ where

(i) X is the sample space,

(ii) α is a fixed σ-field of subsets of X, called the measurable sets or the events, and

(iii) $\mathcal{P} = \{P_\theta | \theta \varepsilon \Omega\}$ is a fixed family of probability measures P_θ on α.

The family $\mathcal{P}$ is indexed by the unknown state of nature (the parameter) θ. The set of all possible values of θ is the parameter space Ω.

By the term statistic we mean a characteristic of the sample x. A general and abstract formulation of the notion of a statistic is that of a mapping of X onto a space Y. Thus, a statistic $T = T(x)$ is an arbitrary function with X as its domain. Every statistic T defines an equivalence relation [$x \sim x'$ if $T(x) = T(x')$] on the sample space X. This leads to a partition of X into equivalent classes of sample points. As we need not distinguish between statistics that induce the same partition of X, it is convenient to think of a statistic T as a partition $\{\pi\}$ of X into a family of mutually exclusive and collectively exhaustive parts π.

The statistic (partition) $T = \{\pi\}$ is said to be wider (larger) than the statistic $T^* = \{\pi^*\}$ if every π is a subset of some π^*–in other words, if every π^* is a union of a number of π's. Given a statistic $T = \{\pi\}$, consider the class of all measurable sets (members of α) that are unions of some π's. They constitute a sub-σ-field (sub-field) of α and is denoted by α_T. We call α_T the sub-field induced by T. If T is wider than T* then $\alpha_T \supset \alpha_{T^*}$.

An abstract and very general formulation of the notion of sufficient statistic is the following :

Definition : The statistic T is sufficient $[\alpha, \mathcal{P}]$ if, corresponding to every realvalued bounded, α-measurable function f, there exists an α_T-measurable f^* such that for all $B \varepsilon \alpha_T$ and $\theta \varepsilon \Omega$

$$\int_B f \, dP_\theta \equiv \int_B f^* \, dP_\theta .$$

The notion of sufficiency has been studied in great details in statistical literature. In the particular case where the family $\mathcal{P}$ of probability measures is dominated by a σ-finite measure λ, we have the following factorization theorem of fundamental importance.

Theorem : Let $p_\theta = dP_\theta/d\lambda$ be a fixed version of the Radon-Nikodym derivative of P_θ wrt λ. A necessary and sufficient condition for the sufficiency of the statistic T is that there exists, for each $\theta \in \Omega$, an α_T-measurable function g_θ and a fixed α-measurable function h such that, for each $\theta \in \Omega$,

$$p_\theta(x) = g_\theta(x)h(x) \text{ aew } [\lambda] .$$

In a dominated set-up, most of the properties of sufficient statistics flow from the above factorization theorem. For example, if T is sufficient then any statistic T* that is wider (larger) than T is also sufficient. Again, with a separability condition on $\mathcal{P}$, it is true that there exists a sufficient statistic T which is essentially smaller (narrower) than every other sufficient statistic T*. Such a sufficient statistic is called the minimal (or least) sufficient statistic.

That neither of the above two propositions need hold for general undominated set-up has been exhibited by Burkholder (1961) and Pitcher (1957). Consider the following two examples.

Example 1 : Let X be the real line, α the σ-field of Borel sets and $\mathcal{P}$ the class of all discrete two-point probability distributions P_θ on the line that are symmetric about the origin. [That is, the entire mass P_θ is equally distributed over the two points $-\theta$ and θ, where $\theta > 0$.] Let E be a non-Borel set that excludes the origin but is symmetric about it. Let $T(x) = |x|$ and let

$$T^*(x) = \begin{cases} |x| & \text{if } x \in E \\ x & \text{if } x \notin E . \end{cases}$$

Clearly, T* is wider than T. However, in this example, T is sufficient but T* is not.

Example 2 : Let X,α and E be as in the previous example and let $\mathcal{P} = \{P_\theta\}$ be defined as follows. If $\theta \in E$ then the whole mass of P_θ is equally distributed over the points $-\theta$ and θ. If $\theta \notin E$, then P_θ is degenerate at θ. In this example, there does not exist a minimal sufficient statistic.

In each of the above two examples, we are dealing with a family of measures each member of which is discrete. In example 1, each measure has its entire mass concentrated at two points only; in example 2, each measure has its entire mass distributed over at most two points. True, we are dealing, in each case, with an undominated family of measures. But that is not where the real trouble lies. In these examples, our difficulties stem from our artificially restricting ourselves to Borel sets only. If in the above two examples we take α to be the class of all subsets, then we do not have to face the above kind of anomalous situations. The natural domain of definition of discrete measures is the σ-field of all subsets. In sample survey theory, we need not consider non-discrete probability measures. By a discrete model we mean the following.

Definition : The statistical model (X, α, P_θ), $\theta \epsilon \Omega$, is called a discrete model if

(i) each P_θ is a discrete measure,

(ii) α is the class of all subsets of X, and

(iii) for each $x \varepsilon X$, there exists a $\theta \varepsilon \Omega$, such that, $P_\theta(\{x\}) > 0$.

[Remark : Condition (iii) only ensures that we do not entangle ourselves with possibilities that have zero probabilities for each possible value of the parameter θ . Condition (ii) ensures that all sets and functions are measurable.]

We, henceforth, deal with discrete models only. A discrete model is undominated if and only if X is uncountable.

3. Sufficiency in Discrete Models

Let (X, α, P_θ), $\theta \varepsilon \Omega$, be a discrete model. For each $x \varepsilon X$ let

$$\Omega_x = \{\theta | P_\theta(x) > 0\}.$$

[We, henceforth, write $P_\theta(x)$ for $P_\theta(\{x\})$.] The set Ω_x is the set of parameter points that are consistent with the observation (sample point) x. No Ω_x is vacuous.

For discrete models, the minimal sufficient statistic always exist and is uniquely defined as follows. Consider the binary relation on X : "$x \sim x'$ if $\Omega_x = \Omega_{x'}$ and $P_\theta(x)/P_\theta(x')$ is a constant in θ for all $\theta \varepsilon \Omega_x = \Omega_{x'}$" .

The above is an equivalence relation on X. The partition (statistic) induced by the equivalence relation is the minimal (least) sufficient statistic. This is an easy consequence of the following factorization theorem (Basu and Ghosh, 1967).

Theorem : If (X, α, P_θ), $\theta \epsilon \Omega$, be a discrete model, then a necessary and sufficient condition for a statistic (partition) $T = \{\pi\}$ to be sufficient is that there exists a function g on X such that, for all $\theta \varepsilon \Omega$ and $x \varepsilon X$,

$$P_\theta(x) \equiv g(x)P_\theta(\pi_x),$$

where π_x is that part of the partition $\{\pi\}$ that contains x.

The above factorization theorem is a direct and easy consequence of the definitions of sufficient statistics and discrete models (as stated in Section 2).

If $T = \{\pi\}$ be a sufficient partition and if g be defined as in the previous theorem, then it follows that $g(x)>0$ for all $x \varepsilon X$ and that, for each π,

$$\sum_{x\varepsilon\pi} g(x) = 1.$$

Each part of a sufficient partition must be countable. What the above factorization theorem is telling us is nothing but the intuitively obvious proposition that $\{\pi\}$ is sufficient if and only if, for each part π, it is true that the conditional distribution of the sample x given π is θ-free. This is indeed the original definition of sufficiency as proposed by Fisher.

Another consequence of the above theorem is that if $T = \{\pi\}$ is a sufficient statistic then any statistic T^* that is wider then T is necessarily sufficient. It also follows that (for discrete models) there exists a one-one correspondence between sufficient statistics (partitions) and sufficient sub-fields (Basu and Ghosh, 1967).

An alternative (but equivalent) way of characterizing the minimal sufficient statistic for a discrete model is the following. For each $x \varepsilon X$ let $L_x(\theta)$ stand for the likelihood function, i.e.

$$L_x(\theta) = \begin{cases} P_\theta(x) & \text{for} \quad \theta\varepsilon\Omega_x \\ 0 & \text{for} \quad \theta\notin\Omega_x. \end{cases}$$

Let us standardize the likelihood function as follows.

$$\bar{L}_x(\theta) = \frac{L_x(\theta)}{\sup_\theta L_x(\theta)}.$$

Consider the mapping

$$x \to \bar{L}_x(\theta),$$

a mapping of X into a class of real-valued functions on Ω. This mapping is the minimal sufficient statistic. [A little reflection would show that the partition (of X) induced by the above mapping is the same as the one induced by the equivalence relation described earlier in this section.]

4. The Sample Survey Models

The principal features of a sample survey situation are as follows. There exists a well-defined population Π — a finite set of distinguishable objects called the (sampling) units. Typically, there exists a list of these units - the so-called sampling frame. Let us list the population as

$$\Pi = (1, 2, 3, \ldots, N).$$

The unit i has an unknown characteristic Y_i. The unknown state of nature is

$$\theta = (Y_1, Y_2, \ldots, Y_N).$$

The statistician has some prior information or knowledge K about θ . This knowledge K is largely of a qualitative and speculative nature. For example, the statistician knows that θ is a member of a well-defined set Ω (the parameter space). He also knows, for each unit i, some characteristic A_i of the unit i. Let us denote this set of known auxiliary characteristics by

$$\mathbf{A} = (A_1, A_2, \ldots A_N).$$

Thus, **A** is a principal component of K. In K is also embedded what the statistician thinks (knows) to be the true relationship between the unknown θ and the known **A**.

It is within the powers of the statistician to find out or "observe" the characteristic Y_i for any chosen unit i. A survey problem arises when the statistician plans to gain further "information" about some function $\tau = \tau(\theta)$ of the parameter θ by observing the Y-characteristics of a set (sequence)

$$\mathbf{i} = (i_1, i_2, \ldots, i_n)$$

of units selected from Π .

Let us denote the observed Y-characteristics by

$$\mathbf{y} = (Y_{i_1}, Y_{i_2}, \ldots, Y_{i_n}).$$

The problem is to make a "suitable" choice of **i** and then to make a "proper" use of the observations x = (**i**, **y**) in conjunction with the prior "knowledge" K to arrive at a "reasonable" "judgement" about τ.

Now, let us examine how probability theory comes into the picture. If we ignore observation errors, then there is no discernable source of randomness in the above general formulation of a survey problem (excepting some very intangible quantities like "belief", "knowledge" etc. which the Bayesians try to formalize as probability.) In any survey situation there are bound to be some observation errors (the so-called non-sampling errors). Unfortunately, in current sample survey research it is not often that we find mention of this source of randomness. It is tacitly assumed that the observation errors are negligible in comparison with the so-called "sampling error". This sampling error is the distinguishing feature of the current sample survey theory. Here is a phenomenon of randomness that is not inherent to the problem but is artificially injected into the problem by the statistician himself. The survey statistician does not lean on probability theory for the purpose of understanding and controlling the mess created by an unavoidable source of randomness or uncertainty (observation errors). He uses his knowledge of probability theory to introduce into the problem a well-understood (fully controlled) element of randomness and seems to derive all his strength (intellectual conviction) from that.

The "sampling error" is the randomness that the statistician injects into the problem by selecting the set (sequence) $\mathbf{i} = (i_1, i_2, \ldots, i_n)$ in a random manner. Given a sampling plan $\mathcal{S}$, for each possible $\mathbf{i}$ there exists a number $p(\mathbf{i})$ which is the probability of ending up with $\mathbf{i}$. Usually, this $p(\mathbf{i})$ does not depend on the parameter θ, although quite often it is made to depend on the auxiliary information **A**. [However, one may consider sequential sampling plans for which $p(\mathbf{i})$ depends on θ. For instance, consider the sampling plan — "Choose unit 1 and observe Y_1. If Y_1 (which we suppose is real valued) is larger than b then choose unit 2, otherwise choose unit N". For this plan $\mathbf{i}$ is either (1, 2) or (1, N) and $p(\mathbf{i})$ depends on θ through Y_1.] Typically, the random choice of $\mathbf{i}$ is made in the statistical laboratory well in advance of the time the observation job is in progress. For such typical sampling plans, the probability $p(\mathbf{i})$ for any possible $\mathbf{i}$ does not depend on θ at all. However, even if we agree to consider sequential sampling plans of the type described within the parentheses before, it is clear that $p(\mathbf{i})$ for such plans can depend on $\theta = (Y_1, Y_2, \ldots, Y_N)$ only through $\mathbf{y} = (Y_{i_1}, Y_{i_2}, \ldots, Y_{i_n})$. As we shall presently see, this remark is important. In the sequel we write $p(\mathbf{i}|\,\theta)$ for $p(\mathbf{i})$.

The sample is $x = (\mathbf{i}, \mathbf{y})$, the set $\mathbf{i}$ together with the observation $\mathbf{y}$. [For some sampling plans — like sampling with replacement — it is more natural to think of $\mathbf{i}$ as a finite sequence of units with repetitions allowed.] The sample space X is the set of all possible samples x.

Now, the sample x, when observed, tells us the exact Y-value of some population units, i.e., tells us about some coordinates of the vector θ. Let Ω_x be the set of parameter points θ that are consistent with a given sample x. If $P_\theta(x)$ be the probability that the sampling plan ends up with sample $x = (\mathbf{i},\mathbf{y})$, then it is clear that

$$P_\theta(x) = \begin{cases} p(\mathbf{i} \mid \theta) & \text{for} \quad \theta \in \Omega_x \\ 0 & \text{otherwise.} \end{cases}$$

Thus, Ω_x is also the set of all parameter points that allot non-zero probabilities to x.

As we have said before, in typical sampling plans $p(\mathbf{i} \mid \theta)$ does not depend on θ. In sequential sampling plans (where the choice of population unit at any stage is made to depend on the observed Y-values of the previously selected units) we have noted before that $p(\mathbf{i} \mid \theta)$ depends on θ through **y**. It is important to observe then that, for any sampling plan,

$$P_\theta(x) = \begin{cases} \text{constant} & \text{for} \quad \theta \in \Omega_x \\ 0 & \text{for} \quad \theta \notin \Omega_x . \end{cases}$$

This leads us to the following general characterization of a sample survey (SS) model.

Definition : The model $(X, \alpha, P_\theta), \theta \in \Omega$ is called an SS-model if the model is discrete and if $P_\theta(x)$ is a constant for all $\theta \in \Omega_x$, where

$$\Omega_x = \{\theta | P_\theta(x) > 0\} .$$

From what we have said in Section 3, it then follows that

Theorem : If (X, α, P_θ), $\theta \in \Omega$ be an SS-model, then the minimal (least) sufficient statistic is the mapping $x \to \Omega_x$.

In typical survey situations, the minimal sufficient statistic (the information core of the sample) is the set of (distinct) population unit-labels that are drawn in the sample together with the corresponding Y-values.

The distinguishing feature of an SS-model is that for every possible sample the likelihood function is flat. That is, for every $x \in X$ the likelihood function $L_x(\theta)$ is zero for all θ outside a set Ω_x and is a constant for $\theta \in \Omega_x$. The following is an example of a non-discrete model with the above feature.

Example 3 : Let $x = (x_1, x_2, \ldots, x_n)$ be n independent observations on a random variable that has a continuous and uniform distribution over the interval $(\theta - 1/2, \theta + 1/2)$, where θ is the parameter $(-\infty < \theta < \infty)$. Let Ω_x be the interval $(m(x), M(x))$ where $m(x) = \max x_i - 1/2$ and $M(x) = \min x_i + 1/2$. Here, $L_x(\theta) = 1$ for all $\theta \in \Omega_x$ and is zero for $\theta \notin \Omega_x$. The mapping $x \to (m(x), M(x)) = \Omega_x$ is the minimal sufficient statistic.

5. The Sufficiency and Likelihood Principles

The twin principles of sufficiency and likelihood both attempt to answer the same question. The likelihood principle, however, goes a great deal further in its assertion.

The question is : "What characteristic of the sample x is relevant for making an inference about the parameter θ ?" In general, the sample x is a very complex entity. Must we take into account the sample x in all its details? Could it be that some characteristics of x are totally irrelevent for making any inference about the state of nature θ ? For instance, if in the observation x we have incorporated the outcome u from a number of tosses of a symmetric coin, then it seems very reasonable to argue that the characteristic u of x is totally irrelevent and must be ignored.

The sufficiency principle is the following. If $T = T(x)$ be a sufficient statistic, then only the characteristic $T(x)$ of x is relevant for inference making. That is, if $T(x) = T(x')$ then the inference about θ should be the same whether the sample is x or x'. The relevant information core of x is then the statistic $T_0(x)$, where T_0 is the minimal sufficient statistic.

The suficiency principle has gained rather wide acceptance. The Neyman-Pearson school of statisticians tend to justify the principle by proving some complete class theorem that tells us that it is not necessary to consider decision rules (inference procedures) that do not depend on x through $T_0(x)$. On the other hand the Bayesians have no objection to the sufficiency principle as they point out that the posterior distribution for the parameter θ—whatever be its prior distribution — depends on x only through the minimal sufficient statistic $T_0(x)$.

As we have stated in Section 3, the mapping $x \to \bar{L}_x(\theta)$, where $\bar{L}_x(\theta)$ is the standardized (modified) likelihood function, is the minimal sufficient statistic. Thus, according to the sufficiency principle, two sample points x and x' are equally informative if

$$\bar{L}_x(\theta) = \bar{L}_{x'}(\theta) \quad \text{for all } \theta.$$

The sufficiency and the likelihood principles do not tell us anything about the nature of the information supplied by x. Whereas the sufficiency principle can compare two possible samples x and x' only when they are points in the same sample space, the likelihood principle can compare them even when they are points in different sample spaces. Consider the following example.

Example 4 : Let θ be the unknown probability of head for a given coin. The following is a list of three different experiments (among the many that one can think of) that one may perform for the purpose of eliciting information about the unknown θ .

E_1 : Toss the coin 5 times

E_2 : Toss the coin until there are 3 heads

E_3 : Toss the coin until there are 2 consecutive heads.

We give below an example x_i of a possible sample point for each experiment E_i ($i = 1, 2, 3$). [H = head, T = tail]

x_1 : H T H H T

x_2 : T T H H H

x_3 : H T T H H

[Note that the sample spaces for the three experiments are different from one another. Also note that x_1 cannot be a sample point for either E_2 or E_3. Similarly, x_2 cannot be a sample point for E_3.]

It is easy to check that the likelihood function for x_i (when it is referred to experiment E_i is $\theta^3(1-\theta)^2$ and this is irrespective of whether i is 1, 2, or 3. The principle of likelihood tells us that sample x_1 for experiment E_1 gives the same information about θ as does sample x_2 for E_2 and sample x_3 for E_3.

From the Bayesian point of view the likelihood principle is almost a truism. The starting point for a Bayesian (in his inference making effort) is a prior probability distribution over the parameter space Ω . Let $q = q(\theta)$ be the prior probability frequency function. Having observed the sample x, the Bayesian uses the likelihood function $\bar{L}_x(\theta)$ to arrive at the posterior distribution

$$q_x^*(\theta) = \frac{q(\theta)\bar{L}_x(\theta)}{\Sigma_\theta q(\theta)\bar{L}_x(\theta)} .$$

To a Bayesian, the role of the sample x is only to change his prior scale of preference (probability distribution) $q = q(\theta)$, for various possible values of θ , to the posterior scale $q^* = q_x^*(\theta)$. And this change is effected through the likelihood function $\bar{L}_x(\theta)$. Possible sample points x and x' (whatever sampling experiments might generate them) are equivalent as long as they induce identical (modified) likelihood functions. The likelihood principle is essentially a Bayesian principle. It is hard to justify the principle under the Neyman-Pearson set-up.

6. Role and Choice of the Sampling Plan

Let $\mathcal{S}$ be the chosen sampling plan and let $x = (\mathbf{i}, \mathbf{y})$ be the data (sample) generated by $\mathcal{S}$. In the matter of analyzing the data, how relevant is the plan $\mathcal{S}$?

If $\mathbf{i} = (i_1, i_2, \ldots, i_n)$ and $\mathbf{y} = (Y_{i_1}, Y_{i_2}, \ldots, Y_{i_n})$, then Ω_x is the set of all $\theta \in \Omega$ whose j-th co-ordinate is $Y_j (j = i_1, i_2, \ldots i_n)$ - the set of θ's that are consistent with the data. Note that Ω_x depends only on x and Ω , it has nothing to do with the plan $\mathcal{S}$. The minimal sufficient statistic is the mapping $x \to \Omega_x$ and the likelihood function $\bar{L}_x(\theta)$ is

$$\overline{L}_x(\theta) = \begin{cases} 1 & \text{for } \theta \in \Omega_x \\ 0 & \text{otherwise} \end{cases}$$

If $q = q(\theta)$ be the Bayesian prior distribution over Ω, then the posterior distribution is

$$q_x^*(\theta) = \frac{q(\theta)\,\overline{L}_x(\theta)}{\Sigma_\theta q(\theta)\,\overline{L}_x(\theta)} = \begin{cases} c(x)\, q(\theta) \text{ for } \theta \in \Omega_x \\ 0 \qquad\qquad \text{otherwise.} \end{cases}$$

The posterior distribution $q_x^*(\theta)$ is nothing but the restriction of q to the set Ω_x. And the plan $\mathcal{S}$ does not enter into the definition of Ω_x. Thus, from the Bayesian (and the likelihood principle) point of view, once the data x is before the statistician, he has nothing to do with the plan $\mathcal{S}$. He does not even need to know what the plan $\mathcal{S}$ was. [This is because, in sample survey situations, the plan $\mathcal{S}$ is an artificial source of randomness. In other statistical situations, where randomess is unavoidable and is an inherent part of the observation process, the statistician has to "understand" the process well enough to be able to arrive at his likelihood function.]

In the Neyman-Pearson type of analysis of the data, the statistician considers not only the data x in hand but also pays a great deal of attention to what other data x' he might have obtained. In other words, he needs to know the model $(X, \alpha, \mathcal{P})$ as well as the sample x. The Bayesian needs to know only the likelihood function $\overline{L}_x(\theta)$, which, in a sample survey situation, is entirely independent of the model (the sampling plan $\mathcal{S}$). The author does not think that any reconcilliation between the two approaches to data analysis is possible.

A majority of statisticians of the Neyman-Pearson school would readily agree to the proposition that the Bayesian analysis of the data is sensible (acceptable) when the following condition holds :

Condition B : It is reasonable to think of the parameter θ as a random variable, and the random process governing θ is at least partially discernable.

However, it is hard to understand how such statisticians reconcile themselves to the contrary positions : (a) Only the data x (the likelihood function) is relevant (for inference making) when condition B holds and (b) the whole sample space X (the model) is relevant when B does not hold. There exists a continuous spectrum of conditions between the extremes of B and not-B. But the shift of emphasis from the sample x to the sample space X is not continuous. [Fisher with his theory of ancillary statistics and choice of reference sets, made a bold but unsuccessful (see Basu, 1964) attempt to bridge the gap between the above polarities in statistical theory.]

It seems to the author that the Bayesian analysis of the data x is very appropriate in sample survey situation. Given the data x = (**i**, **y**) the sampling plan $\mathbb{S}$— the model (X, α, $\mathcal{P}$)— ceases to be of any relevance for inference making about the parameter θ . Given the data x the statistician arrives at his posterior preference scale $q^*_x(\theta)$ for the parameter θ . If $\tau = \tau(\theta)$ be the parameter of interest, then the statistician can compute the marginal posterior distribution $q^*(\tau)$ of the variable τ. The question, "Given x, how much information we have about τ?", can then be answered by first agreeing upon a suitable definition of information. [For example, we may agree to work with the Shannon definition of information or with the posterior variance (or its reciprocal) of τ.]

Given a sample x, we can now tell how good (informative) the sample is. The object of planning a survey should be to end up with a good sample. The term "representative sample" has often been used in sample survey terminology. But no one has cared to give a precise definition of the term. It is implicity taken for granted that the statistician with his biased mind is unable to select a representative sample. So a simplistic solution is sought by turning to an unbiased die (the random number tables). Thus, a deaf and dumb die is supposed to do the job of selecting a "representative sample" better than a trained statistician. It is, however, true that we do not really train our statisticians for the job of selecting and observing survey type data. [In contrast, the medical practitioner is given a much more meaningful training in understanding the many variables and their interrelations in his chosen field of specialization.]

In a Bayesian plan for selecting the sample x, there is no place for the symmetric die. Very little attention has so far been paid to the problem of devising suitable sampling strategies from this point of view. We end this section by describing a Bayesian sampling strategy for the very simple case where the statistician wants to select and observe only one unit. Suppose his prior probability distribution is $q(\theta)$. If he selects unit i and observes Y_i then his posterior (marginal) distribution for τ would be, say, $q^*(\tau \mid i, Y_i)$. Once a suitable definition of "information" is agreed upon, he can use the above distribution to compute the quantity $I(i, Y_i)$— the information about τ gained from the sample (i, Y_i). At the planning stage of the experiment, the statistician does not know the value of Y_i that he is going to observe for the unit i. Let J(i) be the average value of $I(i, Y_i)$ when the averaging is done over all possible values of Y_i (weighted by the prior distribution of Y_i). Thus, J(i) is the "expected" information to be gained from observing unit i. Faced with the problem of deciding which unit i to select (and then observe), the statistician would not be acting unreasonably if he selects the unit i that has maximum J(i). [What if the J(i)'s are all equal? Such would be the case if the prior distribution of $\theta = (Y_1, Y_2, \ldots, Y_N)$ is symmetric in the coordinates. In this situation the statistician is indifferent as to which i is selected for observation. In principle, he cannot object now to a random (with equal or unequal probabilities) selection procedure for i. However, this does not mean that he will be willing to let another person (say, a field investigator) make the choice for him. If for nothing else, a scientist ought to be always on his guard against letting an unknown element enter into the picture.]

Of course, a non-Bayesian would sneer at the arbitrariness inherent in the difinition of J(i). But the procedure described above is certainly more justifiable than our current naive reliance on the symmetric die. Any reasonable Bayesian sampling strategy would have the following characteristics. (a) The sampling plan would usually be sequential. The statistician would continue sampling (one or a few units at a time) until he is satisfied with the information thus obtained or until he reaches the end of his rope (time and cost). His decision to select the units for a particular sampling stage would depend (non-randomly) on the sample obtained in the previous stages. (b) The probability that the statistician would end up observing the units $\mathbf{i} = (i_1, i_2, \ldots, i_n)$ in this order, would depend on $\mathbf{i}$ and the state of nature θ . This probability would be degenerate, i.e., zero for some values of θ and unity for the rest of the values of θ .

Concluding Remarks

(a) Godambe (1966a) noted that the application of the likelihood principle in the sampling situation would mean that the sampling design is irrelevant for data analysis. On page 317 he writes, "One implication of this, as can be seen from (4), is that the inference about θ must not depend on the sampling design even through the probability p($\mathbf{i}$) of the $\mathbf{i}$ that has actually been drawn. In particular, the estimator of τ should not depend on p($\mathbf{i}$) or the sampling design". [In this and in the following quotation the author has taken the liberty of changing some of the notations. This was done for the sake of bringing them in line with the notations used in this article] It is interesting to observe that Godambe immediately shies away from the revolutionary implication of his remark and tries to find some excuses for not applying the likelihood principle in the sampling situation. He writes (p 317, Godambe, 1966a), "In connection with the likelihood principle, it may be further noted that here θ is the parameter and ($\mathbf{i},\mathbf{y}$) is the sample. Thus, possibly there is some kind of relationship between the parametric space and the sample space (when the sample is observed, the parameter cannot remain completely unknown) which forbids the use of the likelihood principle. The relationships between parametric and sample spaces restricting the use of the likelihood principle are referred to by Barnard, Jenkins and Winsten (1962)". In the two 1966 papers referred to here, Godambe tries very hard to justify a particular linear estimator as the only reasonable one for the population total. Godambe's estimator depends on the sampling design. The author finds Godambe's arguments very obscure.

(b) Let us repeat once again that the posterior distribution of τ depends only on the prior distribution q (on Ω) and the sample x = (**i**, **y**). It does not depend on the sampling design $\mathcal{S}$. Thus, any fixed q on Ω would give rise to a Bayes 'estimation procedure' B_q that would tell us how to estimate τ for each possible sample x – no matter what design $\mathcal{S}$ is used to arrive at x. [Note that B_q is well-defined as a functiuon on the union of sample spaces for all designs $\mathcal{S}$]. Now, if we consider B_q in relation to a fixed design $\mathcal{S}$, then it would be classified as an admissible estimator in the sense of Wald. The findings of Godambe (1960) and Joshi (1968) therefore appear to the author as rather obvious in nature. It is so easy to reel off any number of such universally admissible estimation procedures.

CHAPTER XII

ON THE LOGICAL FOUNDATIONS OF SURVEY SAMPLING

1. An Idealization of the Survey Set-up

It is a mathematical necessity that we idealize the real state of affairs and come up with a set of concepts that are simple enough to be incorporated in a mathematical theory. We have only to be careful that the process of idealization does not distort beyond recognition the basic features of a survey set-up, which we list as follows:

(a) There exists a population – a finite collection $\mathcal{P}$ of distinguishable objects. The members of $\mathcal{P}$ are called the (sampling) units. [Outside of survey theory the term population is often used in a rather loose sense. For instance, we often talk of the infinite population of all the heads and tails that may be obtained by repeatedly tossing a particular coin. Again, in performing a Latin-square agricultural experiment the actual yield from a particular plot is conceived of as a sample from a conceptual population of yields from that plot. It is needless to mention that such populations are not real. The existence of a down-to-Earth finite population is a principal characteristic of the survey set-up.]

(b) There exists a sampling frame of reference. By this we mean that the units in $\mathcal{P}$ are not only distinguishable pairwise, but are also observable individually; that is, there exists a list (frame of reference) of the units in $\mathcal{P}$ and it is within the powers of the surveyor to pre-select any particular unit from the list and then observe its characteristics. Let us assume that the units in $\mathcal{P}$ are listed as

$$1, 2, 3, \ldots, N,$$

where N is finite and is known to the surveyor. [We are thus excluding from our survey theory such populations as, for example, the insects of a particular species in a particular area or the set of all color-blind adult males in a particular country. Such populations as above can, of course, be the subject matter of a valid statistical inquiry but the absence of a sampling frame makes it impossible for such populations to be surveyed in the sense we understand the term survey here.]

(c) Corresponding to each unit $j \varepsilon \mathcal{P}$ there exists an unknown quantity Y_j in which the surveyor is interested. The unknown Y_j can be made known by observing (surveying) the unit j. The unknown state of nature is the vector quantity

$$\theta = (Y_1, Y_2, \ldots, Y_N).$$

However, the surveyor's primary interest is in some characteristic (parameter) $\tau = \tau(\theta)$

Editor's Note : This chapter is based on an invited paper presented at a Conference at Waterloo University, Canada in 1970, and published with discussions in the conference volume, Foundations of Statistical Inference, V.P. Godambe and D.A. Sprott (eds.), 1971, Holt : New York. The author's views are elaborated in Chapter XIV.

of the state of nature θ . [Typically, the Y_j's are vector quantities themselves and the surveyor is seeking information about a multiplicity of τ's. However, for the sake of pinpointing our attention to the basic questions that are raised here, we restrict ourselves to the simple case where the Y_j's are scalar quantities (real numbers) and $\tau = \Sigma Y_j$.]

(d) The surveyor has prior knowledge K about the state of nature θ. This knowledge K is a multi-dimensional complex entity and is largely of a qualitative and speculative nature. We consider here the situation where K has at least the following two well-defined components. The surveyor knows the set Ω of all the possible values of the state of nature θ and, for each unit j, (j = 1,2, . . . , N), he has access to a record of some known auxiliary characteristic A_j of j. [Typically, each A_j is a vector quantity. However, in our examples we shall take the A_j's to be real numbers.] The set Ω and the vector

$$\alpha = (A_1, A_2, \ldots, A_N)$$

are the principal measurable components of the surveyor's prior knowledge K. Let us denote the residual past of the knowledge by R and write

$$K = (\Omega, \alpha, R).$$

(e) The purpose of a survey is to gain further knowledge (beyond what we have described as K) about the state of nature θ and, therefore, about the parameter of interest $\tau = \tau(\theta)$. Since the surveyor is supposed to know the set Ω of all the possible values of θ , he knows the set $\mathcal{J}$ of all the possible values of τ . Initially, the surveyor's ignorance about τ is, therefore, spread over the set $\mathcal{J}$. [Later on, we shall quantify this initial spread of ignorance as a prior probability distribution.] In theory, the surveyor can dispel this ignorance and gain complete knowledge by making a total survey (complete enumeration) of $\mathcal{P}$. If he observes the Y-characteristic of every unit j(j = 1,2, . . . , N), then he knows the actual value of $\theta = (Y_1, Y_2, \ldots, Y_N)$ and, therefore, that of $\tau(\theta)$. We are, however, considering the case where a total survey is impracticable. By a survey of the population $\mathcal{P}$ we mean the selection of a (usually small) subset

$$u = (u_1, u_2, \ldots, u_n)$$

of units from $\mathcal{P}$ and then observing the corresponding Y-values

$$y = (Y_{u_1}, Y_{u_2}, \ldots, Y_{u_n})$$

of units in the subset u.

(f) We make the simplifying assumption that there are no non-response and observation errors; that is, the surveyor is able to observe every unit that is in the subset u, and when he observes a particular unit j, he finds the true value of the hitherto unknown Y_j without any error.

(g) The Surveyor's blueprint for the survey is usually a very complicated affair. The survey plan must take care of myriads of details. However, in this article we idealize away most of these details and consider only two facets of the survey project, namely, the sampling plan and the fieldwork. The sampling plan is the part of the project that yields the subset u of $\mathcal{P}$ and fieldwork generates the observations y on members of u. The data (sample) generated by the survey is

$$x = (u, y).$$

For reasons that will be made clear later, it is important to distinguish between the two parts u and y of the data x.

(h) Let $\mathcal{S}$ stand for the sampling plan of the surveyor. The plan (when set in motion) produces a subset u of the population $\mathcal{P} = (1,2, \ldots, N)$. We write $u = (u_1, u_2, \ldots, u_n)$, where $u_1 < u_2 < \ldots < u_n$ are members of $\mathcal{P}$. The fieldwork generates the vector $y = (Y_{u_1}, Y_{u_2}, \ldots, Y_{u_n})$ which we often write as $(y_1, y_2, \ldots, y_n)$. [Occasionally, we shall consider sampling plans that introduce a natural selection order among the units that are selected. For such plans it is more appropriate to think of u, not as a subset of $\mathcal{P}$, but as a finite sequence of elements $u_1, u_2, \ldots, u_n$ drawn from $\mathcal{P}$ in that selection order. In rare instances, the sampling plan may allow the possibility of a particular unit appearing repeatedly in the sequence $(u_1, u_2, \ldots, u_n)$. From the description of the sampling plan $\mathcal{S}$ it will usually be clear if we intend to treat u as a set or a sequence. In either case, we can think of u as a vector $(u_1, u_2, \ldots, u_n)$ and y as the corresponding observation vector $(y_1, y_2, \ldots y_n)$ where $y_i = Y_{u_i}$.]

(i) **Summary :** Our idealized survey set-up consists of the following :

(i) A finite population $\mathcal{P}$ whose members are listed in a sampling frame as 1,2,..., N. Availability of each $j \varepsilon \mathcal{P}$ for observation.

(ii) The unknown state of nature $\theta = (Y_1, Y_2, \ldots, Y_N)$ and the parameter of interest $\tau = \tau(\theta)$.

(iii) The prior knowledge $K = (\Omega, \alpha, R)$.

(iv) Absence of non-response and observation errors.

(v) Choice of a sampling plan $\mathcal{S}$ as part of the survey design.

(vi) Putting the sampling plan $\mathcal{S}$ and the fieldwork into operation, thus arriving at the data (sample)

$$x = \{u = (u_1, u_2, \ldots, u_n), \quad y = (y_1, y_2, \ldots, y_n)\}$$

where $u_i \varepsilon \mathcal{P}$ and $y_i = Y_{u_i} (i = 1, 2, \ldots, n)$.

(vii) Making a proper use of the data x in conjunction with the prior knowledge K to arrive at a reasonable judgment (or decision) related to the parameter τ.

The operational parts of the survey are its design (v), the actual survey (vi) and the data analysis (vii). In this article we are concerned only with the design and the analysis of a survey.

2. Probability in Survey Theory

We posed the survey set-up as a classic problem of inductive inference – a problem of inferring about the whole from observations on only a part. The basic questions are : Which part does one observe? Does the part (actually observed) tell us anything about the whole? and, then the main question, Exactly what does it tell? Let us now examine how probability enters into the picture.

There are three different ways in which probability theory finds its way into the mathematical theory of survey sampling. First, there is the time honored way through a probabilistic model for observation errors. Indeed, this is how probability theory first infiltrated the sacred domain of science. When we observe the Y-value Y_j of unit j, there is bound to be some observation error. In current survey theory we classify this kind of error as non-sampling error. In this article we have idealized away this kind of probability by assuming that there exists no observation error. We have deliberately taken this simplistic view of the survey set-up. The idea is to concentrate our attention on the other two sources of probability.

In current survey theory, the main source of probability is randomization, which is an artificial introduction (through the use of random number tables) of randomness in the sampling plan $\mathcal{S}$. Randomization makes it possible for the surveyor to consider the set (or sequence) u, and therefore the data x = (u, y), as random elements. With an elemet of randomization incorporated in the sampling plan $\mathcal{S}$, the surveyor can consider the space U of all the possible values of the random element u and then the probability distribution p_θ of u over U. [For sampling plans usually discussed in survey textbooks, the probability distribution p_θ of u is uniquely determined (by the plan) and is, therefore, independent of the state of nature θ.] Now, let X be the space of all the possible values of the data (sample) x = (u,y) of which we have already recognized (thanks to randomization) the part u to be a random element. The space X is our sample space. Let P_θ be the probability distribution of x over the sample space X. If T = T(x) is an estimate of τ , then (prior to sampling and fieldwork) we can consider T to be a random variable and speculate about its sampling distribution and its average performance characteristics (as an estimator of τ) in an hypothetical sequence of repeated experimentations. This decision-theoretic approach is not possible unless we regard randomization as the source of probability in survey theory. From the point of view of a frequency-probabilist, there cannot be a statistical theory of surveys without some kind of randomization in the plan $\mathcal{S}$.

Apart from observation errors and randomization, the only other way that probability can sneak into the argument is through a mathematical formalization of what we have described before as the residual part R of the prior knowledge $K = (\Omega, \alpha, R)$. This is the way of a subjective (Bayesian) probabilist. The formalization of R as a prior probability distribution of θ over Ω makes sense only to those who interpret the probability of an event, not as the long range relative frequency of occurrence of the event (in an hypothetical sequence of repetitions of an experiment), but as a formal quantification of the illusive (but nevertheless very real) phenomenon of personal belief in the truth of the event. According to a Bayesian, probability is a mathematical theory of belief and it is with this kind of a probability theory that one should seek to develop the guidelines for inductive behavior in the presence of uncertainty. The purpose of this essay is not to examine the logical basis of Bayesian probability nor to describe how one may arrive at the actual qualification of R as a prior probability distribution of θ over Ω. [Of late, a great deal has been written on the subject. See, for instance, I.R. Savage's delightfully written new book, Statistics : Uncertainty and Behavior.]

Can the two kinds of probability co-exist in our survey theory? This is what we propose to find out.

3. Non-Sequential Sampling Plans and Unbiased Estimation

By a non-sequential sampling plan we mean a plan that involves no fieldwork. If the sampling plan $\mathcal{S}$ is non-sequetial, then the surveyor can (in theory) make the selection of the set (or sequence) u of population units right in the office and then send the field investigators to the units selected in u and thus obtain the observation part y of the data $x = (u, y)$. A great majority of survey theoreticians have so far restricted themselves to non-sequential plans that involve an element of randomization in it. In this section we consider such plans only. The essence of non-sequentialness of a plan $\mathcal{S}$ is that the probability distribution of u does not involve the state of nature θ. Thus, the sampling plan where we continue to draw a unit at a time with equal probabilities and with replacements until we get υ distinct units is a non-sequential plan.

Given $u = (u_1, u_2, ..., u_n)$, the observation part $y = (y_1, y_2, ..., y_n)$, where $y_i = Y_{u_i}$ $(i = 1, 2, ..., n)$, is obtained through the field work and is uniquely determined by the state of nature $\theta = (Y_1, Y_2, ..., Y_N)$. The conditional probability distribution of y given u is degenerate, the point of degeneration depending on θ. That is, for all y, u and θ

$$\text{Prob}(y|\ u, \theta) = 0 \text{ or } 1\ . \tag{3.1}$$

[We are taking the liberty of using the symbols u, y, x and θ both as variables and as particular values of the variables.]

For each sampling plan $\mathcal{S}$ we have the space U of all the possible values of u. The probability distribution p of u over U is θ-free, that is, is uniquely defined by the plan $\mathcal{S}$. The probability distribution p is clearly discrete. There is no loss of generality in assuming that $p(u) > 0$ for all $u \varepsilon U$. If the non-sequential plan is purposive (that is, the plan involves no randomization) then U is a single-point set and the distribution of u is degenerate at that point.

Let X be the sample space, the set of all possible samples (data) $x = (u, y)$ where u is generated by the plan $\mathcal{S}$ and y by the fieldwork. For each $\theta \varepsilon \Omega$, we have a probability distribution P_θ over X. Whatever the plan $\mathcal{S}$, the probability distribution P_θ is necessarily discrete. We write $P_\theta(x)$ or $P_\theta(u, y)$ for the probability of arriving at the data $x = (u, y)$ when θ is the true value of the state of nature. Clearly,

$$P_\theta(x) = P_\theta(u, y) = p(u)\text{Prob}(y|\ u, \theta). \tag{3.2}$$

The surveyor takes a peep at the unknown $\theta = (Y_1, Y_2, ..., Y_N)$ through the sample $x = (u, y)$. Prior to the survey, the surveyor's ignorance about θ was spread over the space Ω . Once the data x is at hand, the surveyor has exact information about some coordinates of the vector θ . These are the coordinates that correspond to the distinct units that are in u. The data x rules out some points in Ω as clearly inadmissible. Let Ω_x be the subset of values of θ that are consistent with the data x. In other words, $\theta \varepsilon \Omega_x$ if $P_\theta(x) > 0$; that is, it is possible to arrive at the data x when θ is the true value of the state of nature. The subset Ω_x of Ω is well-defined for every $x \varepsilon X$. Without any loss of generality we may assume that no Ω_x is vacuous. From (3.1) and (3.2) it follows that the likelihood function $L(\theta)$ is given by the formula

$$L(\theta) = P_\theta(x) = \begin{cases} p(u) & \text{for all } \theta \varepsilon \Omega_x \\ 0 & \text{otherwise.} \end{cases} \tag{3.3}$$

In other words, whatever the data x, the likelihood function $L(\theta)$ is flat (a positive constant) over the set Ω_x and is zero outside Ω_x. This remark holds true for sequential plans also (Basu, 1969b).The importance of the remark will be made clear later on.

A major part of survey theory is concerned with unbiased estimation. A statistic is a characteristic of the sample x. An estimator T = T(x) is a statistic that is well-defined for all $x \varepsilon X$ and is used for estimating a parameter $\tau = \tau(\theta)$. By an unbiased estimator of $\tau(\theta)$ we mean an estimator T that satisfies the identity

$$E(T|\theta) = \sum_{x} T(x)P_\theta(x) \equiv \tau(\theta), \text{ for all } \theta \varepsilon \Omega. \qquad (3.4)$$

Let $w(t, \theta)$ be the loss function. That is, $w(t, \theta)$ stands for the surveyor's assessment of the magnitude of error that he commits when he estimates the parameter $\tau = \tau(\theta)$ by the number t. We assume that

$$w(t, \theta) \geqslant 0 \text{ for all } t \text{ and } \theta,$$

the sign of equality holding only when $t = \tau(\theta)$. The risk function $r_T(\theta)$ associated with the loss function w and the estimator T is then defined as the expected loss

$$r_T(\theta) = E[w(T,\theta)|\theta] = \sum_{x} w(T(x), \theta)P_\theta(x). \qquad (3.5)$$

[If the reader is not familiar with the decision-theoretic jargons of loss and risk, he may restrict himself to the particular case where $w(t, \theta)$ is the squared error $(t - \tau(\theta))^2$ and the risk function $r_T(\theta)$ is the variance $V(T|\theta)$ of the unbiased estimator T.] The following theorem proves the non-existence of a uniformly minimum risk (variance) unbiased estimator of τ.

Theorem. Given an unbiased estimator T of τ and an arbitrary (but fixed) point $\theta_0 \varepsilon \Omega$, we can always find an unbiased estimator T_0 (of τ) such that $r_{T_0}(\theta_0) = 0$, that is, T_0 has zero risk at θ_0.

Proof : We find it convenient to write $T(u, y)$ and $P_\theta(u, y)$ for $T(x)$ and $P_\theta(x)$ respectively. It has been noted earlier that the conditional distribution of y given u is degenerate at a point that depends on θ. Let $y_0 = y_0(u)$ be the point of degeneration of y, for given u, when $\theta = \theta_0$. Consider the statistic

$$T_0 = T(u, y) - T(u, y_0) + \tau(\theta_0) \qquad (3.6)$$

The statistic $T(u, y_0)$ is a function of u alone and so its probability distribution, and therefore its expectation are θ-free. Indeed,

$$\begin{aligned} E[T(u, y_0)] &= \sum_{u} T(u, y_0)p(u) \\ &= \sum_{u,y} T(u,y)P_{\theta_0}(u,y) \\ &= E[T(u,y)|\theta_0] \\ &= \tau(\theta_0). \end{aligned}$$

Thus, the statistic T_o as defined in (3.6) is an unbiased estimator of τ. Now, when $\theta = \theta_o$, the statistics $T(u,y)$ and $T(u,y_o)$ are equal with probability one, and so $T_o = \tau(\theta_o)$ with probability one. This proves the assertion that $r_{T_o}(\theta_o) = 0$.

The impossibility of the existence of a uniformly minimum risk unbiased estimator follows at once. For, if such an estimator T exists then $r_T(\theta)$ must be zero for all $\theta \varepsilon \Omega$. That is, whatever the value of the state of nature θ it should be possible to estimate $\tau(\theta)$ without any error (loss) at all. Unless the sampling plan $\mathcal{S}$ is equivalent to a total survey of the population, such a T clearly cannot exist for a parameter τ that depends on all the coordinates of θ. The following two examples will clarify the theorem further.

Example 1. Consider the case of a simple random sample of size one from the population $\mathcal{P} = (1, 2, ..., N)$. The sample is (u, y) where u has a uniform probability distribution over the N integers 1, 2, ..., N and $y = Y_u$. Let the population mean

$$\overline{Y} = \frac{1}{N}(Y_1 + ... + Y_N)$$

be the parameter to be estimated. Clearly, y is an unbiased estimator of $\overline{Y}$. Let $\theta_o = (a_1, a_2, ..., a_N)$ and $\overline{a} = (\Sigma a_j)/N$. The statistic

$$T_o = y - a_u + \overline{a} \tag{3.7}$$

is an unbiased estimator of $\overline{Y}$ with zero risk (variance) when $\theta = \theta_o$. The variance of y is $\Sigma(Y_j - \overline{Y})^2/N$ and that of T_o is $\Sigma(Z_j - \overline{Z})^2/N$ where $Z_j = Y_j - a_j$ (j = 1, 2, ..., N).

Example 2. Let $\mathcal{S}$ be an arbitrary non-sequential sampling plan that allots a positive selection probability to each population unit. That is, the probability Π_j that the unit j appears in the set (or sequence) u is positive for each j(j = 1, 2, ..., N). Since $\mathcal{S}$ is non-sequential, the vector

$$\Pi = (\Pi_1, \Pi_2, ..., \Pi_N)$$

is θ-free. Let Y be the population total ΣY_j. A particular unbiased estimator of Y that has lately attracted a great deal of attention is the so-called Horvitz-Thompson (HT) estimator (relative to the plan $\mathcal{S}$). The HT-estimator is defined as follows. Let $u_1 < u_2 < ... < u_\upsilon$ be the distinct population units that appear in u and let $\hat{y} = (y_1, y_2, ..., y_\upsilon)$ be the corresponding observation vector. Let $p_i = \Pi_{u_i}$ (i = 1, 2, ..., υ). The HT-estimator H is then defined as

$$H = \frac{y_1}{p_1} + ... + \frac{y_\upsilon}{p_\upsilon} \tag{3.8}$$

That H is an unbiased estimator of Y will be clear when we rewrite (3.8) in a different form. Let E_j (j = 1, 2, . . . , N) stand for the event that the plan $\mathcal{S}$ selects unit j, and let I_j be the indicator of the event E_j. That is, I_j = 1 or 0 according as unit j appears in u or not. It is now easy to check that

$$H = \sum_{j=1}^{N} \Pi_j^{-1} I_j Y_j \,. \tag{3.9}$$

That H is an unbiased estimator of Y follows at once from the fact that

$$\begin{aligned} E(I_j) &= \text{Prob}(E_j) \\ &= \Pi_j \quad (j = 1, 2, \ldots, N). \end{aligned}$$

Now, let $\theta_o = (a_1, a_2, \ldots, a_N)$ be a point in Ω that is selected by the surveyor (prior to the survey) and let H_o be defined as

$$H_o = H - \Sigma\, \Pi_j^{-1} I_j a_j + \Sigma a_j \,. \tag{3.10}$$

It is now clear that H_o is an unbiased estimator of Y and that $V(H_o|\theta) = 0$ when $\theta = \theta_o$. Since the variances of H and H_o are continuous functions of θ, it follows that

$$V(H_o|\,\theta) < V(H|\theta) \tag{3.11}$$

for all θ in a certain neighborhood Ω_o of the point θ_o. If the surveyor has the prior knowledge that the true value of θ lies in Ω_o, then the modified Horvitz-Thompson estimator H_o is uniformly better than H. The estimator H_o will look a little more reasonable if we rewrite it as

$$H_o = [\Sigma\, \Pi_j^{-1} I_j(Y_j - a_j)] + \Sigma a_j. \tag{3.12}$$

If (3.9) is a reasonable estimator of $Y = \Sigma Y_j$, then the variable part of the right hand side of (3.12) is an equally reasonable estimator of

$$\Sigma(Y_j - a_j) = Y - \Sigma a_j \,.$$

The strategy of a surveyor who advocates the use of (3.12) [in preference to that of (3.9)] as an estimator of $Y = \Sigma Y_j$ is quite clear. Instead of defining the state of nature as

$$\theta = (Y_1, Y_2, \ldots, Y_N)$$

he is defining it as

$$\theta' = (Y_1 - a_1, Y_2 - a_2, \ldots, Y_N - a_N).$$

Suppose the surveyor has enough prior information about the state of nature, so that by a proper choice of the vector $(a_1, a_2, \ldots, a_N)$ he can make the coordinates of θ' much less variable than that of θ. He is then in a better position to estimate the total of the coordinates of θ' than one who is working with θ. Consider the situation where the surveyor knows in advance that the jth coordinate Y_j of θ lies in a small interval around the number $a_j (j = 1,2, \ldots, N)$. In such a situation the surveyor ought to shift the origin of measurement (for θ) to the point $(a_1, a_2, \ldots, a_N)$ and represent the state of nature as

$$\theta' = (Y_1 - a_1, Y_2 - a_2, \ldots, Y_N - a_N).$$

If the numbers $a_1, a_2, \ldots, a_N$ has a large dispersion, then shifting the origin of measurement to $(a_1, a_2, \ldots, a_N)$ will cut down the variability in the coordinates of the state of nature to a large extent. The effect will be similar to what is usually achieved by stratification.

At this point one may very well raise the questions: Why must the surveyor choose his knowledge vector $(a_1, a_2, \ldots, a_N)$ before the survey? Is it not more reasonable for him to wait until he has the survey data at hand and then take advantage of the additional knowledge gained thereby? Once the data is at hand, the surveyor knows the exact values of the surveyed coordinates of θ. The natural post-survey choice of a_j for any surveyed j is, therefore, Y_j. For a non-surveyed j, the surveyor's best estimate a_j of the unknown Y_j would still be of a speculative nature. If in formula (3.12) we allow the surveyor to insert a post-survey specification of the vector $(a_1, a_2, \ldots a_N)$, then the first part of the right hand side of (3.12) will vanish and the estimator will look like

$$\begin{aligned} H_* &= \Sigma a_j \\ &= (Y_{u_1} + Y_{u_2} + \ldots + Y_{u_\nu}) + \sum_{j \notin u} a_j \qquad (3.13) \\ &= S + S^* , \end{aligned}$$

where S is the sum total of the Y-values of the distinct surveyed units ad S* is the surveyor's post-survey estimate of the total Y-values of the non-surveyed units.

A decision-theorist will surely object to our derivation of formula (3.13) as naive and incompetent. He will point out that we have violated a sacred cannon of inductive behavior, namely, NEVER SELECT THE DECISION RULE AFTER LOOKING AT THE DATA. He will also point out that S* in (3.13) is, as yet, not well-defined (as a function on the sample space X), and he will reject H_* (as an estimator of Y) with the final remark that the whole thing stinks of Bayesianism!

Nevertheless, the fact remains that formula (3.13) points to the very heart of the matter of estimating the population total Y. A survey leads to a complete specification of a part of the population total. This part is the sample total S as defined in (3.13). At the end of the survey the remainder part Y-S is still unknown to the surveyor. If the surveyor insists on putting down T as an estimate of Y, then he is in effect saying that he has reason to believe that T-S is close to Y-S. And then he should give a reasonable justification for his belief. Of course we can write any estimate T in the form

$$T = S + S^*$$

where S is the sample total and $S^* = T - S$. But then, for some T, the part S^* (of T) would appear quite preposterous as an estimate of the unknown part $Y - S$ of Y. The following two examples will make clear the point that we are driving at.

Example 3. The circus owner is planning to ship his 50 adult elephants and so he needs a rough estimate of the total weight of the elephants. As weighing an elephant is a cumbersome process, the owner wants to estimate the total weight by weighing just one elephant. Which elephant should he weigh? So the owner looks back on his records and discovers a list of the elephants' weights taken 3 years ago. He finds that 3 years ago Sambo the middle-sized elephant was the average (in weight) elephant in his herd. He checks with the elephant trainer who reassures him (the owner) that Sambo may still be considered to be the average elephant in the herd. Therefore, the owner plans to weigh Sambo and take 50 y (where y is the present weight of Sambo) as an estimate of the total weight $Y = Y_1 + \dots + Y_{50}$ of the 50 elephants. But the circus statistician is horrified when he learns of the owner's purposive samplings plan. "How can you get an unbiased estimate of Y this way?" protests the statistician. So, together they work out a compromise sampling plan. With the help of a table of random numbers they devise a plan that allots a selection probability of 99/100 to Sambo and equal selection probabilities of 1/4900 to each of the other 49 elephants. Naturally, Sambo is selected and the owner is happy. "How are you going to estimate Y?", asks the statistician. "Why? The estimate ought to be 50y of course," says the owner. "Oh! No! That cannot possibly be right," says the statistician, "I recently read an article in the Annals of Mathematical Statistics where it is proved that the Horvitz-Thompson estimator is the unique hyperadmissible estimator in the class of all generalized polynomial unbiased estimators." "What is the Horivitz-Thompson estimate in this case?" asks the owner, duly impressed. "Since the selection probability for Sambo in our plan was 99/100," says the statistician, "the proper estimate of Y is 100y/99 and not 50y." "And, how would you have estimated Y," inquiress the incredulous owner, "if our sampling plan made us select, say, the big elephant Jumbo?" "According to what I understand of the Horvitz-Thompson estimation method," says the unhappy statistician, "the proper estimate of Y would then have been 4900y, where y is Jumbo's weight." That is how the statistician lost his circus job and perhaps became a teacher of statistics!

Example 4. Sampling with unequal probabilities has been recommended in situations that are less frivolous than the one considered in the previous example but the recommended unbiased estimators for such plans sometimes look hardly less ridiculous than the one just considered. Let us consider the so-called pps (probability proportional to size) plans about which so many research papers have been written in the past 20 years. A pps sampling plan is usually recommended in the following kind of situation. Suppose for each population unit j we have a record of an auxiliary characteristic A_j (the size of j). Also suppose that each A_j is a positive number and that the surveyor has good reason to believe that the ratios

$$A_j = Y_j/A_j \quad (j = 1, 2, ..., N) \tag{3.14}$$

are nearly equal to each other. In this situation it is often recommended that the surveyor adopts the following without replacement pps sampling plan.

Sampling plan. Let $A = \Sigma A_j$ and $P_j = A_j/A$ ($j = 1, 2, ..., N$). Choose a unit (say, u_1) from the population $\mathcal{P} = (1, 2, ..., N)$ following a plan that allots a selection probability P_j to unit j ($j = 1, 2, ..., N$). The selected unit u_1 is then removed from the sampling frame and a second unit (say, u_2) is selected from the remaining $N - 1$ units with probabilities proportional to their sizes (the auxiliary characters A_j). This process is repeated n times so that the surveyor ends up with n distinct units

$$u_1, u_2, \ldots, u_n$$

listed in their natural selection order. After the fieldwork the surveyor has the sample

$$x = \{(u_1, y_1), \ldots, (u_n, y_n)\}$$

where $y_i = Y_{u_i}$ ($i = 1, 2, ..., n$). Let us write p_i for P_{u_i} ($i = 1, 2, ..., n$) and

$$x^* = \{(u_1, y_1), \ldots, (u_{n-1}, y_{n-1})\}$$

for the vector defined by the first $n - 1$ coordinates of x. It is then easy to see that (see Desraj (1968) Theorem 3.13)

$$E\left(\frac{y_n}{p_n}\middle| x^*\right) = \Sigma' \frac{Y_j}{P_j} \cdot \frac{P_j}{1 - p_1 - p_2 - \cdots - p_{n-1}}$$

$$= (\Sigma' Y_j)/(1 - p_1 - \cdots - p_{n-1}),$$

where the summation is carried over all j that are different from $u_1, u_2, \ldots, u_{n-1}$. Since $\Sigma' Y_j = Y - (y_1 + \ldots + y_{n-1})$ it follows at once that

$$E(y_1 + \ldots + y_{n-1} + \frac{y_n}{p_n}(1 - p_1 - \ldots - p_{n-1}) | x^*) = Y. \tag{3.15}$$

Therefore, the unconditional expectation of the lefthand side of (3.15) is also Y. And so we have the so-called Desraj estimator

$$D = y_1 + \ldots + y_{n-1} + \frac{y_n}{p_n}(1 - p_1 - \ldots - p_{n-1}), \tag{3.16}$$

which is an unbiased estimator of Y. Writing S for the sample total $y_1 + \ldots + y_n$ we can rewrite (3.16) as

$$D = S + S^* \tag{3.17}$$

where

$$S^* = \frac{y_n}{p_n}(1 - p_1 - \ldots - p_n).$$

Let us examine the face-validity of S^* as an estimate of Y^*, the total Y-values of the unobserved population units.

Writing $A = \Sigma A_j$, $a_i = A_{u_i}$ $(i = 1, 2, \ldots, n)$ and $A^* = A - a_1 - \ldots - a_n$ (the total A-value of the unobserved units), we have

$$S^* = \frac{y_n}{p_n}(1 - p_1 - \ldots - p_n) \tag{3.18}$$

$$= \frac{y_n}{a_n} A^* \quad (\text{since } p_i = \frac{a_i}{A}).$$

Clearly, S^* would be an exact estimate of Y^* if and only if

$$\frac{y_n}{a_n} = \frac{Y^*}{A^*} = \frac{\Sigma' Y_j}{\Sigma' A_j} \tag{3.19}$$

(the summation is over the unobserved j's).

Now, if the surveyor claims that according to his belief (3.17) is a good estimate of Y, then that claim is equivalent to an assertion on belief in the near equality of the two ratios

$$\frac{y_n}{a_n} \text{ and } \frac{Y^*}{A^*}.$$

What can be the logical basis for such a belief? We started with the assumption that the surveyor has prior knowledge of near equality in the N ratios in (3.14). At the end of the survey, the surveyor has observed exactly n of these ratios and they are

$$\frac{y_1}{a_1}, \frac{y_2}{a_2}, \ldots, \frac{y_n}{a_n}. \tag{3.20}$$

The surveyor is now in a position to check on his initial supposition that the ratios in (3.14) are nearly equal. Suppose he finds that the observed ratios in (3.20) are indeed nearly equal to each other. This will certainly add to the surveyor's conviction that the unobserved ratios A_j (where j is different from $u_1, u_2, \ldots, u_n$) are nearly equal to each other and that they lie within the range of variations of the observed ratios in (3.20). Now, Y^*/A^* is nothing but a weighted average of the unobserved ratios (the weights being the sizes of the corresponding units). It is then natural for the surveyor to estimate Y^*/A^* by some sort of an average of the observed ratios. For instance, he may choose to estimate Y^*/A^* by $(y_1 + \ldots + y_n)/(a_1 + \ldots + a_n)$. This would lead to the following modification of the Desraj estimate (3.17):

$$D_1 = S + \frac{y_1 \cdots + y_n}{a_1 \cdots + a_n} A^* \qquad (3.21)$$
$$= \frac{y_1 + \cdots + y_n}{a_1 + \cdots + a_n} A$$

(and this we recognize at once as the familiar ratio estimate). Alternatively, the surveyor may choose to estimate the ratio Y^*/A^* by the simple average

$$\frac{1}{n}\left(\frac{y_1}{a_1} + \cdots + \frac{y_n}{a_n}\right)$$

of the observed ratios. This will lead to another variation of the Desraj estimate, namely

$$D_2 = (y_1 + \cdots + y_n) + \frac{1}{n}\left(\frac{y_1}{a_1} + \cdots + \frac{y_n}{a_n}\right)A^* . \qquad (3.22)$$

What we are trying to say here is the simple fact that both (3.21) and (3.22) have much greater face validity as estimates of Y than the Desraj estimate (3.17). In the Desraj estimate we are trying to evaluate Y^*/A^* by the nth observed ratio y_n/a_n and are taking no account of the other n - 1 ratios. This is almost as preposterous as the estimate suggested by the circus statistician in the previous example. Suppose the surveyor finds that the n observed ratios y_i/a_i (i = 1. 2. ..., n) are nearly equal alright, but y_n / a_n is the largest of them all. In this situation how can he have any faith in the Desraj estimate

$$D = S + \frac{y_n}{a_n} A^*$$

being nearly equal to Y? [Remember, the factor A^* will usually be a very large number.] Again, what does the surveyor do when he discovers that his initial supposition that the ratios Y_j/A_j (j = 1, 2, ..., N) are nearly equal, was way off the mark? Will it not be ridiculous to use the Desraj estimate in this case? Here we are concerned not with the mathematical property of unbiasedness of an estimator but with the hard-to-define property of face validity of an estimate. An estimate T of the population total Y has little face validity if after we have written T in the form

$$T = S + S^*$$

we are hard put to find a reason why the part S^* should be a good estimate of Y^*.

4. The Label-Set and the Sample Core

We have noted elsewhere that, for a non-sequential sampling plan $\mathcal{S}$, the label part u of the data x = (u, y) is an ancillary statistic; that is, the sampling distribution of the statistic u does not involve the state of nature θ. The sampling distribution of u is uniquely determined by the plan. It is therefore obvious that the label part of the data cannot, by itself, provide any information about θ. Knowing u, we only know the names (labels) of the population units that are selected for observation. [When u is a sequence, we also know the order and the frequency of appearance of each selected unit in u.] With a non-sequential plan, the knowledge of u alone cannot make the surveyor any wiser about θ. The surveyor may, and often does, incorporate his prior knowledge of the auxiliary characters $\alpha = (A_1, A_2, \ldots, A_N)$ in the plan $\mathcal{S}$. But this does not alter the situation a bit. The lable part u of the data x will still be an ancillary statistic.

If the label part u is informationless, then can it be true that the observation part y of the data x = (u, y) contains all the available information about θ? A little reflection will make it abundantly clear that the answer must be an emphatic, no. A great deal of information will be lost if the label part of the data is suppressed. Without the knowledge of u, the surveyor cannot relate the components of the observation vector y to the population units and so he cannot make any use of the auxiliary characters $\alpha = (A_1, A_2, \ldots, A_N)$ and whatever other prior knowledge he may have about the relationship between θ and α.

Let us call a statistic T = T(u, y) label-free if T is a function of y alone. So far, the only label-free estimator that we have come across is the estimator $\bar{y}$ of $\bar{Y}$ in Example 1. If in this case the surveyor has prior knowledge that the true value of θ lies in the vicinity of the point $\theta_0 = (a_1, a_2, \ldots, a_N)$, then he would naturally prefer the estimator (3.7) as an unbiased estimator of $\bar{Y}$. The surveyor can arrive at an estimate like (3.7) only if he has access to the information contained in u. In survey literature, we find several attempts at justifying label-free estimates. But a reasonable case for a label-free estimate can be made only under the assumption of a near complete ignorance in the mind of the surveyor. But, in these days of extreme specialization, who is going to entrust an expensive survey operation in the hands of a very ignorant surveyor?! To remain in survey business, the surveyor has to carefully orient himself to each particular survey situation, gather a lot of auxiliary data $A_1, A_2, \ldots, A_N$ about the population units, and then make intelligent use of such data in the planning of the survey and in the analysis of the survey data. Considerations of label-free estimates are, therefore, of only an academic interest in survey theory.

Let us denote by $\hat{u}$ the set of distinct population units that are selected (for survey) by the sampling plan. The set $\hat{u}$ is a statistic – a characteristic of the sample $x = (u, y)$. We call $\hat{u}$ the label-set and find it convenient to think of $\hat{u}$ as a vector

$$\hat{u} = (\hat{u}_1, \hat{u}_2, \ldots, \hat{u}_\nu),$$

where $\hat{u}_1 < \hat{u}_2 < \ldots < \hat{u}_\nu$ are the ν distinct unit-label that appear in u, arranged in an increasing order of their label values. The observation-vector $\hat{y}$ is then defined as

$$\hat{y} = (\hat{y}_1, \hat{y}_2, \ldots, \hat{y}_\nu),$$

where $\hat{y}_i = Y_{\hat{u}_i} (i = 1, 2, \ldots, \nu)$.

We denote the pair $(\hat{u}, \hat{y})$ by $\hat{x}$ and call it the sample-core. For each sample $x = (u, y)$ we have a well-defined sample-core $\hat{x} = (\hat{u}, \hat{y})$. In the literature the sample-core has been called by other fancy names like order statistic or sampley, etc. [It should be noted that, though we can think of $\hat{u}$ as a subset of $\mathcal{P}$, we cannot think of $\hat{y}$ as a set, because the values in $\hat{y}$ need not be all different. Even if it were possible to think of $\hat{y}$ as a set, it would not be fruitful to do so. For, if $\hat{u}$ and $\hat{y}$ are both conceived as sets, then we have no way to relate a member of $\hat{y}$ to the corresponding label in $\hat{u}$. This is the reason why we prefer to think of the label-set $\hat{u}$ as a vector and of $\hat{y}$ as the corresponding observation-vector.]

The sample core $\hat{x}$ is a statistic. The mapping $x \to \hat{x}$ is usually many-one. For instance, in the pps plan of Example 4, the number ν (of distinct units selected) is the same as n, but, for each value of the label-set $\hat{u}$, there are exactly n! values of u (corresponding to the n! different selection-orders in which the n units might have been selected). Here the mapping $x \to \hat{x}$ is n! to 1.

The sample-core $\hat{x}$ is a sufficient statistic. This means that, given $\hat{x}$, the conditional distribution of the sample x is uniquely determined (does not involve the unobserved part of the state of nature θ). The widely accepted principle of sufficiency tells us that if T be a sufficient statistic then every reasonable estimator (of every parameter τ) ought to be a function of T. Following Fisher we may call an estimator H insufficient if H is not a function of the sufficient statistic T. The Desraj estimator (3.17) of the population total Y is then an insufficient estimate. If we rewrite the Desraj estimate as

$$D = S + \frac{y_n}{a_n} A^*$$

where S is the sample Y - total and A* is the total A - values of the unobserved units, then it is clear that both S and A* are functions of the sample-core $\hat{x} = (\hat{u}, \hat{y})$. [Indeed, S is a function of $\hat{y}$ and A* is a function of $\hat{u}$.] However, y_n/a_n (the ratio corresponding to the last unit drawn in the without replacement pps plan) is not a function of $\hat{x}$. Knowing $\hat{x}$, we only know that the ratio y_n/a_n may have been any one of the n ratios

$$\hat{y}_i/\hat{a}_i \quad (i = 1, 2, \ldots, n)$$

where $\hat{y}_i = Y_{\hat{u}_i}$ and $a_i = A_{\hat{u}_i}$.

If we define λ_i (i = 1, 2, . . . , n) as the conditional probability of u_n/a_n being equal to $\hat{y}_i/\hat{a}_i$, then the λ_i's are well-defined (θ-free) constants, $\Sigma\lambda_i = 1$ and

$$\bar{D} = E(D|\hat{x}) = S + [\sum_1^n \lambda_i \frac{\hat{y}_i}{\hat{a}_i}] A^* . \qquad (4.1)$$

Since D is an unbiased estimator of $Y = \Sigma Y_j$, so also is the estimator $\bar{D}$. From the Rao-Blackwell theorem it follows that (if $n > 1$) the variance of $\bar{D}$ is uniformly smaller than that of D. The estimator $\bar{D}$ has been variously called in the literature, the symmetrized or the un-ordered Desraj estimator. In view of what we explained in the previous section, the symmetrized Desraj estimator (4.1) looks much better than the original Desraj estimator on the score of face-validity. However, the coefficients $\lambda_1, \lambda_2, \ldots, \lambda_n$ in (4.1) are much too complicated to make $\bar{D}$ an acceptable estimator of Y. The estimates (3.21) and (3.22) have about the same face-validity as that of (4.1) and are much simpler to compute. However, (4.1) scores over the other two estimates on the dubious criterion of unbiasedness!

The estimator (4.1) cannot be the only unbiased estimator of Y that is a function of the sample core $\hat{x}$. Consider the estimator

$$\frac{y_1}{p_1} = \frac{y_1}{a_1} A , \qquad (4.2)$$

where y_1 is the Y - value of the first unit that was drawn (by the pps plan of Example 4) and a_1 is the corresponding A-value. Clearly, (4.2) is an insufficient unbiased estimator of Y. The symmetrized version of (4.2) will be

$$\left(\Sigma \mu_i \frac{\hat{y}_i}{\hat{a}_i}\right) A, \tag{4.3}$$

where

$$\mu_i = P\left(\frac{y_1}{a_1} = \frac{\hat{y}_i}{\hat{a}_i} \mid \hat{x}\right) \quad (i = 1, 2, \ldots, n).$$

The estimator (4.3) is unbiased and is a function of $\hat{x}$.

Of late, quite a few papers have been written in which the main idea is the above described method of un-ordering an ordered estimate, that is, making use of the Rao-Blackwell theorem and the sufficiency of the sample core $\hat{x}$. Whatever the sampling plan $\mathcal{S}$ is, the sample-core is always sufficient. Indeed, the sample-core is (in general) the minimum (minimal) sufficient statistic. However, for a non-sequential sampling plan $\mathcal{S}$, the sufficient statistic $\hat{x}$ is never complete. By the incompleteness of $\hat{x}$ we mean the existence of non-trivial functions of $\hat{x}$ whose expectations are identically zero for all possible values of the state of nature θ. This is because (when the plan is non-sequential) the label-set $\hat{u}$ (which is a component of $\hat{x}$) is an ancillary statistic. For every parameter of interest $\tau(\theta)$, there will exist an infinity of unbiased estimators each of which is sufficient in the sense of Fisher (that is, is a function of the minimal sufficient statistic $\hat{x}$.)

5. Linear Estimation in Survey Sampling

During the past years a great many research papers have been written dealing exclusively with the topic of linear estimation of the population mean $\bar{Y}$ or, equivalently, the population total Y. Some confusion has, however, been created by the term linear. An estimator is a function on the sample space X. Unless X is a linear space we cannot, therefore, talk of a linear estimator. In our formulation, X is the space of all samples x = (u, y) and so X is not a linear space. How then are we to reconcile ourselves to the classical statement that, in the case of a simple random sampling plan, the sample mean is the best unbiased linear estimate of the population mean? We have the often quoted contrary assertion from Godambe that in no realistic sampling situation (whatever the plan $\mathcal{S}$) can there exist a best estimator in the class of linear unbiased estimators of the population mean. This section is devoted entirely to the notions of the so-called linear estimates.

Consider first the case of a simple random sampling plan in which a number n (the sample size) is chosen in advance and then a subset of n units is selected from the population $\mathcal{P}$ in such a manner that all the $\binom{N}{n}$ subsets of $\mathcal{P}$ with n elements are alloted equal selection probabilities. Let us suppose that the plan calls for a selection of the n sample units one by one without replacements and with equal probabilities, so that we can list the selected units in their natural selection order as $u_1, u_2, \ldots, u_n$. The label part of the data is then the sequence $u = (u_1, u_2, \ldots, u_n)$ and the observation part is the corresponding observation vector $y = (y_1, y_2, \ldots, y_n)$. Clearly, the y_i's are identically distributed (though not mutually independent) random variables with

$$E(y_i) = (\Sigma Y_j)/N = \overline{Y} \quad (i = 1, 2, \ldots, n).$$

Now, if the surveyor chooses to ignore the lable part u of the data, then he/she can define a linear estimator of $\overline{Y}$ as a linear function

$$T = b_o + b_1 y_1 + b_2 y_2 + \cdots + b_n y_n \tag{5.1}$$

of the observation vector y, where the coefficients $b_o, b_1, \ldots, b_n$ are pre-selected constants. All estimators of the above kind are label-free estimators. Let L be the class of all unbiased estimators of $\overline{Y}$ that are of the type (5.1). In other words, L is the class of all estimators of the type (5.1) with

$$b_o = 0 \text{ and } b_1 + \cdots + b_n = 1. \tag{5.2}$$

The sample mean $\bar{y} = (\Sigma y_i)/n$ is a member of L. The classical assertion that we referred to before is to the effect that, in the class L, there exists a uniformly minimum variance estimator and that is the sample mean $\bar{y}$. This result is well-known and a fairly straightforward proof may be given for the particular case where we define variance as the mean square deviation from the mean. We, however, consider it appropriate to sketch a proof that ties in well with the general spirit of this article.

Consider the sample core $\hat{x} = (\hat{u}, \hat{y})$ where we write the label-set $\hat{u}$ as a sequence $(\hat{u}_1, \ldots, \hat{u}_n)$ with $\hat{u}_1 < \hat{u}_2 < \ldots < \hat{u}_n$ and look upon $\hat{y}$ as the corresponding observation vector $(\hat{y}_1, \ldots, \hat{y}_n)$. Note that the mapping $x \to \hat{x}$ is n! to 1 and that the vector $\hat{y}$ is obtained from the vector y by rearranging its coordinates in an increasing order of their corresponding unit labels. Now, given $\hat{x}$, the conditional distribution of x is equally distributed over the n! possible values of x and so it follows that

$$E(y_i | \hat{x}) = (\Sigma\, \hat{y}_i)/n = \overline{y} \quad (i = 1, 2, \ldots, n). \tag{5.3}$$

Thus, if $T = \Sigma a_i y_i$, with $\Sigma a_i = 1$, is any member of L then from (5.3) it follows that

$$E(T | \hat{x}) = \Sigma\, (a_i \overline{y}) = \overline{y}\,. \tag{5.4}$$

And so from the Rao-Blackwell theorem it follows that $\bar{y}$ is better than T [and this is irrespective of the loss function $w(t, \theta)$ (see § 3) as long as $w(t, \theta)$ is convex (from below) in t for each fixed value of θ]. Observe that, in the class L, the sample mean

$$\bar{y} = (\Sigma\, y_i)/n = (\Sigma\, \hat{y}_i)/n$$

is the only one that is a function of the sample core $\hat{x}$. Every other member of L is insufficient in the sense explained in the earlier section. And so it is no wonder that $\bar{y}$ beats every other member of L in its performance characteristics. The class L - $\{\bar{y}\}$ is certainly not worth any consideration at all.

At this stage one may ask : Why not consider the class of all linear functions of the vector $\hat{y} = (\hat{y}_1, \ldots, \hat{y}_n)$? The snag is that the variables $\hat{y}_1, \ldots, \hat{y}_n$ have very complicated distributions and their expectations are not easy to obtain. For instance, the variable $\hat{y}_1$ can take only the values $Y_1, Y_2, \ldots, Y_{N-n+1}$ and its expectation is a complicated linear function of these N - n + 1 values. It is, therefore, not easy to characterize the class of unbiased estimators of $\overline{Y}$ that are linear functions of the vector $\hat{y}$. In any case, our representation of $\hat{y}$ as a vector is a rather artificial one and it is difficult to see why we should consider linear functions of the vector $\hat{y}$.

Let us look at the problem from another angle. True, the sample space X is not linear, but the parameter space Ω of all the possible values of the state of nature $\theta = (Y_1, \ldots, Y_N)$ is a part of the N-dimensional linear space R_N. A linear function on Ω is a function of the type

$$B_o + B_1 Y_1 + \ldots + B_N Y_N \qquad (5.5)$$

where $B_o, B_1, \ldots, B_N$ are constants. But (5.5) is a linear function of the parameter θ and cannot be conceived of as a statistic. Consider, however, a modification of (5.5) where we replace the coefficient B_j by the variable B_jI_j where I_j is the indicator of the event E_j that the unit j is selected by the sampling plan $\mathcal{S}$ (j = 1, 2, . . . , N). For each set of coefficients $B_o, B_1, \ldots, B_N$ we then have a sort of a linear function [see formula (3.9)]

$$T = B_o + \sum_{I}^{N} B_jI_jY_j \qquad (5.6)$$

on Ω, where the coefficients B_jI_j (j = 1, 2, . . . , N) are random variables. The indicator I_j is a function of the label-set $\hat{u}$ [$I_j(\hat{u}) = 1$ or 0 according as j is a member of $\hat{u}$ or not]. It is easy to recognize T as a statistic — indeed as a function of the sample core $\hat{x} = (\hat{u}, \hat{y})$. Only observe that we may rewrite (5.6) as

$$T = B_o + \sum_{1}^{n} b_i\hat{y}_i \qquad (5.7)$$

where $b_i = B_{\hat{u}_i}$ (i = 1, 2, . . . , n). Let us repeat once again that T is not a linear function on the sample space X, but that we may stretch our imagination a little bit to conceive of T as a random linear function on Ω with coefficients that are determined by the label-set $\hat{u}$. If T is defined as in (5.6) then

$$E(T) = B_o + \Sigma B_j \Pi_j Y_j \tag{5.8}$$

where $\Pi_j = E(I_j) = P(E_j)$. And so T is an unbiased estimator of $\bar{Y}$ if and only if [we are assuming that each $\Pi_j > 0$ and that Ω does not lie in a subspace (of R_N) of dimension lower than N]

$$B_o = 0 \text{ and } B_j = (N\Pi_j)^{-1} \ (j = 1, 2, \ldots, N). \tag{5.9}$$

If we define a linear estimator as in (5.6), then it follows that the Horvitz-Thompson estimator [see (3.8) and (3.9)] is the only unbiased linear estimator of $\bar{Y}$. Following Godambe we, therefore, take one step further and define the class of linear estimators in the following manner :

Definition. Let $\beta_o, \beta_1, \ldots, \beta_N$ be well-defined functions of the label-set $\hat{u}$. By a generalized linear estimator T we mean a statistic that may be represented as

$$T = \beta_o + \sum_1^N \beta_j I_j Y_j . \tag{5.10}$$

Note that the β_j's and I_j's are functions of $\hat{u}$ and that it is only the observed Y_j's that really enter into the definition of T. We may rewrite T in the alternative form

$$T = \beta_o + \sum_1^n \beta_{\hat{u}_i} \hat{y}_i \tag{5.11}$$

and thus recognize it as a function of the sample core $\hat{x}$.

The generalized linear estimator T [as defined in (5.10)] is an unbiased estimator of $\bar{Y}$ if and only if

$$E(\beta_o) = 0 \text{ and } E(\beta_j I_j) = N^{-1} \text{ for all } j . \tag{5.12}$$

Let us denote by $\mathcal{L}$ the class of generalized linear unbiased estimators of $\bar{Y}$. If each $\Pi_j > 0$, then $\mathcal{L}$ is never vacuous, for we have already recognized the Horvitz-Thompson estimator

$$H = \Sigma(N \Pi_j)^{-1} I_j Y_j \tag{5.13}$$

as a member of $\mathcal{L}$. If $\theta_o = (a_1, a_2, \ldots, a_N)$ be a fixed point in Ω and we define H_o as

$$H_o = \Sigma(N \Pi_j)^{-1} I_j(Y_j - a_j) + (\Sigma a_j)/N \tag{5.14}$$

then H_o is a member of $\mathcal{L}$ and has zero risk (variance) when $\theta = \theta_o$. It follows that in the class $\mathcal{L}$ of generalized linear unbiased estimators of $\bar{Y}$ there cannot exist a best (uniformly minimum risk) estimator. This then is the celebrated Godambe assertion that we referred to in the opening paragraph of this section.

If we go back to the case of simple random sampling and compare the two classes L and $\mathcal{L}$ [defined in (5.2) and (5.12) respectively] then we shall observe that the two classes have precisely one member in common, namely the sample mean

$$\bar{y} = (\sum_1^n y_i)/n = \sum_1^N (n^{-1} I_j Y_j).$$

The Godambe class $\mathcal{L}$ of generalized linear estimators is not an extension of the class L. The two classes L and $\mathcal{L}$ are essentially different in character and scope. Thus, the classical assertion that the sample mean is the best linear unbiased estimate of the population mean and Godambe's denial that no such best linear unbiased estimate can ever exist are both true (each rather trivially) in their separate contexts.

Following Hanurav, we may extend the Godambe class of linear estimators by defining a linear estimate as

$$T^* = \beta_o^* + \Sigma \beta_j^* I_j Y_j \tag{5.15}$$

where β_o^*, β_1^*, . . . , β_N^* are well-defined functions of u — the label part of the data x = (u, y). The only difference between (5.10) and (5.15) is that in the former the β's are functions of $\hat{u}$, whereas in the latter, the β*'s are functions of u. Once we remember that the I_j's are functions of the label-set $\hat{u}$, it follows at once that

$$E(T^* | \hat{x}) = \beta_o + \Sigma \beta_j I_j Y_j \tag{5.16}$$

where

$$\beta_j = E(\beta_j^* | \hat{x}) = E(\beta_j^* | \hat{u})$$

is a function of the label-set $\hat{u}$ (j = 0, 1, 2, . . . , N). Thus, the conditional expectation of each T*, given the sufficient statistic (sample-core) $\hat{x}$, is a T as defined in (5.10). From the Rao-Blackwell theorem it then follows that for each estimator of type (5.15) we can find an estimator of type (5.10) with a performance characteristic that is at least as good as (uniformly) that of the former. From the decision theoretic point of view the extension of the class (5.10) by the class (5.15) is, therefore, sort of vacuous.

6. Homogeneity, Necessary Bestness and Hyper-Admissibility

During the past few years, altogether much too much has been written on the subject of linear estimates of the population total Y. The original sin was that of Horvitz and Thompson who in 1952 sought to give a classification of linear estimates of Y. The tremendous paper-writing pressure of the past decade has taken care of the rest. For a plan $\mathcal{S}$ that requires that the n sample units be drawn one at a time, without replacements, and with equal or unequal probabilities, Horvitz and Thompson called an estimator T to be of T_1-type if T be of the form (5.1) with $b_0 = 0$, where $b_1, b_2, \ldots, b_n$ are prefixed constants and $y_1, y_2, \ldots, y_n$ are the n observed Y-values in their natural selection order. An estimator of the type (5.6) with $B_0 = 0$ was classified as a T_2-type estimator. By a T_3-type estimator, Horvitz and Thompson meant an estimator $T = \beta S$, where S is the sample total and $\beta = \beta(\hat{u})$ is an arbitrary function of the lable-set $\hat{u}$. That is, a T_3-type estimator is of the form (5.10) with $\beta_0 = 0$ and $\beta_1 = \beta_2 = \ldots = \beta_N$. Prabhu Ajgaonkar (1965) combined the features of the T_2 and T_3 type estimators to define his T_5-type (someone else must have defined the T_4-type!) estimators as estimators of the type (5.10) with

$$\beta_0 = 0 \text{ and } \beta_j = \beta B_j \quad (j = 1,2, \ldots, N) \tag{6.1}$$

where β is a function of $\hat{u}$ and $B_1, B_2, \ldots, B_N$ are pre-fixed constants. With the exception of the T_1-type estimators, all the other types are subclasses of the Godambe class of linear homogeneous estimators, that is, estimators of the type (5.10) with $\beta_0 = \beta_0(\hat{u}) \equiv 0$ for all values of $\hat{u}$. Let us denote the Godambe class of linear homogeneous unbiased estimators of Y by $\mathcal{L}_0$. The rest of this section is devoted to a study of the class $\mathcal{L}_0$.

The class $\mathcal{L}_0$ is the class of all estimators of the form

$$T = \Sigma \beta_j I_j Y_j , \tag{6.2}$$

where β_j is a function of the label-set $\hat{u}$, I_j is the indicator of the event $j \varepsilon \hat{u}$ and

$$E(\beta_j I_j) = 1 \quad (j = 1, 2, \ldots, N). \tag{6.3}$$

Let us count the degrees of freedom that we have in setting up an estimator in $\mathcal{L}_0$. Let U be the set of all the possible values (given a plan $\mathcal{S}$) of the label-set $\hat{u}$ and let U_j be the subset of those $\hat{u}$'s that include the unit j. [The event E_j that $j \varepsilon \hat{u}$ is then the same as $\hat{u} \varepsilon U_j$.] Let m_j be the number of members in the set U_j. We are assuming that no U_j is vacuous; that is, no E_j is an impossible event; that is, $\Pi_j = P(E_j) > 0$ for all j. Thus,

$$m = \Sigma m_j \geqslant N. \tag{6.4}$$

For defining a T in $\mathcal{L}_0$ we need to define the N functions $\beta_1, \beta_2, \ldots, \beta_N$ on U. Since $I_j = I_j(\hat{u}) = 0$ for all $\hat{u} \notin U_j$, it is clear that we really need to define β_j on the set U_j only $(j = 1, 2, \ldots, N)$. [The values of β_j outside the set U_j have no bearing on the statistic T as defined in (6.2).] Thus, we can think of each β_j as an m_j-dimensional vector. Now (6.3) is, in reality, a linear restriction on the m_j-dimensional vector β_j. We, therefore, have $m_j - 1$ degrees of freedom in our choice of the function (vector) β_j and so we have in all

$$\Sigma(m_j - 1) = m - N \tag{6.5}$$

degrees of freedom in our selection of a T in $\mathcal{L}_0$. We may visualize $\mathcal{L}_0$ as an m - N dimensional surface (plane) in the m-dimensional Euclidean space R_m.

Let us stop for a moment to consider the extreme (and rather trivial) situation where m = N, that is, $m_j = 1$ for all j. This is the case of a unicluster (the terminology is Hanurav's) sampling plan, that is, a plan $\mathcal{S}$ that partitions the population $\mathcal{P}$ into a number of mutually exclusive and collectively exhaustive parts and then selects just one of these parts as the label-set $\hat{u}$. In this case we have no degree of freedom in the selection of a T; that is, the class $\mathcal{L}_0$ is a one point set consisting only of the Horvitz-Thompson estimator

$$T_0 = \Sigma\, \Pi_j^{-1}\, I_j Y_j\ . \tag{6.6}$$

Let us return to the non-trivial case where m > N. As we remarked before, a member T in $\mathcal{L}_0$ is then determined by our choice of $(\beta_1, \beta_2, \ldots, \beta_N)$ which we may look upon as an m-dimensional vector lying in an m - N dimensional plane. The problem is to choose a T in $\mathcal{L}_0$ that has minimum variance. Now, if T be as in (6.2) then

$$V(T) = \sum_j V(\beta_j I_j) Y_j^2 + 2 \sum_{j<k} \mathrm{Cov}(\beta_j I_j,\ \beta_k I_k) Y_j Y_k \tag{6.7}$$

which depends on the state of nature $\theta = (Y_1, Y_2, \ldots, Y_N)$. For each $\theta \in \Omega$, it is then clear that V(T) is a (positive semi-definite) quardratic form in the m-dimensional vector $(\beta_1, \beta_2, \ldots, \beta_N)$. For each θ in Ω, there clearly exists a choice of the vector $(\beta_1, \beta_2, \ldots, \beta_N)$ that minimizes (6.7). Except in some very special situations (with Ω a very small set), there cannot exist a choice of $(\beta_1, \beta_2, \ldots, \beta_N)$ that will minimize (6.7) uniformly for all $\theta \in \Omega$. In the class $\mathcal{L}_0$ of all linear homogeneous unbiased estimators of Y there does not exist a uniformly minimum variance unbiased estimator (Godambe, 1955a).

So the search was on for some other performance criterion that would uphold some estimator as the best in the class $\mathcal{L}_0$ (or in some other smaller or larger class). Of late two rather curious such criteria have been proposed for consideration. They are (a) Ajgaonkar's criterion of necessary bestness and (b) Hanurav's criterion of hyper-admissibility. Let us first consider necessary bestness, the curiouser of the two criteria.

"In order to choose a serviceable estimator from the practical point," writes Ajgaonkar (1965, p. 638), "we propose the following criterion of the necessary best estimator."

Definition (Ajgaonkar). Between two unbiased estimators T and T' (of the population total Y) with variances

$$V(T) = \Sigma a_j Y_j^2 + 2 \sum_{j<k} a_{jk} Y_j Y_k$$

and

$$V(T') = \Sigma b_j Y_j^2 + 2 \sum_{j<k} b_{jk} Y_j Y_k$$

the estimator T is necessary better than T' if $a_j \leqslant b_j$ for all j. The estimator T (in the class C) is necessary best in C if it is necessary better than every other estimator in C.

From (6.7) and the above definition it then follows that the estimator $T = \Sigma\beta_j I_j Y_j$ is necessary best in the class $\mathcal{L}_0$ if and only if $V(\beta_j I_j)$ is uniformly minimum for all j. From the Schwarz inequality we have

$$V(\beta_j I_j)V(I_j) \geqslant [\mathrm{Cov}(\beta_j I_j, I_j)]^2 = [E(\beta_j I_j^2) - E(\beta_j I_j)E(I_j)]^2. \tag{6.8}$$

Since $I_j^2 = I_j$, $E(I_j) = \Pi_j$, $V(I_j) = \Pi_j(1 - \Pi_j)$ and $E(\beta_j I_j) = 1$ for all j, we at once have

$$V(\beta_j I_j) \geqslant (1 - \Pi_j)/\Pi_j \quad (j = 1,2, \ldots, N). \tag{6.9}$$

The sign of equality holds for all j in (6.9) if we select

$$\beta_j = \Pi_j^{-1} \quad (j = 1, 2, \ldots, N),$$

that is, if T is the Horvitz-Thompson estimator. Thus, in $\mathcal{L}_0$ there exists a unique necessary best estimator and that is the Horvitz-Thompson estimator (5.22). [Ajgaonkar (1965) gave a very complicated looking proof of the necessary bestness of (6.6) in the subclass of T_5-type estimators as defined in (6.1), and for a particular class of sampling plans. The present proof is a simplification of a proof suggested by Hege (1967).]

But why necessary bestness? It is hard to figure out how Ajgaonkar stumbled across this curious name and definition. Let us hazard a guess. We begin with a most unrealistic assumption that the space Ω contains points of the type

$$(0, \ldots, 0, Y_j, \ldots, 0),$$

that is, vectors with only one non-zero coordinate $Y_j (j = 1, 2, \ldots, N)$, and let Ω_0 be the subset of all points of the above kind. For a typical $\theta \in \Omega_0$ the variance of $T = \Sigma\beta_j I_j Y_j$ is equal to

$$V(\beta_j I_j)Y_j^2 \text{ (for some j and } Y_j).$$

Hence, if we restrict our attention to the subset Ω_0 of Ω, the necessary best estimator in $\mathcal{L}_0$ is also the uniformly minimum variance estimator. The Horvitz-Thompson estimator has uniformly minimum variance (in $\mathcal{L}_0$) over the subset Ω_0.

Let us now consider the hyper-admissibility thesis of Hanurav (1968). Hyper-admissibility as the name suggests, is a strengthening of the decision-theoretic notion of admissibility. In order not to draw the attention of the reader away from the present context, let us define admissibility in the narrow framework of unbiased point estimation (of the population total Y) with variance as the risk function. Let T_0 and T_1 be unbiased estimators of Y.

Definition. T_0 is uniformly better than T_1 if

$$V(T_0) \leqslant (T_1) \text{ for all } \theta \in \Omega$$

with the strict sign of inequality holding for at least one $\theta \in \Omega$.

Let C be a class of unbiased estimators of Y. We tacitly assume that C is a convex class, that is, when T_0 and T_1 are both members of C then so also is $(T_0 + T_1)/2$. For instance, the class $\mathcal{L}_0$ is convex.

Definition. $T_0 \in C$ is admissible in C if there does not exist a $T_1 \in C$ that is uniformly better than T_0.

If T_0 is admissible in C, then for any alternative $T_1 \in C$ it must be true that T_1 is not uniformly better than T_0; that is, either

$$\text{(a)} \quad V(T_0) \equiv V(T_1) \quad \text{for all } \theta \in \Omega \text{, or}$$

$$\text{(b)} \quad V(T_0) < V(T_1) \quad \text{for at least one } \theta \in \Omega.$$

Now, in view of the admissibility of T_0 and the convexity of C, the alternative (a) is impossible. Suppose (a) holds. Consider the estimator

$$T_* = (T_0 + T_1)/2$$

and observe that

$$V(T_*) = \frac{1}{4}\{V(T_0) + V(T_1)\} + \frac{1}{2}\rho\sqrt{V(T_0)V(T_1)}$$

$$= V(T_0)(1 + \rho)/2$$

$$\leqslant V(T_0),$$

where ρ is the correlation coefficient between T_0 and T_1. Since T_0 is admissible, it follows that $V(T_*) \equiv V(T_0)$ for all θ , i.e., $\rho \equiv 1$ for all θ . Therefore, $T_0 = a + bT_1$. Since, T_0 and T_1 are both unbiased estimators of Y, it follows that $a = 0$ and $b = 1$. And this contradicts the initial supposition that T_0 and T_1 are different estimators.

Thus, in our present context, we may redefine admissibility as

Definition (Hanurav) : $T_o \varepsilon C$ is admissible in C if, for any other $T_1 \varepsilon C$, it is true that

$$V(T_o) < V(T_1)$$

for at least one value of θ , say θ_{o1}, in Ω . [The point θ_{o1} will usually depend on T_o and T_1.]

It is clear that the admissibility of an estimator T_o depends on two things, namely, (1) the extent of the class C that T_o is referred to and (2) the extent of the space Ω in which θ is supposed to lie. The smaller the class C and the larger the space Ω , the easier it is to establish the admissibility of a T_o in C. A little while ago we noted that, in the class $\mathcal{L}_o$, the Horvitz-Thompson estimator (6.6) is the only one that has uniformly minimum variance over the set Ω_o of all points θ with only one non-zero coordinate. If we are allowed to make the unrealistic assumption that $\Omega \supset \Omega_o$, then the admissibility of (6.6) in $\mathcal{L}_o$ follows at once. Godambe and Joshi (1965) proved the admissibility of (6.6) in the wider class of all unbiased estimators of Y, under the very unrealistic assumption that $\Omega = R_N$. As we have noted earlier [see (3.10) and (3.11)], the Horvitz-Thompson estimator is no longer admissible [even in the small class $\mathcal{L}$ of all linear unbiased estimators of Y) if it is known that Ω is a small neighborhood of a point $\theta_o = (a_1, a_2, ..., a_N)$.

Hanurav sought to strengthen the notion of admissibility as follows. Following Godambe, he made the unrealistic assumption that $\Omega = R_N$ and then defined a 'principal hyper-surface (phs)' of Ω as a linear subspace of all points $\theta = (Y_1, Y_2, ..., Y_N)$ with

$$Y_{j_1} = Y_{j_2} = \ldots = Y_{j_k} = 0$$

where $0 \leqslant k < N$ and $(j_i, ..., j_k)$ is a subset of (1, 2, . . ., N). [The whole space Ω corresponds to the case k = 0. There are $2^N - 1$ phs's of Ω.] Let Ω^* be a typical phs in Ω . And let C be a class of unbiased estimators of Y.

Definition (Hanurav) : $T_o \varepsilon$ C is hyper-admissible in C if, for every phs $\Omega^* \subset \Omega$, it is true that T_o is admissible in C when we restrict θ to Ω^*.

It follows at once that the H-T estimator $T_o = \Sigma \Pi_j^{-1} I_j Y_j$ is the unique hyper-admissible estimator in $\mathcal{L}_o$. Suppose $T = \Sigma \beta_j I_j Y_j$ is hyper-admissible in $\mathcal{L}_o$. Consider the phs Ω_j^* of all points θ with $Y_i = 0$ for all $i \neq j$. For a typical $\theta \varepsilon \Omega^*$

$$V(T) = V(\beta_j I_j) Y_j^2$$

and this [as we have noted in (6.9)] is greater than

$$V(T_o) = \Pi_j^{-1}(1 - \Pi_j)\, Y_j^2$$

unless $\beta_j = \beta_j(\hat{u}) = \Pi_j^{-1}$. Thus, the admissibility of T in each phs Ω_j^* implies that $T = T_o$. That T_o is hyper-admissible, that is, is admissible on each phs, is equally trivial. Let Ω^* be a typical phs and let $T^* = \Sigma\beta_j^* I_j Y_j$ be a member of $\mathcal{L}_o$ such that

$$V(T^*) \leqslant V(T_o) \quad \text{for all} \quad \theta \in \Omega^*. \tag{6.10}$$

For each one-dimensional phs $\Omega_j^* \in \Omega^*$, we must have the sign of equality in (6.10) for all $\theta \in \Omega_j^*$, and so it follows that $\beta_j^* = \Pi_j^{-1}$ for each j such that $\Omega_j^* \in \Omega^*$. Therefore, the sign of equality holds in (6.10) for all $\theta \in \Omega^*$. In other words, it is impossible to find an estimator T^* in $\mathcal{L}_o$ that is uniformly better than T_o in the phs Ω^*; that is, T_o is admissible (in the class $\mathcal{L}_o$) when we restrict θ to Ω^*.

In the context of the class $\mathcal{L}_o$, the twin criteria of necessary bestness and hyper-admissibility are mathematically equivalent. Before we proceed to examine the logical basis of the criterion of hyper-admissibility, let us point out a curious error committed by Hanurav (1968, p. 626). In his relation (3.2) Hanurav mistakenly asserts that T_o is hyper-admissible (in C) if and only if, for every alternative $T_1 \in C$ and every phs $\Omega^* \subset \Omega$, we can find a point $\theta_{o1} \in \Omega^*$ such that

$$V(T_o | \theta = \theta_{o1}) < V(T_1 | \theta = \theta_{o1}). \tag{6.11}$$

We give an example to contradict the above assertion. Consider T_o and T_1 where T_o is as in (6.6) and

$$T_1 = \beta_1 I_1 Y_1 + \sum_{j=2}^{N} \Pi_j^{-1} I_j Y_j \text{ with } E(\beta_1 I_1) = 1.$$

In the phs Ω^* of all θ's with $Y_1 = 0$, it is clear that

$$V(T_o) \equiv V(T_1).$$

So in Ω^* we cannot find a point θ_{o1} satisfying (6.11) and this in spite of T_o being hyper-admissible in $\mathcal{L}_o$ and T_1 being an alternative member of $\mathcal{L}_o$.

The main result of Hanurav is to the effect that, for any nonunicluster sampling plan $\mathcal{S}$, the H-T estimator (6.6) is the unique hyper-admissible estimator in the class $\mathcal{M}^*$ of all polynomial unbiased estimators of Y. A quadratic estimator of Y is a statistic T of the form

$$T = \beta_o + \Sigma\beta_j I_j Y_j + \Sigma\beta_{jk} I_{jk} Y_j Y_k \tag{6.12}$$

where $I_{jk} = I_j I_k$ is the indicator of the event that both j and k are in the label set $\hat{u}$, and the β's are functions of $\hat{u}$ with the (unbiasedness) conditions

$$E(\beta_o) = 0,\ E(\beta_j I_j) \equiv 1,\ E(\beta_{jk} I_j I_k) \equiv 0 \quad \text{for all } j \text{ and } k.$$

A polynomial estimator is similarly defined.

Now, let us examine the logical content of the hyper-admissibility criterion. Let $\mathcal{P}^*$ be an arbitrary but fixed subset (subpopulation) of the population $\mathcal{P}$ and let Y* be the total Y-value of the units in $\mathcal{P}^*$; this is,

$$Y^* = \sum_{j \in \mathcal{P}^*} Y_j \ . \qquad (6.12)$$

Suppose, along with an estimate of Y, the surveyor also needs to estimate the parameter Y*. Once the surveyor has decided upon an estimator $T = T(\hat{u}, \hat{y})$ for Y, he may choose to derive an estimate T* for Y* in the following manner. Recall that $\hat{y}$ is the vector $(\hat{y}_1, \hat{y}_2, \ldots, \hat{y}_\nu)$ of the (observed) Y-values of the ν distinct units $\hat{u}_1\ \hat{u}_2 \ldots \hat{u}_\nu$ in the label set $\hat{u}$. Define y* as the vector

$$y^* = (y_1^*, y_2^*, \ldots, y_\nu^*),$$

where y_i^* is y_i or zero according as $\hat{u}_i$ is or is not a member of $\mathcal{P}^*$. In other words, we derive y* by substituting by zeros those coordinates of the observation vector $\hat{y}$ that corresponds to units that are outside the sub-population $\mathcal{P}^*$. Now define

$$T^* = T(\hat{u}, y^*). \qquad (6.13)$$

If T is an unbiased estimator of Y, then it is almost a truism that T* is an unbiased estimator of Y*. If T is the linear homogenous estimator $\Sigma\beta_j I_j Y_j$, then T* is the estimator $\Sigma^* \beta_j I_j Y_j$, where the summation Σ* extends over all j that belong to $\mathcal{P}^*$. In particular, if $\mathcal{P}^*$ is the single member subpopulation consisting of the unit j alone, then the H-T estimator $T_o = \Sigma \Pi_j^{-1} I_j Y_j$ gives rise to the estimator

$$T_o^* = \Pi_j^{-1} I_j Y_j = \begin{cases} \Pi_j^{-1}\ Y_j & \text{if } j \text{ is surveyed} \\ 0 & \text{otherwise} \end{cases} \qquad (6.14)$$

for the parameter $Y^* = Y_j$.

The estimate (6.14) is similar to the one considered in example 3 of Section 3 and is, of course, utterly ridiculous. But in the makebelieve world of mathematicians, we are allowed to make any supposition. Let us pretend that when a surveyor estimates Y by T, he naturally commits himself to estimating each of the $2^N - 1$ subtotals Y* by the corresponding derived estimate T*. Given a class C = { T} of estimators of Y, let us consider, for each subtotal Y*, the class C* = {T*} of derived estimators of Y*. The estimator $T_0 \varepsilon$ C is hyper-admissible in C if, for each subtotal Y*, the derived estimator T_0^* is admissible in C*. According to Hanurav, if the sampling plan $\mathcal{S}$ is nonuniclaster, then given any linear (or polynomial) unbiased estimator T that is different from the H-T estimator, he can always find another unbiased estimator T_1 and a subtotal Y* such that the derived estimator T_1^* (for Y*) is uniformly better than the derived estimator T*.

7. Linear Invariance

We have idealized away many of the mathematically intractable features of the survey operation. But even with our oversimplified mathematical framework, the dimension N of the state of nature θ will usually run into several hundred thousands. It is clear that we are dealing with a most complex inference situation. A typical survey operation is an essentially non-repeatable, once in a lifetime affair. The surveyor, who is a specialist in the particular survey area, plans the survey, collects and analyzes the huge survey data and then arrives at his estimates of the various parameters of interest. Why does he need to consult a mathematician? How can the deductive processes of mathematics be of any use to the surveyor in his purely inductive inference making efforts? The author suspects that the answer lies in the general concensus among the scientific community that the mathematicians are the true watchdogs of rationalism. It may well be argued that this great reverence for mathematicians, this identification of rationalism with deduction, this over-eagerness to put every argument (be it in the realm of economics, psychology, survey theory, even philosophy) in the mold of pure deduction have done more harm than good to the general growth of knowledge. True, a good mathematician, having sharpened his mind with constant exercises in deductive reasoning, will often be able to comb out many a tangle created by unclear thinking on the part of the scientist. But new tangles are created by our over-eagerness to force a mathematical model for a situation that is essentially non-mathematical in nature. We close the essay with one more example of such a tangle in survey theory.

When the surveyor calls upon a decision-theorist (let us abbreviate the name to DT) to audit his survey work, the DT does not attempt to evaluate the thought process by which the surveyor arrived at his estimate T (for, say, the population total Y) from the data x. Indeed, the DT denies the very existence of a rational thought process that may lead us from the particular data x to the estimate T. [So far we have been freely using the two terms estimate and estimator and did not care to distinguish between them. But the whole controversy that is now raging in survey theory may be summarized as the difference between the estimate and the estimator. To the surveyor, the parameter Y is an unknown variable and the estimate T is a constant suggested by the data x at hand. The DT thinks of Y as an unknown constant and looks upon T as a random variable — a function on the sample space X.] As the DT cannot evaluate the estimate T, he proceeds to force an estimator out of the surveyor. For this he needs the sampling plan $\mathcal{S}$ to be randomized and, preferably, non-sequential. Once the DT has figured out the space X of all the possible data x (that the surveyor might have obtained from the survey), he would ask the surveyor to answer the impossible question of how he would have estimated Y for each x in X. If the function T(x) is very complicated (as it would usually be) then that would be the end of the DT's audit. The estimator T(x) better be simple enough so that the DT can evaluate the risk function — the average performance characteristics of the estimator. But before the risk function is evaluated, the DT would like to know the surveyor's loss function which again better be a simple one. As the DT cannot answer the question: How good (rational) is the estimate T?, he evades the issue and proceeds to answer what he thinks to be a nearly equivalent question: How good is the average performance characteristic of the estimator T ?

Instead of looking at the average performance characteristics of T, the DT may try to evaluate the estimator T by examining it directly as a function on X. A criterion that is frequently used for such direct evaluation of the estimator T is the criterion of linear invariance. The DT tries to find out if the surveyor's estimate T of the population total depends in some way on the scale in which the population values (the state of nature θ) are measured. With a linear shift (change of origin and scale) in the measurement scale for the population values, the state of nature $\theta = (Y_1, Y_2, \ldots, Y_N)$ will be shifted to $\theta' = (a + bY_1, a + bY_2, \ldots, a + bY_N)$ and the parameter Y will be shifted to $Y' = Na + bY$. With the same shift in the measurement scale the data

$$x = \{(u_1, u_2, \ldots, u_n), (y_1, y_2, \ldots, y_n)\}$$

will appear as

$$x' = \{(u_1, u_2, \ldots, u_n), (a + by_1, \ldots, a + by_n)\} \tag{7.1}$$

Since x and x' represent the same data (in two different scales) it is natural to require that they lead to the same estimate (in the two scales) of the population total. This leads us to the following

Definition. The estimator $T = T(x)$ is origin and scale invariant if

$$T(x') \equiv Na + bT(x) \tag{7.2}$$

for all x, a, and b > 0, where x' is defined as in (7.1). We call T scale invariant if the above identity holds with a = 0.

One reason why there is so much interest (see Section 6) in linear homogeneous estimators is that they are supposed to be scale invariant. [As we shall presently point out, the above supposition is true only under some qualifications.] The Horvitz-Thompson estimator $T_o = \Sigma \Pi_j^{-1} I_j Y_j$ is clearly scale invariant. It will be origin invariant only if

$$\Sigma \Pi_j^{-1} I_j \equiv N \tag{7.3}$$

for all samples. Since the expected value of the left hand side is clearly equal to N, we may restate the identity (7.3) as $V[\Sigma \Pi_j^{-1} I_j] = 0$ or equivalently

$$N^2 = E[\Sigma \Pi_j^{-1} I_j]^2$$

$$= \Sigma \Pi_j^{-1} + \sum_{j \neq k} [\Pi_{jk}/(\Pi_j \Pi_k)] \tag{7.4}$$

where Π_{jk} is the probability that both j and k are in the sample. In the case of simple random sampling with sample size n, it is clear that $\Pi_j = n/N$ for all j and so the H-T estimator reduces to the simple origin and scale invariant estimator

$$G = (N\,S)/n \tag{7.5}$$

where S is the sample total.

So far mathematicians have generally avoided the non-homogeneous linear estimators of the type

$$\beta_o + \Sigma \beta_j I_j Y_j \tag{7.6}$$

in the mistaken belief that such estimators cannot possibly be scale invariant. It is tacitly assumed that any function $\beta = \beta(\hat{u})$ of the label-set $\hat{u}$ is necessarily scale-free; that is, the value of $\beta(\hat{u})$ depends only on $\hat{u}$ and not on the scale in which the population values are measured. That this need not be so is seen as follows. Suppose the surveyor defines β as

$$\beta(\hat{u}) = \Sigma I_j a_j \tag{7.7}$$

where $\theta_o = (a_1, a_2, \ldots, a_N)$ is a pre-selected fixed point in the space Ω. The function β is clearly scale invariant. That is, if the surveyor is told that, in the new measurement scale, each of the population values is to be multiplied by the scaling factor b, then he (the surveyor) will automatically represent the point θ_o as $(ba_1, ba_2, \ldots, ba_n)$ and re-compute $\beta(\hat{u})$ as $b\beta(\hat{u})$. Let us look back on the modified H-T estimator

$$H_o = \Sigma \Pi_j^{-1}(Y_j - a_j) + \Sigma a_j \tag{7.8}$$

that we had considered earlier in (3.12), where $\theta_o = (a_1, a_2, \ldots, a_N)$ is a pre-selected fixed point in Ω. A surveyor using (7.8) as his estimating formula for Y can never be accused of violating the canon of linear invariance. [We are not saying that H_o is a respectable or a reasonable estimator of Y. We are only saying that, apart from being an unbiased estimator of Y with zero variance when $\theta = \theta_o$, the estimator H_o is origin and scale invariant.] It has been repeatedly asserted by Godambe (see either of his 1968 papers) that, in the class of all estimators that are functions of the label-set $\hat{u}$ and the sample total S, the estimator G = (NS)/n (where n is the number of units in $\hat{u}$) is the unique origin and scale invariant one. However, observe that if $\beta = \beta(\hat{u})$ is any scale-free function of $\hat{u}$ and

$$\beta_o(\hat{u}) = \Sigma a_j - \beta(\hat{u}) \Sigma I_j a_j ,$$

where $\theta_o = (a_1, a_2, \ldots, a_N)$ is a fixed point in Ω, then the estimator

$$G_o = \beta_o + \beta S = \Sigma a_j + \beta \, \Sigma I_j (Y_j - a_j) \tag{7.9}$$

is an origin and scale invariant function of $\hat{u}$ and S.

CHAPTER XIII

ON THE LOGICAL FOUNDATIONS OF SURVEY SAMPLING DISCUSSIONS

Editor's Note : The preceding article was formally presented at a symposium on Foundations of Statistical Inference held at Waterloo University, Canada in March/April 1970 and was followed by discussions. Professor Jerzy Neyman was in the chair. The discussants were G. A. Barnard, V. P. Godambe, J. Hájek, J. C. Koop and R. Royall. At the actual presentation, Basu gave a very brief outline of his essay and went on to make a number of critical comments which he said will be fully presented in the second part of the essay. We present here a brief outline of these comments followed by the discussions and the author's reply.

D. Basu :

The sample core $\hat{x} = (\hat{u}, \hat{y})$ is always a sufficient statistic. In general, it is minimal sufficient. The sufficiency principle tells us to ignore as irrelevant all details of the sample $x = (u, y)$ that are not contained in the sample core $\hat{x}$. The likelihood principle tells us much more. The (normalized) likelihood function is the indicator of the set Ω_x of all parameter points θ that are consistent with the sample x. The set Ω_x depends on the sample x oly through the sample-core $\hat{x}$. Given $\hat{x}$, the set Ω_x has nothing to do with the sampling plan $\mathcal{S}$. And this is true even for sequential sampling plans. For one who believes in the likelihood principle (as all Bayesians do) the sampling plan is no longer relevant at the data analysis stage.

A major part of the current survey sampling theory was dismissed by D. Basu as totally irrelevant. He stressed the need for science oriented, down to earth data analysis. The survey problem was posed as a problem of extrapolation from the observed part of the population to the unobserved part. An analogy was drawn between the problem of estimating the population total ΣY_j and the classical problem of numerical integration. In the latter the problem is to 'estimate' the value of the integral $\int_a^b Y(u)du$ by 'surveying' the function Y(u) at a number of 'selected points' $u_1, u_2, \ldots, u_n$. Which points to select and how many of them, are problems of 'design'. Which integration formula to use and how to assess the 'error' of estimation, are problems of 'analysis'. True, it is possible to set up a statistical theory of numerical integration by forcing an element of randomness in the choice of the points. But, how many numerical analysts will be willing to go along with such a theory?

The mere artifact of randomization cannot generate any information that is not there already. However, in survey practice, situations will occasionally arise where it will be necessary to insist upon a random sample. But this will be only to safeguard against some unknown biases. In no situation, is it possible to make any sense of unequal probability sampling.

The inner consistency of the Bayesian point of view is granted. However, the analysis of the survey data need not be fully Bayesian. Indeed, who can be a true Bayesian and live with thousands of parameters? According to the author, survey statistics is more an art than a science.

DISCUSSIONS

G. A. Barnard :

First, a point of detail. Dr. Basu suggested, in the unwritten part of his paper, a method of estimation using an assistant and dividing the data into two parts, D_1 and D_2. He said that no estimate of error would be available. I simply want to point out that an error estimate could be obtained, in an obvious way, if Dr. Basu has a twin, $Basu^1$, and his assistant also has a twin, $assistant^1$. Then the data are divided into four sets, D_1, D'_1, D_2 and D'_2.

Second, a general point. Dr. Basu and others here are concerned particularly with the problems which arise when it is necessary to make use of the additional information or prior knowledge α . In many sample survey situations this prior knowledge of individuals is negligible (at least for a large part of the population under discussion) and in this case the classical procedures, in particular the Horvitz-Thompson estimators, apply in a sensible manner. It is important that we should not appear in this conference to be casting doubt on procedures which experience has shown to be highly effective in many practical situations.

The problem of combining external information with that from the sample is in general difficult to solve. For instance, in ordinary (distribution-free) least squares theory, the additional information that one or more of the unknown parameters has a bounded range makes the usual justifications inapplicable and no general theory appears to be possible, though it is easy to see what we should do in some particular cases.

With Dr. Basu's elephants, a realistic procedure (on the data he has given) would seem to be to think of the measurement to be made on one elephant as providing some estimate of how the animals have put on weight, or lost it, during the past three years. The circus owner should be able to give good advice on how elephants grow, but in the absence of this it would seem plausible to assume that the heaviest elephant, being fully mature, will have gained nothing, and that the percentage growth will be a linear function of weight three years ago. It would then be wise to select an elephant somewhat lighter than Sambo for weighing. The estimation procedure is clear.

Evidently, as Dr. Basu suggests, no purely mathematical theory is ever likely to be able to account for an estimation procedure such as that suggested. But I do not think this implies that all mathematical theories are time wasting in this context. A judicious balance is necessary. In particular, as I have said, we should not throw overboard the classical theory, or the work of Godambe, Horvitz, Thompson and others, just because we can envisage situations where these results would clearly not be applicable.

V. P. Godambe :

Professor Basu has given a very interesting presentation of some ideas in survey sampling theory which many of us have been contemplating for some time. I find it difficult however to agree with him in one respect. The likelihood principle, which does not permit the use of the sampling distribution generated by randomization for inference purposes, is unacceptable to me in relation to survey sampling. It seems as though the likelihood principle has different implications for two intrinsically similar situations: for the coin tossing experiment the likelihood principle allows the use of binomial distributions while infering about the binomial parameter but if the experiment is replaced by one of the drawing balls from a bag containing black and white balls, the likelihood principle does not allow the use of corresponding binomial (or hypergeometric if sampling is without replacement) distribution to infer the unknown proportion of white balls in the bag.

Professor Basu's comments on HT-estimator (example of weighing elephant and so on) are humourous and I wonder if he wants us to take them at all seriously. The comments fail to take into account the fact that the inclusion probabilities involved in HT-estimator are inseparably tied to the prior knowledge represented or approximated by a class of (indeed a very very wide one) prior distributions (Godambe, J. Roy. Statist. Soc., 1955) on the parametric space. I believe the only way of making sense of sampling practice and theory is through studying the frequency properties implied by the distributions generated by different modes of randomization (that is, different sampling designs) of the estimators obtained on the basis of the considerations of prior knowledge; of course one should also study the implications of reversing the role of frequency properties and prior knowledge (reference: section 7, Godambe's and Thompson's Symposium paper).

At the end of his paper Professor Basu comments on "origin and scale invariant estimator" in my paper, "Bayesian Sufficiency in Survey-Sampling", Ann. Inst.

Stat. Math., 1968. My assertion about the uniqueness in the paper is certainly true. Basu's comments suggest a different type of invariance which is already discussed in our (Godambe and Thompson) symposium paper (section 3).

J. Hájek :

Professor Basu and myself both like the likelihood function connected with sampling from finite populations, but for opposite reasons. He likes it to support the likelihood principle in sample surveys, and I like it to discredit this principle by showing its consequences in the same area. We both are wrong, because the probabilities of selection of samples are in a vague sense dependent on the unknown parameter, because they depend on the same prior facts (prior means and expectations, etc.) that have influenced the values under issue. Consequently, we do not have exactly the situation assumed in applications of the likelihood and conditionality principles. Of course this dependence of parameter and sample strategy is hard to formalize mathematically. My recognition of this dependence is due to a discussion I had recently with Professor Rubin on the conditionality principle.

As to the Horvitz-Thompson estimate, its usefulness is increased in connection with ratio estimation. For example, if the probabilities of inclusion are π_i and we expect the Y_i's to be proportionate to A_i, then we should use the estimate

$$\left(\sum_{i=1}^{N} A_i\right) \frac{\sum_{i \varepsilon s} Y_i/\pi_i}{\sum_{i \varepsilon s} A_i/\pi_i},$$

which would save the statistician's circus job. This estimate is not unbiased but the bias is small, and the idea of unbiasedness is useful only to the extent that greatly biased estimates are poor no matter what other properties they have.

J. C. Koop :

Professor Basu's essay is very stimulating and sometimes also provocative.

Regarding Sambo, I find the choice of selection probability for him (equal to 99/100) rather unwise in the face of the existence of a list of elephants' weights taken three years ago in the owner's possession. Sambo, we are told, was a middle-sized elephant, and knowing the existence of Jumbo in the herd, it might have been wiser to choose the selection probabilities directly proportional to the respective weights of the elephants according to the available records. The reason being that if the elephants grew such that their present weights are directly proportional to their weights three years ago, then the variance of the estimate (equal to the selected elephant's weight divided by its selection probability), is zero. The circus statistician ought to have known better, and one should not be surprised that he was fired!

I am in complete agreement with him that the label of each unit in a sample (or in my terminology, the identity of a unit) cannot be discarded on the ground that it does not provide information. His discussion on this important point is very clear and can be read with profit.

However, I am somewhat surprised at his lack of appreciation for the basic ideas contained in Horvitz and Thompson's path-breaking paper of 1952 as evidenced by the following statement in section 6 of his paper : "During the past few years, altogether too much has been written on the subject of linear estimators of the population total Y. The original sin was that of Horvitz and Thompson who in 1952 sought to give a classification of linear estimates of Y. The tremendous paper-writing pressure of the past decade has taken care of the rest." These two writers constructed three linear estimators, each depending on one of the following three basic features of what I subsequently termed as the axioms sample formation in selecting units one at a time, namely, (i) the order of appearance of a unit in a sample, (ii) the presence or absence of a unit in the sample and (iii) the identity of the sample itself. Sample survey theorists have since benefited from their work. I for one felt in 1956 that the various types of estimators in the literature of that time needed classification and starting with these three features of sample formation, showed that $2^3-1=7$ types or classes of linear estimators, $T_1, T_2, \ldots, T_7$ were possible for one-stage sampling, three of which were those of Horvitz and Thompson. Godambe in his fundamental paper of 1955 found what I subsequently classified as the T_5-type of estimator, which should certainly not be attributed to Ajgaonkar, whose work began much later. In the process of this classification, it was found that an estimator given in the early pages of Sukhatme's text book of 1954 is of the T_4-type, that is, an estimator where the coefficients attached to the variate-values (observations in Basu's terminology) depended on the identity of the unit (label) and the order of appearance of the unit. Among other things, all this work was described in a thesis accepted by the North Carolina State University in 1957 and published in its Institute of Statistics Mimeo Series as No. 296, in 1961. Subsequently in 1963 I revised some of this work and amplified some of its ramifications in a paper in Metrika, Vol. 7(2) and (3).

One may ask what is the use of recognizing the three features of sample formation? In the context of the real world of sample surveys it must be said that they have physical meaning, which has some bearing on how an estimator may be constructed. Equally important, they point to the information supplied by the sample even before (field) observations on its members are made. In discussing Dr. C. R. Rao's excellent paper, I constructed a class of estimators where two of the features of sample formation were used, viz., (ii) which is equivalent to recognizing the identity of the distinct units (labels) and (iii) the identity of the sample itself, to show that an estimator of this class can have smaller M.S.E. than the U.M.V. estimator, derived through an appeal to the principles of maximum likelihood, sufficiency and completeness, carried over almost bodily from classical estimation theory, thus bringing into question the extent of relevance of these principles in estimation theory for sample surveys of a finite universe. (It must be stressed that this does not detract from Dr. Rao's valuable paper which I interpret as a probe to uncover the difficulties of the subject.)

R. Royall :

Although I agree with much of what is said in this paper, I must take exception to one fundamental point. In section 2 Professor Basu states that : "From the point of view of a frequency probabilist, there cannot be a statistical theory of surveys without some kind of randomization in the plan $\mathcal{S}$"

"Apart from observation errors and randomization, the only other way that probability can sneak into the argument is through a mathematical formalization of what we have described before as the residual part R of the prior knowledge, $K = (\Omega, \alpha, R)$. This is the way of a subjective (Bayesian) probabilist. The formalization of R as a prior probability distribution of θ over Ω makes sense only to those who interpret the probability of an event, not as the long range relative frequency of occurrence of the event (in a hypothetical sequence of repetitions of an experiment), but as a formal quantification of the ... phenomenon of personal belief in the truth of the event."

It seems frequently to be true that at some time before the values $y_1, y_2, \ldots, y_N$ are fixed it is natural and generally acceptable to consider these numbers as values, to be realized, of random variables $Y_1, Y_2, \ldots, Y_N$. For instance, these might be the numbers of babies born in each of the N hospitals in the state during the next month. What particular values will appear is uncertain, and this uncertainty can be described probabilistically. Although subjectivists would presumably accept these statements, in many finite populations such models are precisely as objective as those used everyday by frequentists. If such a model is appropriate before the y's are realized, it seems to be equally appropriate after they are fixed but unobserved. If a fair coin is flipped, the probability that it will fall beads is one half; if the coin was flipped five minutes ago, but the outcome has not yet been observed, my statement that the probability of heads is one half is no less objective now than it was six minutes ago. The state of uncertainty is not transformed from objective to subjective by the single fact that the event which determines the outcome has already occurred.

It can be argued that since the event has already occurred, the outcome should be treated as a fixed but unknown constant (so that now the probability of heads is one if the fixed but unknown outcome is heads and otherwise is zero). Such an argument leads back to the conventional model but rests on an unduly restrictive notion of the scope of objective probability theory.

The probability of one half for heads arises from my failure to notice that the coin is slightly warped. It can be argued that all probability models for real phenomena are likewise conditioned on personal knowledge and should therefore be called subjective. Be that as it may, (i) many statisticians do not consider themselves to be subjectivists and (ii) super-population models are frequently as objective as any other probability models used in applied statistics. Since such models, in conjunction with non-Bayesian statistical tools, can be extremely useful in practice as well as in theory, it seems to me to be a mistake to insist that they are available only to subjective Bayesians without pointing out that in this context the term applies to essentially all practicing statisticians.

AUTHOR'S REPLY

Professor Koop and Professor Godambe seem to think that the real difficulty in the elephant problem lies in the 'unrealistic' sampling plan - a plan that is 'not related' to the background knowledge. I always thought that the real purpose of a sampling plan is to get a good representative sample. If the owner knows how to relate the present weight of the representative elephant Sambo to the total weight of his fifty elephants, then he ought to go ahead and select Sambo. Why does he need a randomized sampling plan? Professor Koop wants to allot larger selection probability to Jumbo, the large elephant. Does he really prefer to have Jumbo rather than Sambo in his sample? I think Professor Koop is actually indifferent as to which elephant he selects for weighing. He knows more about the circus elephants than the circus owner. He 'knows' that the 50 ratios of the present and past weights of the elephants are nearly equal. Therefore, he has made up his mind that the ratio estimate is a good one irrespective of which elephant is selected. But he is not prepared to go all the way with me and assert the goodness of the ratio estimate irrespective of the selection plan. Professor Koop needs to allot unequal selection probabilities (proportional to their known past weights) to the 50 elephants so that he can mystify his non-statistical customers with the assertion that his estimate is then an unbiased one. As a scientist he has been trained to make a show of objectivity. May I ask what Professor Koop would do if the elephant trainer informs him that Jumbo (the big elephant) is on hunger strike for the past 10 days? Should he not try to avoid selecting Jumbo? He should, because now he does not know how to relate the present weight of Jumbo to the total weight of the 50 elephants.

In survey literature, we often come across the term representative sample. But to my knowledge the term has never been properly defined. At one time it used to be generally believed that the simple random sampling plan yields a representative sample. However, the difficulty with this naive sampling plan was soon recognized and so surveyors turned to stratification and other devices (like ratio and regression estimation) to exploit their background information about a specific survey problem. It is not easy to understand how surveyors got messed up with the idea of unequal probability sampling. I think it started with the idea of making the ratio estimate look unbiased. Thus Lahiri devised his method of using the random number tables in such a manner that the probability of selecting a particular sample set of units is proportional to the total 'size' of the units. This plan made the ratio estimate look 'good'. The flood-gate of unequal probability sampling was then opened and a surprisingly large number of learned papers have been published on the subject. What is even more surprising is that no one seems to worry about the fact that the surveyor can allot only one set of selection probabilities $\pi_1, \pi_2, \ldots, \pi_N$, but that he has usually to estimate a vast number of different population totals. For each particular population total the surveyor may be able to find an appropriate ratio (or regression) estimate. But how can he possibly make all these different ratio estimates look 'good'?

Of late, a great deal has been written about the Horvitz-Thompson estimate. A little while ago Professor Rao proved an optimum property of the method. But to me the H-T estimate looks particularly curious. Here is a method of estimation that sort of contradicts itself by alloting weights to the selected units that are inversely proportional to their selection probabilities. The smaller the selection probability of a unit, that is, the greater the desire to avoid selecting the unit, the larger the weight that it carries when selected.

The question that Professor Hájek raised in the first part of his comments is exceedingly important and is one that,at one time, had given me a great deal of trouble. As Professor Hájek admitted, the question is hard to formulate and is even harder to answer. In the second part of my essay, I shall discuss the problem in greater detail. To-day, let us try to understand the difficulty in the context of the circus elephants. Suppose the surveyor (the owner) selects three elephants u_1, u_2 and u_3

with probabilities proportional to their past weights (and, say, with replacements) so that the data is $x = [(u_1, y_1), (u_2, y_2), (u_3, y_3)]$. In this case, the selection probability of the labels $u = (u_1, u_2, u_3)$ depends on the past weights of the 50 elephants and, therefore, also depends on their present weights - the state of nature $\theta = (Y_1, Y_2, \ldots, Y_{50})$. If the selection probability of u depends on θ , then the very fact of its selection gives the surveyor some information about θ . Should the surveyor ignore this fact and act as if he always wanted to select this set of labels u and analyze the data x on that basis? This is precisely what I am advising the surveyor to do and this is what Professor Hájek thinks to be an error. But let us stop and think for a moment. Does the information that u is selected tell the surveyor anything (about θ) that the surveyor did not know already? When the question is phrased this way, one will be forced to admit that there is no real difference between the above plan and a simple random sampling plan. Indeed, the important point that I am trying to make is this, that even when the sampling plan is sequential, the relevant thing is the data generated by the plan and the likelihood function (which depends only on the data and has nothing whatsoever to do with the plan).

However, contrast the above sampling plan with a plan where the owner asks the elephant trainer to give him the names of three elephants that come first to his mind. If (u_1, u_2, u_3) are the three elephants that are selected by the above plan, then the surveyor does not really know how he got the labels (u_1, u_2, u_3) and so he cannot analyze the data x. Could it be that the three elephants were refusing to eat for some time and that is why they were on the trainer's mind at the time? If the owner must depend on the trainer for the names and present weights of three sample elephants, and if he does not have the sampling frame (so that he cannot select the labels himself), then he may be well advised to instruct the trainer to select the three sample labels at random. Randomness is a devil no doubt, but this is a devil that we understand and have learnt to live with. It is easier to trust a known devil than an unknown saint !

The second point raised by Professor Hájek is easier to deal with. If the surveyor knows that the ratios of the present and past weights of the elephants are nearly equal, then why does he not use the ratio estimate itself? I do not see any particular merit in the estimate suggested by Professor Hájek.

Now, let us turn to Professor Godambe's objection to the likelihood principle in the context of survey sampling. It will be easier for us to understand Godambe's point if we examine the following example. In a class there are 100 students. An unknown number τ of these students have visited the musical show Hair. Suppose we draw a simple random sample of 20 students are record for each student, not his/her name, but only whether he/she has seen Hair. The likelihood is then a neat (hypergeometric) function involving only the parameter of interest τ . Godambe likes this likelihood function. However, if we had also recorded the name of each of the selected students, then the likelihood function would have been a lot messier. It would no longer have been a direct function of τ , but would have been a function of the state of nature $\theta = (Y_1, Y_2, \ldots, Y_N)$, where Y_j is 1 or 0 according as the student j has or has not seen Hair. Godambe does not know how to make any sense of this likelihood function. My advice to Professor Godambe will be this: "If the names (labels) are 'not informative', if there is no way that you can relate the labels to the state of nature θ , then do not make trouble for yourself by incorporating the labels in your data". After all, isn't this what we are doing all the time? When we toss a coin several times to determine the extent of its bias, do we record for each toss the exact time of the day or the face that was up when the coin was stationary on the thumb? We throw out such details from our data in the belief that they are not relevant (informative). Statistics is both a science and an art. It is impossible to rationalize everything that we do in statistics. These days we are hearing a lot of a new expression - rationality of type II. It is this second kind of rationality that will guide a surveyor in the matter of selection of his sample and the recording of his data.

The final remark of Professor Godambe seems to suggest that he has not quite understood what I said in the last paragraph of my essay. It is simply this that the constants in the estimating formula of the surveyor need not be regarded (indeed, they should not be) as pure numbers like π and e. The estimating formula (estimator) that the surveyor chooses surely depends on the particular inference situation. If the mathematician wishes to find out how the estimator behaves in the altered situation where the population values are measured in a different scale, he should first ascertain from the surveyor whether he (the surveyor) would like to adjust the constants in his formula to fit the new scale. When the surveyor is given this freedom, then it is no longer true that $G = NS/n$ is the only linearly invariant estimator in the class of all estimators that depend only on the label-set and the sample total. The Godambe assertion holds true only in the context of a severely restricted choice.

If I have understood Professor Royall correctly, then he claims that his super-population models for the parameter $\theta = (Y_1, Y_2, \ldots, Y_N)$ are non-Bayesian in the sense that such models do have objective frequency interpretations. His contention about the tossed coin in the closed palm is somewhat misleading. Let us examine a typical super-population model in which the Y_j's are assumed to be independent random variables with means αA_j and variances βA_j^{γ}, where A_j is a known auxiliary character of unit j and α, β, γ are known (or unknown) constants ($j = 1, 2, \ldots, N$). To me, such a model looks exactly like a Bayesian formalization of the surveyor's background knowledge or information. Certainly, there is nothing objective about the above model. Indeed, is any probability model objective? When a scientist makes a probability assumption about the observable X, he is supposed to be very objective about it. But as soon as he makes a similar statement about the state of nature θ he is charged with the unmentionable crime of subjectivity. Mr. Chairman, you have always been telling us that the ultimate decision is an 'act of will' on the part of the decision (inference) maker. Isn't it equally true that the choice of the probability model for the observable X is also an act of will on the part of the statistician? Equally subjective is the choice of the 'performance characteristics'. A true scientist has to be subjective. Indeed, he is expected to draw on all his accumulated wisdom in the field of his specialization. My own subjective assessment of the present day controversy on objectivity in science and statistics is this that the whole thing is only a matter of semantics.

If we define mathematics as the art and science of deductive reasoning – an effort at deducing theorems from a set of basic postulates, using only the three laws of logic – then statistics (the art and science of induction) is essentially anti-mathematics. A mathematical theory of statistics is, therefore, a logical impossibility!

CHAPTER XIV

RELEVANCE OF RANDOMIZATION IN DATA ANALYSIS

0. Notes

Editor's Note: This chapter is based on an invited paper presented at a Conference on Survey Sampling held in 1977 at the University of North Carolina, Chapel Hill, N. Carolina. The paper was published with discussions in the conference volume: Survey Sampling and Measurement, N. K. Namboodiri (ed.), Academic Press, 267-339. We did not include the discussions in this book because one of the discussants did not agree to the inclusion.

Author's Note : While presenting my Waterloo Symposium essay on survey sampling (Chapters XII & XIII), I made the forthright statement that under no circumstances it is possible to make any sense of unequal probability sampling. Notwithstanding several learned papers written on the subject by my good friend V. P. Godambe and others, I still cannot make any sense of it. Let us look at the celebrated πps sampling plan which was especially designed to make the Horvitz-Thomson (H-T) estimate look good.

As we have seen in the elephant example [Ex. 3, Ch. XII], the H-T estimate looks ridiculous when the inclusion probabilities π_1, π_2 π_N are far from being nearly proportional to the unknown population values $Y_1, Y_2, \ldots, Y_N$. Also, with a variable sample size plan the H-T estimate will usually look quite ridiculous. So how can we devise a sampling plan with fixed sample size n such that the π_j's are nearly proportional to the unknown Y_j's? Suppose we have a set of known auxilliary characters X_j, $j = 1, 2, \ldots, N$, such that there are good a priori reasons to believe that the ratios $r_j = Y_j/X_j$ are nearly equal. Then, why not try to make the π_j's exactly proportional to the X_j's and therefore nearly proportional to the Y_j's? What a ridiculously difficult proposition! Has anyone written a computer programme for this? Just imagine feeding the computer with several thousand X-values! Computational difficulties apart, are there any logical compulsions for this unequal probability sampling plan? Consider the simplest possible case of the present kind [All Y's and X's are positive numerals.]

Example : The population consists of only two units 1, 2, the parameters are Y_1, Y_2, the known auxilliary values are X_1, X_2 with the ratios $r_j = Y_j/X_j$, $j = 1, 2$, supposedly nearly equal. We have to estimate $Y = Y_1 + Y_2$ with a single sample drawn from the population of two. The H-T strategy will be to select unit j with probability $\pi_j = X_j/(X_1 + X_2)$, j = 1, 2, and then estimate Y as

$$Y_1/\pi_1 = Y_1 + r_1 X_2 = Y + X_2(r_1 - r_2)$$

or

$$Y_2/\pi_2 = Y_2 + r_2 X_1 = Y + X_1 (r_2 - r_1)$$

depending on whether the sample unit is 1 or 2. Clearly it does not make any sense in this case to let Chance play a part in determining the sample. If $X_1 > X_2$ and we are concerned only with the magnitude of the absolute error of estimation, then the strategy of selecting unit 1 (with probability one) and then estimating Y as $Y_1 + r_1 X_2$ is a superior one. The H-T strategy is not admissible!

In Section 8 of this essay I made a review of many critical remarks that I heard over the years on my neo-Bayesian thesis on survey sampling. Even though Remark 10 happens to be my own, it bothers me even today. As I said there, I do not think that it is realistic to ask for a well-defined theory of survey sampling. The problem is too complex and too varied from case to case. I have no clear-cut prescription for the planning of a survey. Apart from saying that we ought to hold the data as fixed and speculate about the parameters I have indeed very little else to offer.

At the Chapel Hill Conference I heard in amazement a colleague from North Carolina declare that the H-T method was quite adequate for all his data analysis needs. Is it not true that the H-T estimate is always unbiased? Has it not been established by Godambe that the H-T strategy is admissible? Therefore, my esteemed colleague would estimate each one of his 30,000 parameters of interest by the corresponding H-T estimate! I wish I had something as simple to offer.

RELEVANCE OF RANDOMIZATION IN DATA ANALYSIS

1. Introduction

This essay is a natural sequel to an earlier one (Basu, 1971) presented at a symposium held in Waterloo, Ontario in March/April 1970. The writing of this essay was promised and its content foreshadowed in the Waterloo essay. While presenting that essay, I made a number of off-the-cuff remarks challenging the logic of the so-called randomization analysis of data. Here I propose to give a further account of my views on the question of data analysis. The time lag of over seven years between the two essays is only a measure of my diffidence on the important question of the relevance of randomization at the data analysis stage.

I begin on a light-hearted note by giving an account of a brief conversation I had with Professor Jerzy Neyman at a breakfast table in a Waterloo hotel in April 1970. If I remember correctly, the conversation took place on the morning following my spirited presentation of the Waterloo essay. The dialogue ran approximately along the following lines :

Basu : Professor Neyman, I greatly admire your mathematical theory of statistics. All your concepts are so well defined and the theory is filled with so many beautiful theorems. But a big gap exists in your theory.

Neyman : Well, shall we hear about it?

Basu : The theory is concerned more with the sample space than with the particular sample at hand, more with inference-making behaviors and the average performance characteristics of such behaviors than with the problem of ascertaining what particular inference is best warranted by what particular data. Your theory does not seem to recognize the fact that different samples, even though they may be in the same sample space, may differ vastly in their information contents.

Neyman : Let us have an example.

Basu : Here is an oversimplified survey-type example :

Example 1.1 Let us suppose our population consists of 100 units. With each unit is associated an unknown number. Let $Y_1, Y_2, ..., Y_{100}$ be the 100 unknown numbers and let $Y = \Sigma Y_j$ be the parameter of interest that we have to estimate. Let us further suppose that we have the background information that one (but we do not know which one) of the 100 numbers is very large, say, of the order of 10^{10}, and that all the other 99 numbers lie between 0 and 1. If $\mathcal{S}_n$ denotes the experiment of drawing a simple random sample of size n from the population, then I have to agree with you that $\mathcal{S}_{25}$ is a more informative experiment than $\mathcal{S}_5$. But suppose you have drawn a simple random sample of size 25 and you know that I have drawn another such sample of size 5. You find that all your 25 sample numbers are small and that I am beaming with pleasure. You have to concede then that your sample is worthless for the purpose of estimating Y and that I have hit the jackpot.

With a characteristic benevolent smile, Professor Neyman told me :

Neyman : Basu, you amaze me! When you make a point, you do it with so much force and such passion!

That was the end of the dialogue. Clearly, Professor Neyman wanted to enjoy his breakfast and was in no mood to be drawn into a controversy at that time. But I have often wondered about what might have been Professor Neyman's defense had he cared to take up my challenge. Would he have lightly dismissed the example as one that has no practical import? Or, would he have gently chided me for talking about such terrible sampling plans as $\mathcal{S}_5$ and $\mathcal{S}_{25}$ in the context of that kind of background information?

Please note that I am concerned at the moment not with the question of how to plan a survey or experiment but with that of how to analyze the data generated by a survey or experiment. The point that I was trying to drive home with my pathological example is that the information content of the sample generated by a survey or experimental design (however well planned the design may be) depends largely on the sample itself, that a lot of sample-to-sample variation in the information content is inevitable. The question of how to analyze data can be answered only after a careful examination of the data itself.

Of course, it may be argued that, once I have clearly specified what kind of analysis would be appropriate for what data (that may be generated by a particular survey or experiment), my overall inference-making behavior vis-a-vis that particular survey or experiment gets well defined as a decision function.My behavior pattern can therefore be assessed in terms of some average performance characteristics and then compared with other hypothetical behavior patterns. There is more than one snag in this argument.

In most survey situations, the data collected would be so vast and complicated that it would take all my time and mental energy to figure out how best to analyze the particular data obtained. It would be entirely unrealistic for anyone to suggest that I could give an honest answer to the question of how I would analyze all the different samples that might be generated by the survey design. Ask an impossible question and you get an unrealistic answer. Even if I could define my inference-making behavior as a decision function, the function, as a rule, would be so complicated that it would be virtually impossible to work out its average performance characteristics. Finally, it is not clear what statistical intuition guides us through the mathematical process of averaging the "loss" over all possible samples. As I said before, the sample may be good, indifferent, or downright poor in its information content. When I get a good sample, I thank my stars in the hope that my inference-making performance will be good in that situation. Why shall I permit anyone to cast a shodow of doubt on my performance by pointing out the possibility of a poor sample that might have been but was not? On the other hand, when I get a relatively uninformative sample, I should either try to get more information or make a clean breast of the fact and do the best I can in that situation.

Let us look back on our pathological Example 1.1. How is one to analyze the sample generated by the design $\mathcal{S}_{25}$ (a simple random sample of size 25)? Let $y = (y_1, y_2, \ldots, y_{25})$ denote the 25 sample Y values. In a sense we well understand, the statistic $T = 100\bar{y}$ is a "design-unbiased" estimator of $Y = \Sigma Y_j$. But is it not true that the corresponding inference-making behavior is always terribly biased? If y contains the large Y value, then we know that T overestimates Y by a factor of nearly 4. On the other hand, if y fails to contain the large Y value, then T clearly underestimates Y by an astronomical figure. We have no option in this case but to give up the traditional survey criterion of design-unbiasedness and to face the problem of estimation of Y fairly and squarely. Let Λ stand for the large Y value (in the population of 100 units) which to start with we knew to be a number of the order of 10^{10}. Let $Y_s = Y - \Lambda$ be the total of the 99 small Y values. If the sample y contains the large Y value, then our estimate $\hat{Y}$ of Y will perhaps look like

$$\hat{Y} = \Lambda + \hat{Y}_s ,$$

where $\hat{Y}_s$ is some reasonable estimate of Y_s based on the 24 small Y values in the sample. Perhaps the estimate $\hat{Y}_s = 99\bar{y}_s$, where $\bar{y}_s$ is the mean of the 24 small Y values in the sample, will appear reasonable to many of us in this case. On the other hand, if y fails to include Λ, then the sample is of very poor quality, and we should either insist on more data or make a poor job of estimation by using the formula $\hat{Y} = 10^{10} + 99\bar{y}$. The estimator

$$\hat{Y} = \begin{cases} \Lambda + 99\bar{y}_s & \text{if } y \text{ includes } \Lambda, \\ 10^{10} + 99\bar{y} & \text{if } y \text{ fails to include } \Lambda, \end{cases}$$

is not design-unbiased, but who cares? And then, who wants to judge $\hat{Y}$ by averaging its performance over all possible samples each of which is either very good or very poor?

R.A. Fisher was aware of the problem of recognizability of good and bad samples. With his theory of ancillary statistics, he attempted a partial solution of the problem in non-Bayesian terms. (Refer to Basu [1964] for a detailed description of this method.) In this pathological example of ours, good and bad samples are easily identified in terms of the indicator I_E of the event E that the sample y includes the large Y value Λ. Since I_E is an ancillary statistic, one may want to invoke the conditionality argument of Fisher to analyze the data after conditioning it by the observed value of I_E. We shall return to the conditionality argument in a subsequent article.

The point that I was trying to make to Professor Neyman seven years ago could very well have been made in a less sensational fashion with the following realistic survey example.

Example 1.2 We are making an opinion survey among the community of students on a large university campus. We obtain a sampling frame from the Registrar's office and then draw a simple random sample of, say, 100 students. A large body of data is then generated by interviewing (surveying) the 100 sample students. When we carefully scrutinize the data, we discover that our sample contains a disproportionately large number of females and that the response pattern of the females to the questions related to the smoking of marijuana differs markedly from the sample group of males. This postrecognition of the fact that the sample does not truly represent the population in relation to the marijuana questions is certainly going to affect our analysis of the data with reference to the marijuana parameters.

A critic may try to point out that our trouble in this instance really stems from the fact of our initial carelessness in not stratifying the population into males and females. But, it might not have been possible for us to draw samples separately from the subpopulation of the male and female students simply because we did not have separate sampling frames for them. In any case, the hindsight about the fruitful device of prestratification of the population into males and females may have come to us only after we examined the data relative to the marijuana questions. When we turn to the Vietnam issue, we may find that the sample contains far too many foreign students; when we examine the data in relation to the student-power issue, we may find that the sample contains a disproportionate number of undergraduates; and so on.

The survey design is not the only determinant of the quality of the data produced by the survey. The principal determinant of how a particular datum ought to be analyzed is the datum itself. The key concept in survey theory ought to be the notion of poststratification. Example 1.2 highlights the need for poststratification. Depending on what parameter we are trying to estimate, the datum itself will usually suggest how we need to stratify it. Poststratifications of the data in numerous ways should be recognized as an essential process of data analysis.

2. Likelihood

The likelihood principle has an inexorable logic of its own. (For a careful discussion on this topic, see Basu [1975].) Yet, the principle has often been characterized as either irrelevant or inoperative in the context of survey sampling and experimental designs (see, e.g., Kempthorne and Folks, 1971, p. 296). Let us take a close look at the question in terms of yet another survey-type example.

Example 2.1 An urn contains 100 tickets that are known to be numbered consecutively as $\theta + 1, \theta + 2, \ldots, \theta + 100$, where θ (the parameter) is an unknown number in the set $\boldsymbol{\Theta}$ of all integers. Suppose we set in motion a particular sampling plan $\mathcal{S}$, thus generating a set of 10 sample tickets bearing the numbers $y_1, y_2, \ldots, y_{10}$ that are recorded in their natural selection order, if any. How should we analyze the data?

Here $x = (y_1, y_2, \ldots, y_{10})$ is the sample. Let m and M be, respectively, the minimum and the maximum sample values. In order to fix our ideas, let us first consider the case where $\mathcal{S}$ is the traditional plan of drawing 10 tickets one by one with equal probabilities and without replacements. It is now easy to check that the likelihood function generated by the data is

$$L(\theta|x) = \begin{cases} p & \text{if } M - 100 \leqslant \theta \leqslant m - 1, \\ 0 & \text{otherwise,} \end{cases} \tag{1}$$

where the constant p is equal to (90!)/(100!). Writing J_x for the set of integers in the interval $[M - 100, m - 1]$ and $J_x(.)$ for the indicator of the set J_x, we can rewrite (1) as

$$L(\theta|x) = pJ_x(\theta). \tag{2}$$

The preceding representation of the likelihood function leads us to the conclusions :

(a) Since $J_x(\theta)$ depends on the sample x only through the statistic (m, M), it follows that (m, M) is a sufficient statistic in the usual sense. Indeed (m, M) is minimum sufficient.

(b) The sample x rules out as logically impossible all values of θ that fall outside J_x. It cuts down our extent of ignorance about θ from the parameter space $\boldsymbol{\Theta}$ to the set J_x.

(c) The sample x lends equal likelihood support to all the parameter points in J_x.

(d) The length $d_x = 99 - (M - m)$ of the interval $J_x = [M - 100, m - 1]$ may be taken as a reasonable measure of the quality of the data; the smaller d_x is, the better is the sample x. The best possible sample is the one for which $M - m = 99$. In this case we are able to identify the value of θ without the possibility of any error. Observe that $\mathcal{S}_2$, a simple random sampling plan with sample size 2, may yield a sample that is perfect in the preceding sense, whereas a much more extensive survey plan, say $\mathcal{S}_{25}$, may very well fail to do so.

We are at last ready to make the very crucial observation that the preceding conclusions hold true irrespective of the nature of the survey plan $\mathcal{S}$ that produced the sample $x = (y_1, y_2, \ldots, y_{10})$. The likelihood function $L(\theta|x)$ will have the form (2) for any sampling plan $\mathcal{S}$ that we choose. Observe that the factor $J_x(\theta)$ has nothing to do with the plan $\mathcal{S}$ and that the constant factor of proportionality p did not enter into any of the arguments (a) - (d). It is this factor p that depends on $\mathcal{S}$ and also on x in some cases. It will be useful to check the correctness of the preceding assertion for each of the following sampling plans :

$\mathcal{S}_{(1)}$ Continue sampling one at a time, without replacement, and with equal probabilities until the sample range $M - m$ exceeds 50.

$\mathcal{S}_{(2)}$ Draw a sample of size 5 one at a time, without replacement, and with equal probabilities. If the sample mean exceeds 20, draw a further sample of size 5 in the same manner; otherwise stop sampling.

$\mathcal{S}_{(3)}$ Continue sampling one at a time with rerplacement and with equal probabilities until the same number is drawn twice, with the condition that we stop anyway if the first 10 draws are all different and the sample range is at least 25.

What is the likelihood function generated by the sample x = (17, 24, 40, 5, 16, 37, 19, 26, 10, 62) under each of three sampling plans? With such a sample x, we have $m = 5$ and $M = 62$ and so the interval J_x is [- 38, 4]. Irrespective of the sampling plan $\mathcal{S}$ (as long as x is a possible sample for the plan), the data would tell us unequivocally that the true value of θ must lie in the interval J_x and would lend equal likelihood support to each point in J_x. The likelihood function can always be represented in the form (2), where $J_x(\theta)$ does not depend on the plan $\mathcal{S}$. The factor p is $(90!)/(100!)$ in the cases of $\mathcal{S}_{(1)}$ and $\mathcal{S}_{(2)}$, but is 100^{-10} in the case of $\mathcal{S}_{(3)}$.

Thus, irrespective of the nature of the sampling plan $\mathcal{S}$, the likelihood function is flat over the set $J_x = [M - 100, m - 1]$. The fact that all parameter points in the set J_x are equally supported by the data does not mean that we are dealing here with a case of "uninformative" likelihood as I often find suggested by some of my esteemed colleagues. True, we do not have a well-defined maximum likelihood estimate of θ (unless J_x is a one-point set), but that does not mean that we cannot analyze the data in terms of the likelihood function.

In the present case, a Bayesian analysis of the data is quite straightforward and simple. The Bayesian will match the likelihood function L with his prior probability distribution ζ on Θ, arrive at his posterior distribution ζ_x^*, and then use this posterior distribution to justify his inference-making on θ. Since the likelihood function L is flat over the set J_x and is zero outside, it is clear that ζ_x^* is the normalized restriction of ζ to J_x. Since J_x is defined entirely in terms of the statistic (m, M), it is clear that the Bayesian analysis of the data does not depend on the nature of the sampling plan $\mathcal{S}$. At the data analysis stage, the Bayesian can as well forget about the sampling plan!

As I have said elsewhere (Basu, 1969), a typical feature of the survey sampling problem is that we invariably end up with a likelihood function that is flat and design-free. Example 2.1 illustrates this fact beautifully and that is why we have labored with this example at such length. In one important respect, however, this example is not symptomatic of our idealized survey setup. Being an example of an urn model, it is typical of a survey situation where the population is essentially unlabeled, for example, the population of blue whales in the antarctic region. We do not have a sampling frame of reference for the 100 tickets in the urn. As the tickets do not have any presurvey identities, the question of selecting a particular ticket for observing its Y value simply does not arise. This precludes the possibility of using any of the more sophisticated survey designs such as stratified, multistage, systematic, pps, and πps design. Faced with such an unlabeled urn problem, we can at best shake the urn (in the case of the blue whales, even that is beyond us!], blindfold ourselves, pull out several units from the urn, and pretend that we have in effect obtained a simple random sample from the population. With urn models like this, our choice of a sampling plan is essentially restricted to equal probability sampling, with or without replacement and with some simple or fancy stopping rules as illustrated earlier.

3. A Survey Sampling Model

In the Waterloo essay (Basu, 1971), I idealized away most of the troublesome and mathematically intractable features of a large-scale survey operation by characterizing a typical survey setup in the following simplistic terms :

(a) There exists a well-defined population, a finite collection P of distinguishable objects called units.

(b) The units in P are not only distinguishable pairwise but are also identifiable individually. This means that we have a sampling frame of reference (a list of the units in P) which enables us to preselect any particular unit in P for the purpose of observing (surveying) its characteristics. Let us suppose that P is listed as 1, 2, 3, . . . , N.

(c) Corresponding to each $j \in P$, there exists an unknown quantity (possibly vector-valued) Y_j. Let us write $\omega = (Y_1, Y_2, \ldots, Y_N)$ and call ω the universal parameter. The set Ω of all the possible (a priori, that is) values of ω is then the parameter space.

(d) Our main concern is with survey setups when we have a lot of prior information about the universal parameter $\omega = (Y_1, Y_2, \ldots, Y_N)$. Let us suppose that the main ingredient of this large body of prior information comes in the form of a knowledge vector $A = (X_1, X_2, \ldots, X_N)$, where X_j is a known auxilliary characteristic (possibly vector-valued) of unit $j \in P$. All our characterizations of the prior distribution or superpopulation model ζ for the parameter ω will be made in terms of such a knowledge vector A.

(e) We make the simplifying assumption that there are no nonresponse or nonsampling (observational, that is) errors. This means that when we choose to select a particular unit j for the determination of its Y value Y_j, we are always able to locate that particular j and then determine the corresponding Y_j without any observational error.

(f) By a sampling plan or design $\mathcal{S}$, we mean a well-defined body of rules following which we can arrive at a subset s of the population $P = \{1, 2, \ldots, N\}$. We shall call s the sample label-set and shall often characterize it as $s = (i_1, i_2, \ldots, i_n)$, where $i_1 < i_2 < \ldots < i_n$ are a set of distinct elements of P listed in increasing order of their unit indices or labels. We call n the sample size.

(g) Following a sampling plan $\mathcal{S}$, the surveyor selects the label-set $s \subset P$. This selection of s can usually (though, not always) be carried out before the fieldwork. By fieldwork we mean that part of the survey operation that determines the Y value Y_i for each unit $i \in s$. Clearly, fieldwork is the major part of a typical survey operation.

While I recognize the immense complexity of a true-to-life survey problem, I believe it would be quite an achievement if we can sort out the major controversial issues related to survey sampling in terms of the preceding simplistic model.

4. Why Randomize?

By randomization we mean the injection of a fully controlled element of randomness in the selection process of the label-set s. Why is it necessary to randomize?

It is easy to overwhelm the question by a series of counterquestions. What is the alternative to a randomized choice of s? Purposive selection? How can you justify a purposive selection of s? How can you claim to be objective in your scientific methods if you refuse to randomize? With purposive selection, how can you have a sample space? Without a sample space, how can you make a statistical analysis of data? Without randomization, how can you have an unbiased estimate of any population characteristic? And so on.

At this stage, it will be useful if we ask the preceding question and counterquestions in the context of an extremely simple survey-type example. For the sake of pinpointing our whole attention on the theoretical issues, I find it imperative to make all my examples as simple as possible.

Example 4.1 The population is $\{1,2, \ldots, N\}$. Concerning the universal parameter $\omega = (Y_1, Y_2, \ldots, Y_N)$, we have the background information that each Y_j is either 0 or 1 and that $Y_1 \leqslant Y_2 \leqslant \ldots \leqslant Y_N$. The parameter of interest is $\theta = \Sigma Y_j$, that is, the number of ones among the Y_j's. Let us imagine a mechanical device that produces N items (units) on a particular day. The item j produced by the machine is either defective ($Y_j = 1$) or nondefective ($Y_j = 0$). Once the machine produces a defective item, it continues to do so for the rest of the day. At the end of the day, we need to estimate the number of defective items in the lot of N items produced by the machine. The unit j is the jth item produced by the machine.

In this example, the universal parameter ω is completely determined by the parameter of interest θ. In other words, we do not have any nuisance parameter. Let $\mathcal{S}$ be an arbitrary sampling plan. Let $s = (i_1, i_2, \ldots, i_n)$, with $i_1 < i_2 < \ldots < i_n$, be the selected label-set, and let $y = (y_1, y_2, \ldots, y_n)$, where $y_k = Y_{i_k}$, be the observation vector. Our sample is then $x = (s, y)$. What does our sample x tell us about the parameter θ?

Observe that, irrespective of the plan $\mathcal{S}$, some samples $x = (s, y)$ are perfect in the sense that they give us full information about θ. For instance, if $i_1 = 1$ and $y_1 = 1$, that is, if we discover that $Y_1 = 1$, then we know definitely that every $Y_j = 1$ and so also $\theta = N$. Generally speaking, if s contains two consecutive labels, say v and v + 1 such that $Y_v = 0$ and $Y_{v+1} = 1$, then we shall know definitely that $\theta = N - v$.

Let us define $v = v(x)$ to be the largest $i \in s$ such that $Y_i = 0$; if $Y_i = 1$

for all $i \in s$, then define $v = 0$. Likewise, define $w = w(x)$ to be the smallest $i \in s$ such that $Y_i = 1$; if $Y_i = 0$ for all $i \in s$, then define $w = N + 1$. Observe that, irrespective of the survey plan, $v \leqslant w - 1$ for all possible samples **x**, the sign of equality holding only if the sample x is perfect in the sense of being fully informative about θ. Writing J_x for the set of integers in the closed interval $[N - w + 1, N - v]$, it is easy to check that the data x tell us definitely that θ must lie in the set J_x.

It J_x is a one-point set, then we have complete information about θ. Otherwise, we are still left with a measure of doubt about the whereabouts of θ. **What** is this measure? Has the sampling plan $\mathcal{S}$ got anything to do with it?

It would be useful to do a little exercise. Suppose $N = 100$. Think of the data $x = \{(17, 24, 40, 73), (0, 0, 1, 1)\}$. Now think of any sampling plan $\mathcal{S}$ (let it be sequential, stratified, pps, purposive, whatever you like) that could possibly give rise to the preceding sample. Verify that the likelihood function

$$L(\theta) = \text{Prob}(x|\theta, \mathcal{S})$$

is flat over the set $J_x = [61, 62, \ldots, 76]$ and is zero outside. It will then be easy to recognize the fact that the preceding is true for all N, x, and $\mathcal{S}$.

As I have said earlier in the context of Example 2.1, a Bayesian will be very pleased with the simple look of the likelihood function and compute his posterior measure of doubt about θ as the normalized restriction of his prior measure to the set J_x. Since the set J_x is defined entirely in terms of the data x, the Bayesian analysis of the data will be independent of the nature of the sampling plan $\mathcal{S}$.

As in the case of Example 2.1, a statistician schooled in the theory of randomization analysis of data will usually be lost with our data $x = \{(17, 24, 40, 73); (0, 0, 1, 1)\}$. This is because he is not used to thinking of $s = (17, 24, 40, 73)$ as the unique label-set generated by the experiment $\mathcal{S}$. He is trained to look upon s as a variable point in a set S that is determined by $\mathcal{S}$, and so he needs to find out the probability distribution of s over S. Without an element of randomization in the plan $\mathcal{S}$, our conventional survey sampler will not have a roomy space S (well furnished with a probability distribution) to make him feel comfortable. On the other hand, if the plan $\mathcal{S}$ is too complicated, then that will make him uncomfortable again because he will not be able to figure out the probability distribution of s over S. And even when the plan $\mathcal{S}$ is simple enough (say, a simple random sampling plan with sample size 4), his analysis of the data x will suffer from his nonrecognition of the fact that the particular data obtained may have good, moderate, or poor information content.

In the present case, how should we plan to choose a sample of size 4 from the population of N units? To me the simple random sampling plan looks utterly ridiculous. If you are very ignorant about θ, that is, if all the values of θ in the parameter space $\Theta = \{0, 1, 2, \ldots, N\}$ look about equally plausible to you, then why not sample in the following sequential and purposive manner? First, select the unit label [N/2] and observe its Y-value y_1. Depending on whether $y_1 = 1$ or 0, select the second sample label as [N/4] or [3N/4]. If $y_1 = 1$ and $y_2 = 1$, then your third sample label will be [7N/8], and so on. This four-step sequential sampling plan will surely decrease the extent of your ignorance about θ from the interval [0, N] to an interval about one-sixteenth its original size. Although this sampling plan is purposive, in the beginning you would not know what label-set s you are going to end up with. The space S over which s varies is fairly extensive. However, relative to a fixed θ, there is only one label-set s that is possible; that is, the sampling distribution of s is degenerate for each θ.

In the case of this example, it is easy to see that, relative to any given prior distribution ζ of θ, the optimum sampling plan would be sequential and nonrandomized (purposive). Admittedly, it is not easy to justify randomization in this case. However, it would be unfair to deny the usefulness of randomization in survey and experimental designs on the basis of a single pathological counterexample. Let us reexamine the question of randomization.

The counterquestion "How can you justify purposive sampling?" has a lot of force in it. It is only in transparently simple cases, such as the preceding one, that one can give a clear-cut argument in favor of a particular purposive plan. In a true-to-life survey situation, it is very difficult to sell the idea of a fully purposive plan. The very purpose of a purposive plan is rooted in the scientific intuition and knowledge of a surveyor. No two surveyors are likely to agree on the choice of their survey plans. The choice of a purposive plan will make a scientist vulnerable to all kinds of open and veiled criticisms. A way out of the dilemma is to make the plan very purposive, but to leave a tiny bit of randomization in the plan; for inatance, draw a systematic sample with a random start or make a very extensive stratification of the population and then draw a sample of size 1 from each stratum!

The rationale of a fully purposive survey plan will usually be so involved that it would be almost impossible for the surveyor to spell it out clearly for the benefit of anyone other than himself. As a result, hardly anyone else (other than the surveyor) will have a clear understanding of all the factors that contributed to the selection of the label-set s and, therefore, of the data $x = (s, y)$ itself. Without such an understanding, how can anyone check on the validity of the conclusions drawn by the surveyor from his data? How can I disagree with a scientist if I myself cannot analyze his data?

It is thus a clear imperative that the surveyor fully describe his survey plan and carefully explain all the considerations that led to the particular plan. And this inhibits the choice of a purposive plan. The possible criticism that the surveyor's chosen plan was not the optimum one (even with respect to his own background information) may not cast any doubt on his conclusions as long as the critic can analyze his (the surveyor's) data. No wonder, therefore, that all of us choose the path of least resistance and try to incorporate an element of randomness in the survey plan.

There are situations where the surveyor has to protect himself against unknown human frailties and biases by incorporating an element of randomness in his selection process. As an example, consider the case of the huge socioeconomic survey of rural households that is continually carried out by the National Sample Survey Organization (NSSO) of India. There are over half a million villages in India. The NSSO does not have a sampling frame of the rural households (about 70 million of them) in India. An investigator is sent to a sample village to get a list of the households from the village Chowkidar (watchman), to verify the correctness of the list, to select a sample of, say, five households from the list, and then to spend a great deal of time eliciting a lot of information from each of the five sample households. The NSSO cannot trust its investigators in the matter of selecting a representative sample of five households from each sample village. What if an investigator chooses the five most prosperous looking households, or the five nearest to the village Chowkidar's residence? So each investigator is given a sealed envelope containing a page of a random number table and is instructed to use the page in a particular fashion for sample selection after listing the households in alphabetical order according to the names of the household heads. A sample selection process of this kind is easy to comprehend and describe, and it is not so easy to criticize data so generated on the pretext that it cannot be analyzed.

This is about as far as I am willing to go along with the principle of randomization. However, I find that many of my esteemed colleagues have wholeheartedly committed themselves to a much stronger version of the principle. It is with this principle of randomization analysis of data that we are going to be concerned in the rest of the essay.

5. Randomization Analysis of Data

Let us argue within the framework of the simplistic survey model that we described in Section 3. Indeed, we are going to restrict our attention to the simpler situation where the unknown Y_j and the known X_j are real numbers ($j = 1, 2, \ldots, N$) and where the parameter of interest is $Y = \Sigma Y_j$. Let $\mathcal{S}$ be the survey design and $x = (s, y)$ the sample, where $s = \{i_1, i_2, \ldots, i_n\}$, with $i_1 < i_2 < \ldots < i_n$, is the label set, and $y = (y_1, y_2, \ldots, y_n)$, with $y_k = Y_{i_k}$, is the observation vector. The set $\mathcal{X}$ of all the values of x that were possible a priori (i.e., before $\mathcal{S}$ was set in motion) is the sample space. An estimator T is a map of the space $\mathcal{X}$ into the range space $\mathcal{Y}$ of Y.

Our traditional survey theory (hallowed with the names of such authorities as Hansen Mahalanobis, and Yates) cannot get off the ground unless an element of artificial (fully controlled, that is) randomization is injected into the design $\mathcal{S}$. In this theory, the sample x is regarded as a random element that varies over the sample space $\mathcal{X}$ in accordance with a probability law P_ω that is well defined in terms of the universal parameter $\omega = (Y_1, Y_2, \ldots, Y_N)$ and, of course, the design $\mathcal{S}$. An estimator $T = T(x)$ of Y is regarded as a random variable with range $\mathcal{Y}$. If $E(T \mid \omega) = Y$ for all $\omega \varepsilon \Omega$, then T is a design-unbiased (i.e., unbiased relative to design $\mathcal{S}$) estimator of Y. The particular estimate T(x) of Y that corresponds to the data x at hand is then regarded as free of any sampling bias. If the variance $V(T \mid \omega)$ of the unbiased estimator T can be shown to be "small," then the estimate T(x) is supposed to be "close" to the true value of Y. Finally, if the estimator can be shown to be, in some sense, optimum among a class $\mathcal{J}$ of unbiased estimators of Y, then that fact is proudly put forward as a sort of a proof of the objectivity of the surveyor in the choice of the estimator T.

The statistical intuition behind the traditional theory of survey sampling is rooted in the value-loaded notion of unbiased estimation. Along with the concept of significance test and that of confidence interval, the concept of unbiased estimate is one of the three most widely used, most controversial, and, according to me, most misleading notions of statistics.

The traditional survey strategy is to so design the survey that there exists a "good" design-unbiased estimate of Y. With a purposive sampling plan it is not possible to attain the unbiasedness objective. So the surveyor carefully introduces an element of randomization in the survey plan $\mathcal{S}$ and then finds his good unbiased estimate $\hat{Y}$ of Y. Next the surveyor finds a good (and, if possible, unbiased) estimate $\hat{\sigma}^2$ of the variance of $\hat{Y}$ and calls $\hat{\sigma}$ the standard error (another value loaded expression!) of $\hat{Y}$. The surveyor then puts forward the pair $(\hat{Y}, \hat{\sigma})$ as the end-product of the data analysis with the suggestion that almost all of the available information in the data x about the parameter of interest Y is summarized in the estimate $\hat{Y}$ and its standard error $\hat{\sigma}$.

The most attractive feature of the preceding line of argument is that it is not based on any speculative probabilistic supposition such as $u_1, u_2, \ldots, u_n$ are i.i.d. normal variables or the regression of u on v is linear. The statistical argument is "Oh! so nonparametric." Indeed, the only way probability enters into the argument is through the surveyor's well-planned and fully controlled use of the randomization artifact. It is this nonparametric quality of the randomization analysis of data that many statisticians find most irresistable.

The randomization argument is charming, but is it relevant? Inherent in the argument is the supposition that the analysis of a data x =(s, y) must be firmly based on the sampling design $\mathcal{S}$ that produced it. We propose to reexamine this major premise of the randomization argument.

6. Randomization and Information

With the survey data x = (s, y) before us, let us write

$$Y_{obs} = \Sigma\, y_i = \sum_{j\varepsilon s} Y_j \qquad \text{and} \quad Y^* = Y - Y_{obs} = \sum_{j\notin s} Y_j \,.$$

The postsurvey anslysis of the data must begin with the obvious fact that, irrespective of the survey plan $\mathcal{S}$, the data gives us the exact value of the observed part Y_{obs} of the population total Y. The whole purpose of the data analysis is to extract the whole of the relevant information in the data about the yet unobserved part Y* of Y. But is the data informative about the part of the population that is yet unobserved? To answer the question in the negative is to repudiate the basic premise of all survey theory. Yet, I have heard it said that, if the plan $\mathcal{S}$ is purposive, that is, if there is only one possible label set s, then the data (s, y) cannot possibly give us any information about any unobserved Y_j. So naturally I ask the

Question : Why does the artifact of randomization make the data informative about every Y_j?

Answer : With randomization allowed, we can so plan the survey as to ensure that every unit $j \varepsilon P$ is endowed with a nonzero probability of being included in the sample label-set s. So even though a particular j may fail to be selected, the fact could never be denied that it might have been. The statistical purpose of randomization is to make the survey experiment informative about every Y_j and, therefore, about $Y = \Sigma Y_j$.

Question : Could you be a little more explicit about what you mean by your experiment being informative about every Y_j.

Answer : For one thing, we can find an unbiased estimator for each Y_j. We have only to find the selection probability π_j for a given j and then define T_j as

$$T_j = \begin{cases} Y_j/\pi_j & \text{if } j \in s \\ 0 & \text{if } j \notin s. \end{cases}$$

Actually, we need to find π_j only if $j \in s$. Now observe that T_j is an unbiased estimator of Y_j.

Question: Is T_j a respectable estimator of Y_j?

Answer: Yes; estimators of this kind were considered by Horvitz and Thompson (1952). The total of T_j over all j is the celebrated Horvitz-Thompson estimator

$$\hat{Y}_{HT} = \sum_{j \in s} Y_j / \pi_j \,. \tag{3}$$

Several optimum properties of $\hat{Y}_{HT}$ as an estimator of Y have been established by Godambe (1955), Hanurav (1968), Rao (1971), and others.

Question: Generally speaking, what is your definition of a statistical experiment $\mathcal{E}$ that is informative about an unknown state of nature θ ?

Answer: An experiment $\mathcal{E}$ is informative about θ if we can recognize a statistic T (a map of the sample space of $\mathcal{E}$ into another space) whose sampling distribution manifestly depends on θ and, maybe, on some other parameters as well.

Question: Are all the possible outcomes of an informative experiment equally informative?

Answer: Now you have really pulled a fast one. I do not know what you are really talking about. Maybe, you should check with Professor Neyman about this.

So we are back on square one! Before taking up the all-important question of information in the data, let us briefly look at T_j as an informant on Y_j. In a technical sense, T_j is indeed a design-unbiased estimator of Y_j. But with the data $x = (s,y)$ before us, is it not obvious that $T_j(x)$ must be a terribly biased estimate of Y_j irrespective of what x is? When $j \notin s$, the T_j estimate of Y_j is zero. And when $j \in s$, that is, when we have full information on Y_j, the T_j estimate of Y_j is Y_j/π_j. Remember, π_j is usually a very small number!

7. Information in Data

At the data analysis stage, notions such as sampling distribution or design-unbiasedness are hardly relevant. What is really relevant is the likelihood function

$$L(\omega) = L(\omega|x, \mathcal{S}) = \text{Prob}(x|\omega, \mathcal{S})$$

generated by the data - the sample x and the survey design $\mathcal{S}$. Every Bayesian will wholeheartedly agree with this proposition. Even an ardent non-Bayesian like R.A. Fisher found the proposition almost self-evident and used to think of the likelihood function as the only bridge that links the observed sample to the unknown parameter.

In 1967, being somewhat bewildered and bemused by a long list of mostly unreadable papers on survey sampling that appeared in the Annals of Mathematical Statistics and Sankhyā, I resolved to settle to my own satisfaction the following two related questions : In a survey setup, what is the smallest statistic that summarizes in itself all the information in the full data? And, what is all the information in the data? The results of that investigation are carefully laid out by Basu and Ghosh (1967) and Basu (1969). In short, they may be described as follows.

Let $\mathcal{S}$ be an arbitrary sampling plan. The plan may be randomized or purposive, sequential, or nonsequential. For any well-defined sampling plan leading to a sample outcome, we should always be able to work out the corresponding likelihood function. The outcome of $\mathcal{S}$ need not be recorded in the summary form of $x = (s, y)$. Let x' be the form in which the sample outcome of $\mathcal{S}$ is actually recorded. We only suppose that x' is recorded in sufficient detail, so that we can reduce x' to the summary form x if we wish. We call x the sample core.

Let $\omega = (Y_1, Y_2, \ldots, Y_N)$ be the universal parameter and let Ω be the parameter space. The parameter space Ω can be quite arbitrary. For a given outcome x' of $\mathcal{S}$, let $\Omega_{x'}$ be the set of parameter points in Ω that are consistent with the observation x'. In other words, if $x = (s, y)$, where $s = (i_1, i_2, \ldots, i_n)$ and $y = (y_1, y_2, \ldots, y_n)$, is the sample core of x', then

$$\Omega_{x'} = \{ \omega : \omega \in \Omega \text{ and } Y_{i_k} = y_k \; (k = 1, 2, \ldots n)\}.$$

It is clear that the subset $\Omega_{x'}$ of Ω depends on the sample x' only through its core x. Let us, therefore, designate $\Omega_{x'}$ by Ω_x. Observe that once the sample core x is before us, we can determine Ω_x without any reference to the survey design $\mathcal{S}$. It is easy to deduce (see Basu, 1969) that, irrespective of the nature of the design $\mathcal{S}$, the probability that $\mathcal{S}$ gives rise to the sample x' is

$$\text{Prob}(x'|\omega,\mathcal{S}) = \begin{cases} q & \text{if } \omega \in \Omega_x, \\ 0 & \text{if } \omega \notin \Omega_x, \end{cases}$$

where $q = q(x',\mathcal{S})$ depends only on x' and the design $\mathcal{S}$. Writing I_x for the indicator of the set Ω_x, we can now rewrite the preceding as

$$L(\omega|x', \mathcal{S}) = q(x',\mathcal{S})I_x(\omega), \tag{4}$$

where L is the likelihood function determined by the data.

Equation (4) holds the key to both questions that I asked earlier. First, it follows at once that, in the context of a fixed sampling plan $\mathcal{S}$, however complex the plan may be and no matter how the sample x' is recorded, the sample core x is always sufficient. There can never be any loss of information if we insist on recording the outcome of the survey operation as x = (s, y), where s is the subset of population units that are sampled and y the corresponding observation vector. (That is why we have been representing the sample in this form from the very beginning.) Usually, the sample core x will be the minimum sufficient statistic. But if we look back on Example 4.1, it would be apparent that x may contain some redundant information. The minimum sufficient statistic is the map $x' \to \Omega_{x'}(= \Omega_x)$ from the sample space $\mathcal{X}'$ of $\mathcal{S}$ to a class of subsets of the parameter space Ω. (See Basu and Ghosh [1967] for some measure-theoretic difficulties that may arise when $\mathcal{X}'$ and Ω are both suposed to be uncountable and how to overcome such difficulties.)

Second, if we again look at (4) and ask the question "What is all the information in the data?" then we see at once that the relevant part of the likelihood function $L(\omega|x', \mathcal{S})$ is the factor

$$I_x(\omega) = \begin{cases} 1 & \text{if } \omega \in \Omega_x , \\ 0 & \text{otherwise.} \end{cases}$$

Thus, given the data (x, $\mathcal{S}$), all parameter points in the set Ω_x have equal likelihood support and all points outside Ω_x have zero likelihood support (i.e., are rejected outright by the data). As we have noted earlier, given the sample x, we can determine Ω_x without any reference to the design $\mathcal{S}$. Thus, at the data analysis stage, we have no need to concern ourselves with the exact nature of the design $\mathcal{S}$.

As I see it, the question "What do the data tell us?" is a misguided one. The data (x, $\mathcal{S}$) are a representation of an experience. It is for us to interpret that experience in the light of our other experiences. The right question to ask is "How should we make a beginning with the complex process of interpretation of particular data?" My Bayesian answer to the question is "Begin with the likelihood function generated by the data."

The likelihood function is only the beginning and must not be regarded as an end in itself. In our survey situation the function is flat over the set Ω_x and is zero outside. This flatness of the likelihood appears to be a matter of great concern to many non-Bayesians. But this very flatness makes a Bayesian happy, because the mathematics of his data analysis becomes very simple in theory. The Bayesian's posterior distribution for ω is now the normalized restriction of his prior to the set Ω_x. Since Ω_x does not depend on $\mathcal{S}$, the Bayesian can take in his stride any sampling plan $\mathcal{S}$, however complex it may be. Indeed, at the data analysis stage he need not even care to ascertain the exact nature of the plan.

Many eyebows were raised when I made the last remark in the opening section of Basu (1969). The remark was misinterpreted to mean that I can analyze the data (x, .) without knowing what kind of $\mathcal{S}$ filled the blank spot. Of course, that cannot be true. As I have said earlier, no one can analyze a sample x without a clear understanding of how that sample was generated. That is why I must refuse to analyze a sample x that was purposively selected by another person unless I fully understand the rationale of that purposive selection. If, however, I know that the plan $\mathcal{S}$ is one of the set $\{\mathcal{S}_1, \mathcal{S}_2, \ldots, \mathcal{S}_k\}$, every one of which I fully understand, then my Bayesian analysis of the data $(x, \mathcal{S})$ will not depend on the exact nature of $\mathcal{S}$. In this case I can reduce the data $(x, \mathcal{S})$ to the sample x.

8. A Critical Review

Over the years, I have heard many critical remarks on this neo-Bayesian thesis on survey sampling and have indeed suffered from a lot of self-doubt myself. We conclude this chapter with a partial list of such remarks and self-doubts. Immediately following a Remark, the paragraphs marked as (a), (b), etc., will summarize my current thinking on the subject.

Remark 1: It is a mistake to regard the individual Y values as parameters. The fact that the Y_j's are unknown (before the survey) does not mean that they can be regarded as parameters. Is it not a fact that in most textbooks on statistics, a parameter is defined as an unknown characteristic of an unknown (probability) distribution? If F(t) is the proportion of Y values that do not exceed t, then in F we have an unknown cdf. The population mean $\bar{Y}$ is a bonafide parameter because it is the mean of the distribution defined by F. Similarly, the maximum Y value is a parameter. But how can you regard, say, Y_1 as a characteristic of F?

1(a) The remark has a lot of force in it when we are dealing with an unlabeled population such as, say, the blue whales in the antartic region. But here we are talking about a case where each unit j has a presurvey identity. Therefore, the Y value Y_j of the jth unit is a well-defined but unknown state of nature and so is a parameter according to my book.

1(b) Just before the commencement of a football (or cricket) match, an important parameter is the outcome of the impending coin-tossing experiment. Once the experiment is over, all the uncertainties (in the mind of a typical spectator) about this parameter are removed, but the spectator is still uncertain about his parameter of interest, namely, the outcome of the game. In my survey setup, the situation is exactly analogous. In the beginning, I was uncertain about the N parameters, $Y_1, Y_2, \ldots, Y_N$. After the survey, I definitely know every Y_j that correspond to $j \in s$. The state of my uncertainty about the parameter of interest $Y = \Sigma Y_j$ is altered but not removed.

Remark 2 : It is misleading to regard $Y_1, Y_2, \ldots, Y_N$ as parameters. It is much more realistic to think of them as already realized but yet unobserved values of N random variables $\eta_1, \eta_2, \ldots, \eta_N$. This will allow us to build a realistic and useful statistical model ζ (the so-clled superpopulation model) for $\eta_1, \eta_2, \ldots \eta_N$. Relative to such a model ζ, the problem of survey design is to make an optimum choice of the label-set s. Once such a choice is made, the problem of estimating $Y = \Sigma Y_j$ becomes a problem of predicting the unobserved part

$$Y^* = \sum_{j \notin s} Y_j$$

of Y in terms of the observed Y values $\{Y_j : j \in s\}$.

2(a) There is a large measure of agreement between the Bayesian and the preceding prediction approach of Royall (1971). Royall's analysis of survey data is free of the survey design $\mathcal{S}$. Like a Bayesian, Royall would analyze the data $(x, \mathcal{S})$, where $x = (s, y)$, by ignoring $\mathcal{S}$ and then looking upon the label-set s as the one and only one with which he need concern himself. However, Royall would look upon $y = \{Y_j : j \in s]$ as a random vector and would try to find a ζ-unbiased predictor of Y^* that is optimum in some reasonable sense. Royall has built for himself what he considers to be a comfortable halfway house between the Mahalanobis-Yates and the full-fledged Bayesian approach to survey sampling.

2(b) From my remark in 1(b) it should be clear that a Bayesian does not make a distinction between a random variable and a parameter. A parameter (random variable) is any unknown entity that is within reach of human speculation. Before the survey, all the Y values were parameters or random variables. After the survey, some of the parameters become known, but the rest are still random variables. It is not necessary to make a distinction between a random variable η_1 and its realized but still unobserved value Y_1. Thus, it is redundant to hypothesize the existence of the N random variables $\eta_1, \eta_2, \ldots, \eta_N$. The Y_j themselves are random variables. A Bayesian would readily agree that the problem of analyzing survey data is a prediction problem.

Remark 3 : J. Hájek as a discussant on Basu (1971) made in essence the following remark : Let $\mathcal{S}$ be a typical randomized survey design such as pps, πps, etc., where the design probability p(s) of the label-set s is made to depend on some auxiliary characters $X_1, X_2, \ldots, X_N$ that are related to the universal parameter $\omega = (Y_1, Y_2, \ldots, Y_N)$. Hence, the knowledge that a particular set s is selected by the design $\mathcal{S}$ will give us some indirect information on ω. Therefore, Basu as a discussant on Rao (1971) committed a logical error when he invoked the conditionality principle to suggest that, at the data analysis stage, one should hold the chosen s as fixed and not take into account any other possible label-sets s that might have been.

3(a) It is true that p(s) depends directly on $X_1, X_2, \ldots, X_N$ and, therefore, indirectly on ω. However, since we already know $X_1, X_2, \ldots, X_N$, we cannot say that from the knowledge of s we get some additional information on ω via $(X_1, X_2, \ldots, X_N)$. In the case of a non-sequential survey design such as pps, πps, etc., the label-part s of the sample x = (s, y) is truly an ancillary statistic.

Remark 4: For the likelihood principle to be operative, we need a sample space $\mathcal{X}$, a parameter space Ω, and a kernel function $p = p(x \mid \omega)$ mapping $\mathcal{X} \times \Omega$ into the half-line $[0, \infty)$. In problems of survey sampling and experimental designs, we have just one probability distribution defined by the randomization scheme of the survey or experimental design. How can we have the trinity $(\mathcal{X}, \Omega, p)$ of abstractions in such cases?

4(a) This can be done, and this is precisely what I achieved in Basu (1969) in the context of survey designs. The analogous case of experimental designs will be discussed in a subsequent essay.

4(b) Consider, for example, the very simple case where the population $P = \{1, 2, 3\}$ and where $\omega = (Y_1, Y_2, Y_3)$ denotes the unknown Y values. Let $\mathcal{S}$ be the simple random sampling plan of size 1. The sample is x = (i, y), where i is the selected unit and y the corresponding Y value. What are $\mathcal{X}$, Ω, and p in this case?

4(c) We must define the parameter space Ω first. Now define $\mathcal{X}_1$ to be the set of all possible samples of the type (1, y); that is, $\mathcal{X}_1$ is the set of all pairs (1, y) such that there exists an ω in Ω with $Y_1 = y$. The sets $\mathcal{X}_2$ and $\mathcal{X}_3$ are similarly defined. The full sample space is $\mathcal{X} = \mathcal{X}_1 \cup \mathcal{X}_2 \cup \mathcal{X}_3$.

4(d) For a fixed x = (i, y), define Ω_x to be the set of all ω in Ω such that $Y_i = y$. Let I_x denote the indicator of the set Ω_x. For our simple random sampling plan $\mathcal{S}$, the kernel function p is

$$p(x \mid \omega) = \frac{1}{3} I_x(\omega).$$

For a more complicated sampling plan, say, a pps plan, the factor $\frac{1}{3}$ will be replaced by a factor like q(i).

Remark 5 : Let us go back to the simple example in 4(b). For each $\omega = (Y_1, Y_2, Y_3)$, let us define $\mathcal{X}_\omega$ as the set $\{(1, Y_1), (2, Y_2), (3, Y_3)\}$ and call it the (conditional) sample space, given ω . Instead of defining, Basu-fashion, the sample space as $\mathcal{X} = \bigcup\{ \mathcal{X}_\omega : \omega \epsilon \Omega\}$, it is better to define the sample space $\mathcal{X}_\omega$ for each ω separately (Godambe, personal communication).

5(a) There is no real advantage in defining the sample space as a collection $\{ \mathcal{X}_\omega\}$ of conditional sample spaces. In my representation of the sample space as $\mathcal{X} = \bigcup \mathcal{X}_\omega$ the set $\mathcal{X}_\omega$ is the carrier of the measure P_ω (on $\mathcal{X}$) that is indexed by ω . The only consequence of representing the sample space as the collection $\{ \mathcal{X}_\omega\}$ will be that the theory of surveys would appear to fall outside of the mainstream of statistical theory, which it does not.

Remark 6 : Kolmogorov (1933) made the correct fundamental distinction between a "zero probability event" (null set) and "logically impossible event" (empty set). Kolmogorov's setup had only probability measure. However, the same fundamental distinction must be observed, when a single measure is replaced by a family of measures, to resolve some difficulties of statistical logic that otherwise arise. This resolution enabled Godambe and Thompson (1971) to establish some fiducial distributions on sound statistical footings (Godambe, personal communication).

6(a) What Godambe is saying in effect is that anomalies and paradoxes would arise if logically impossible events are included in the class of events (measurable sets) as events with zero probability. Kolmogorov (1933) was interested in "nonatomic" measures on uncountable sets. Therefore, he had to make the logical distinction between impossible events and zero-probability events. In survey theory, all of our probability distributions are discrete. Therefore, we need not distinguish between the two notions. Whenever a subset E of $\mathcal{X}$ has zero P_ω -measure we may regard E as logically impossible with respect to ω and vice versa.

6(b) I have come to the conclusion that the logical difficulties mentioned by Godambe are illusory. And I find the Godambe-Thompson (1971) thesis on fiducial distributions in survey theory utterly incomprehensible.

Remark 7 : In survey sampling, some of the coordinates of the parameter ω are observed, but the rest are not. Essentially based on the fact of nonobservance of some coordinates of ω, it is now asserted that the data x make all parameter points in the set Ω_x equally likely. There must be a logical fallacy in the argument. The assertion that two parameter points are equally likely can be made on the basis of knowledge but never on account of ignorance or lack of observation.

7(a) The trouble lies in some of us taking the value-loaded word "likelihood" too seriously. The statement that the likelihood function is flat over the set Ω_x

should be interpreted to mean that the (inanimate and unintelligent) data x equally support all parameter points in Ω_x. It does not mean that the surveyor should consider all points in Ω_x to be equally likely.

Remark 8 : The fact of a flat likelihood renders any discussion of the maximum of the likelihood null and void. This, also, is the graveyard of the likelihood principle, because if any statistician claims that there is not partial evidence on the set of unobserved Y values, he should visit a psychiatrist or leave statistics. (This is Oscar Kempthorne [at his inimitable best] as a discussant on Rao [1971].)

8(a) The length that even reputable statisticians are sometimes willing to travel in search of a maximum likelihood estimate is really amazing. In the case of the simple example considered in 4(b), if we ignore the label-part i of the sample x = (i, y) and thus reduce the sample x to y, then it is clear that Prob($y|\omega$) is equal to 1 for ω = (y, y, y). Thus, if (y, y, y)$\varepsilon\Omega$ for all y, then the maximum likelihood estimate of the population mean $\bar{Y}$ is well defined provided we reduce the data to y, and is y itself. [The case of a simple random sample of size n is analogous.] It is, however, not generally recognized that y will still be the maximum lkikelihood estimate of $\bar{Y}$, in the preceding restricted sense, even when the plan $\mathcal{S}$ does not allot equal selection probabilities to the three population units!

8(b) During the past several years a lot of debate has taken place on the relevance and/or informativness of the label-set s as part of the data. In my simplistic survey setup, where I have assumed the existence of a set of auxiliary values $X_1, X_2, \ldots, X_N$, the labels are of course always informative and can never be suppressed from the data.

8(c) I have heard it said that the full likelihood function, being flat over the set Ω_x, is uninformative about the unobserved Y values. Indeed, it is this kind of assertion against which Kempthorne was reacting so violently. Once he recognizes that the likelihood is only an intermediate step and not the end product of data analysis, Kempthorne's discomfiture with the likelihood principle will perhaps disappear. Or, will it? (The following is an estimate of what my good friend Kemp might say at this stage.)

Remark 9 : The likelihood principle is only the thin end of your wedge. You are trying to sell me the whole Bayesian package. How can I act like a Bayesian when I do not recognize subjective speculations on uncertainties and utilities as valid scientific methods. There are three kinds of probabilities that statisticians are writing about these days :

(a) Personal probabilities : These are unalloyed and unashamed subjective speculations on uncertainties.

(b) Model probabilities : These conditional probability speculations on the observables involve some subjective elements and a lot of mathematical opportunism.

(c) Randomization probabilities : Ah! These are the only kinds of probability that really exist. Do not ask me to buy a new theory of statistics based on the non-existent probabilities of the kind (a) and (b). I have enough trouble already with the Neyman-Pearson-Wald kind of statistics that is based on (b). By the way, can I interest you in a theory of statistics that is based on (c) alone?

9(a) Let us quote directly from de Finetti's (1970) preface to his two-volume treatise Theory of Probability :

My thesis, paradoxically, and a little provocatively, but nontheless genuinely, is simply this :

Probability does not Exist.

> The abandonment of superstitious beliefs about the existence of Phlogiston, the Cosmic Ether, Absolute Space and Time, . . . or Fairies and Witches, was an essential step along the road to scientific thinking. Probability, too, if regarded as something endowed with some kind of objective existence, is no less a misleading misconception, an illusory attempt to exteriorize or materialize our true probabilistic beliefs.

9(b) What about randomization probabilities? I believe, when de Finetti wrote the preceding stirring remark, he was not thinking of randomization probabilities. If pressed hard on the matter, de Finetti would probably give a grudging recognition of objectivity to this kind of probability, but would perhaps insist on adding the rider that this kind of probability has no relevance at the data analysis stage.

9(c) How can probabilities of type (a) and (b) be useful if they are not real? Similar questions were asked about irrational, negative, and imaginary numbers when they were first introduced into mathematics. Data analysis is a speculative process, a mind interacting with a given data. Probability theory should be looked upon as a guideline for the modes of thought and behavior of human mind when faced with uncertainty. I strongly suspect that the Bayesian guideline, in terms of subjective probabilities of the type (a) and (b), is more reliable than the objective randomization probabilities of Oscar Kempthorne. Whether a mode of thinking that is manifestly superior to the subjective probability mode can be devised for data analysis is a question that can be answered only with a large measure of uncertainty at this moment.

Remark 10 : The inner consistency of the Bayesian point of view is granted. However, the analysis of the survey data need not be fully Bayesian. Indeed, who can be a true Bayesian and live with thousands of parameters? (Basu, 1971, p. 234).

10(a) In a typical large-scale survey situation, the population size N runs into hundreds of thousands, the dimension of the unknown Y-character Y_j for unit j runs into scores, and the dimension of the available auxiliary vector X_j for unit j may also run into dozens. How can we then entertain the thought of calling $\omega = (Y_1, Y_2, \ldots, Y_N)$ the universal parameter and then making a full-fledged Bayesian analysis of the data in terms of a prior and the likelihood function? Our representation of the likelihood function in the pretty, flat, and design-free form of (4) is viable only in some simplistic survey-type problems that I have exemplified in this and in some earlier essays. The primary purpose of this representation was to highlight the simple fact that the information content of the data $(x, \mathcal{S})$ depends only on the sample x.

10(b) The Bayesian as a surveyor must make all kinds of compromises with his theory. In the beginning, he may not have a clear perception of how the unknown multidimensional Y values are related to the known multidimensional X values. His choice of the sample units must, therefore, be based on some ad hoc decisions taken on the basis of some vague, ill-defined suppositions. To avoid public criticisms and to substitute some unknown possible selection biases by a well-understood random selection process, he may even agree to introduce an element of randomization in the plan $\mathcal{S}$.

10(c) Once the data are generated by a large-scale survey, the speculative process of exploratory data analysis begins. I cannot put this enormous speculative process into the straight jacket of a theory. I happen to believe that data analysis is more an art form than a scientific method. In these days of powerful computers, it is possible to analyze the same data over and over again in many different ways. It is this repeated process that will lead us to the underlying relationships between the Y_j and the X_j vectors, if any. We may need to poststratify the particular data in many different ways in order to find out how best to extrapolate from the observed part of the population to the yet unobserved part.

10(d) I have heard it said : "Bayesianism is like a bar of soap. It is a good cleansing agent for the Fisher-Neyman theory, but in the process of cleaning up, it will disappear itself." I do not know if that is going to happen or not. But if the accumulated knowledge and technology of mankind finally sweeps away the Bayesian methods from large-scale survey theory, I strongly believe that the Bayesian wisdom-at the data analysis stage, hold the sample fixed and speculate about the parameters-will linger on.

CHAPTER XV

THE FISHER RANDOMIZATION TEST

0. Notes

Editor's Note : This chapter is based on an invited talk given at the SREB Summer Research Conference in Statistics at Arkadelphia, Arkansas in 1978. A revised version of the talk was then published with invited discussions in the Jl. Am. Statist. Assoc., 75, 575-595.

Author's Note : The Fisher Randomization Test is one of the two supporting pillars of the theory of randomization analysis of experimental data, the other one being the celebrated case of the Lady Tasting Tea. My arguments against the Randomization Test hold equally true against Fisher's Tea Tasting analysis. Let us see how.

The Experiment : The Lady is presented with eight cups of tea neatly arranged in a row. The cups and the tea preparations are identical, excepting that in four of the cups tea is poured in first and in the remaining four cups milk is poured first. Writing T for tea-first and M for milk-first, the experimental layout is a sequence x = MTTM...M of four T's and four M's. The particular layout x is an equal probability random selection from the full set of 70 such sequences. The Lady knows about the experimental setup but is kept ignorant of the particular x. She is asked to identify each of the eight cups as an M-cup or a T-cup. With the information that there are four cups of each kind, her response will naturally be a sequence y = mttm...m of four m's and four t's. The pair (x, y) constitutes the data. The problem is to test whether the Lady can really discriminate between the two modes of tea preparation. Consider therefore the null-hypothesis H_0 that the Lady has no power of discrimination.

We have the data, but where is the parameter? What is the sample space? And the model? In view of the experimental randomization, the x-part of the data may be regarded (before the experiment) as a random variable uniformly distributed oiver the 70 point set. But what about the response poart y? Even under the null-hypothesis H_0, we do not know how to set up a probability model for y.

Let T be the number of matches between the x and the y sequences. Suppose T = 8 for the particular data (x, y), that is, the Lady correctly identified all the 8 cups. How should we measure the strength of this particular evidence against the null-hypothesis H_0? My friend Kemp can never cease to marvel at the following ingenious argument of Sir Ronald.

The Tea Tasting Argument : Hold the response y as fixed and regard x as a variable over the 70 point set of all possible experimental layouts. If H_0 is true then the Lady's choice of y is not influenced by (is independent of) the layout x. Therefore, the conditional distribution of x, given y and H_0, may well be regarded as uniform, thanks to randomization, over the 70 point set. The statistic T seems all right as a test criterion. Therefore, Fisher computes the significance level (SL) or P-value of the data (x, y) with T = 8 as

$$P(T \geqslant 8 | y, H_0) = 1/70.$$

Questions: Why randomize? Was it because we wanted to keep the Lady in the dark about the actual layout? But then, why did we have to tell the Lady that there were exactly four cups of each kind in the layout and that all the 70 choices were equally likely? Why couldn't we choose just any haphazard looking layout and keep the lady uninformed about the choice? But then, how could we compute the SL? Instead of randomizing over the full 70 point set, coundn't we randomize over a smaller, say, 10 point set of haphazard arrangements? How can we explain that in that case the same data (x, y) with T = 8 will be associated with an SL of 1/10? Why are we holding the Lady's response y as fixed and playing this probability game with the ancillary statistic x?

We asked all the above questions and more for the Fisher Randomization Test. The two supporting pillars of the complex theory of randomization analysis of data are based on essentially the same argument. What I pointed out in this article is that the argument is not robust at all. You slightly alter the circumstances and the argument falls to pieces.

THE FISHER RANDOMIZATION TEST

1. Introduction

Randomization is widely recognized as a basic principle of statistical experimentation. Yet we find no satisfactory answer to the question, Why randomize ? In a previous paper (Basu 1978b)the question was examined from the point of view of survey statistics. In this article we take an uninhibited frontal view of a part of the randomization methodology generally known as the Fisher randomization test.

R.A. Fisher's classic text The Design of Experiments (DE) is the principal source of inspiration for a mode of data interpretation that may be characterized as randomization analysis of data. In Chapter III of DE, while discussing Galton's analysis of a Darwin experiment with 15 pairs of self-fertilized and cross-fertilized seeds, Fisher cursorily mentioned how one can take advantage of the physical act of randomization to make a test of significance that needs no assumption of normality for the error terms. This idea of Fisher's was immediately generalized by Pitman (1937) and then pushed to its natural boundary by Kempthorne (1952) and many others. Two variants of the Fisher randomization test are discussed in this article. The variant that is discussed in Section 4 may be regarded as the forerunner of all nonparametric tests. The original variant that is discussed in Section 6 may be regarded as one of the two supporting pillars (the other one being the famous case of the "lady tasting tea") of the complex theory of randomization analysis of experimental data. In between the two sections, I have inserted a section entitled: "Did Fisher Change His Mind?" I speculate that in 1956 Fisher had lost a great deal of his early enthusiasm for randomization analysis.

Whether Fisher changed his mind is not the present issue. What I am asking is whether, in the specific instances discussed in this article, it makes sense to compute a significance level (P value) in the manner of the Fisher randomization test. Can any evidential meaning be attached to a P value so computed?

Let us postpone the debate on significance testing in general and nonparametric tests in particular. Let us keep the issue sharply in focus and ask, can the Fisher randomization test pass the test of common sense?

2. Randomization

Let us define randomization as the incorporation of a fully controlled bit of randomness in the process of data generation. Randomization is usually carried out in the manner of items 1 and 2.

1. Prerandomization. This is the most common form of randomization. As the name suggests, the data generation process begins with a fully controlled randomization exercise that determines the actual experimental (or observational) layout. Typical examples are random allocation of treatments in experimental designs and random selection of units in survey sampling. Along with replication and local control (blocking), prerandomization was characterized by Fisher (1960) as one of the three basic principles of statistical experimentation.

2. Postrandomization. Abraham Wald (1950) was one of the earliest to consider this kind of randomization as a statistical tool. After data x has been obtained, postrandomization is the generation of a further random entity y whose randomness characteristics may depend on x but are completely known to the randomizer. The statistician's conclusions or decisions are then based on the extended data (x, y). The average performance characteristics of a postrandomized decision rule δ are evaluated by taking into account all possible values of (x, y). With postrandomization, the statistician has a wider choice of attainable risk functions.

3. Unrecorded randomization. Occasionally, randomization is allowed to enter into the experimental process in a form quite different from the forms 1 and 2 discussed. For instance, in a randomized-response survey subjects may be instructed to respond to the question "Did you truthfully report your gross income in your 1977 tax return?" in the following manner. Each subject tosses a supposedly unbiased coin twice and then answers the question with a "Yes" if the coin yields two heads, with a "No" if the coin yields two tails, or with a truthful "Yes" or "No" if the coin yields a head and a tail. In this data generation process the statistician may prerandomize to choose his or her sample subjects but has no control over the response randomization done by the subjects. The statistician can only speculate about the outcomes of the response randomizations but cannot observe them. It may be argued that response randomization need not be classified as a form of experimental randomization. We shall not discuss this kind of randomization in this article. Warner (1965) proposed this kind of survey technique for eliminating evasive-answer bias.

3. Two Fisher Principles

As we said in the introduction, our primary concern is the so-called randomization analysis of data generated by a statistical experiment that has a large measure of prerandomization incorporated in it. It will, however, be useful to clear the deck with a short discussion of postrandomization and the two sides of the sufficiency principle.

Postrandomization injects into the data an element whose randomness characteristics are fully controlled by the experimenter. Let x be the initial data (sample) and let y be the postrandomized variable whose probability distribution, given x, depends only on x. In terms of the extended sample (x, y), the statistic x is sufficient and, as Fisher would put it, summarizes in itself the whole of the relevant information available in (x, y). To incorporate y in the inference-making process will be a violation of

The sufficiency principle : If T is a sufficient statistic then any conclusion that can be validly drawn from a statistical analysis of the data ought to depend on the data only through the statistic T.

In accordance with the sufficiency principle the data should be reduced to the minimal sufficient statistic. Not to reduce the data to the minimal sufficient statistic is to keep open the possibility of being influenced by irrelevant data characteristics such as, say, a postrandomization variable. In this connection it is interesting to read Fisher's (1956, pp. 96-98) comments on a postrandomization test proposed by Bartlett.

According to Fisher, a principal difference between the deductive and the inductive modes of inference is that in the former case valid conclusions (theorems) can be drawn from a partial use of the data (the primary postulates), whereas in the latter case no conclusion can be validly drawn from an examination of only a part of the relevant information core of the data. Fisher was quite concerned with the fact that the maximum likelihood estimator is not always a sufficient statistic. This led him to the conditionality principle and the celebrated recovery-of-ancillary-information method. The Fisher concern about using the whole of the relevant information in the data may be loosely stated as

The insufficiency principle : If the statistic T_1 is not sufficient then an inference making procedure that depends on the data only through T_1 is insufficient, that is, lacking in substance.

It is not at all surprising, therefore, that Fisher took a rather dim view of nonparametric methods, especially those that make use of only the rank-order statistics. We shall revert to this theme with a Fisher quotation in Section 5.

4. The Fisher Randomization test

In Chapter III (Sec. 21) of DE, Fisher introduced his randomization test in the following terms : "In these discussions it seems to have escaped recognition that the physical act of randomization, which, as has been shown, is necessary for the validity of any test of significance, affords the means, in respect of any particular body of data, of examining the wider hypothesis in which no normality of distribution is implied." Fisher then gave a brief description of his randomization test as an alternative to the Student's t test. In this section we consider a popular variant of the test that may be regarded as the original permutation test. This is how the test is described in Kempthorne and Folks (1971, p. 342).

Let $x_1, x_2, ..., x_n$ be n independent observations on a random variable x. The problem is to test the null hypothesis H_0 that $\mu = E(x) = 0$. Under the parametric model that x is normally distributed, the test is usually carried out in terms of the studentized sum $T = \Sigma x_i$. Under the wider hypothesis (nonparametric model) that the distribution of x is continuous and is symmetric about mean μ the null hypothesis H_0 may be tested in terms of the criterion $T = \Sigma x_i$ as follows :

Write $\delta_i = \text{sgn } x_i$, $i = 1,2, \ldots, n$; that is, δ_i is -1 or 1 according as x_i is negative or positive. Note that

$$T = \Sigma x_i = \Sigma |x_i| \delta_i$$

and that the sample $(x_1, x_2, ..., x_n)$ may split into the two parts

$$(|x_1|, |x_2|, \ldots, |x_n|) \text{ and } (\delta_1, \delta_2, \ldots, \delta_n).$$

Making the standard pretense that we are dealing with random variables and not particular observations, we recognize at once that the $|x_i|$'s are iid and that so also are the δ_i's. Under the null hypothesis, the two parts of the sample are stochastically independent and each δ_i is uniformly distributed over the two-point set $\{-1, 1\}$.

The distribution of the test criterion T is not well defined under the null hypothesis. If we fix the $|x_i|$'s at their observed values and regard the δ_i 's as random variables, however, then the conditional null distribution of T gets well-defined Although the actual computation may become somewhat tedious, the conditional probability

$$\Pr(T \geq t_1 | H_0, |x_1|, |x_2|, \ldots, |x_n|)$$

can be worked out. Thus, we can carry out one-sided or two-sided tail area tests in terms of the conditional null distribution of T.

The conditional test just described bears the distinctive hallmark of Sir Ronald. It was Fisher who amazed and mystified the statistical world with his sensational 2 × 2 conditional test of independence, and it was he who taught us how to set up a conditional test for the equality of two Poisson means.

In order to find the attained significance level of data vis à vis a null hypothesis H_0, we have to search for an appropriate test criterion T and then refer it to an appropriate sample space (the reference set) for determining the tail-area probability under the null hypothesis. In the present case the criterion is the sample total T and the reference set is the set of all samples of the type

$$(\pm|x_1|, \pm|x_2|, \ldots, \pm|x_n|) ,$$

where $|x_1|, |x_2|, \ldots, |x_n|$ are fixed at their observed values. Before we turn the searchlight of careful scrutiny on this mystifying conditional test, it will be useful to compare it with two familiar nonparametric tests of the null hypothesis $\mu = 0$. (See Kempthorne and Folks 1971, pp. 340-345.)

The sign test : Choose as the test criterion the number S of positive signs among $\delta_i = \text{sgn } x_i$, $i = 1, 2, \ldots, n$. The null distriution of S is bin $(n, \frac{1}{2})$. One-sided or two-sided tests can then be made in terms of S.

The Wilcoxon signed-rank test : Instead of the sample total $T = \Sigma|x_i| \delta_i$, choose the statistic $W = \Sigma r_i \delta_i$ as the test criterion, where r_i is the rank of $|x_i|$ among $|x_1|, |x_2|, \ldots, |x_n|$. Observe that the range of variation of W is the set of alternate integers in the interval $[-n(n + 1)/2, n(n + 1)/2]$. Under the null hypothesis H_0, the two vectors $(r_1, r_2, \ldots, r_n)$ and $(\delta_1, \delta_2, \ldots, \delta_n)$ are stochastically independent and the δ_i's are iid ± 1 variables with equal probabilities. The conditional distribution of W, given $(r_1, r_2, \ldots, r_n)$, can, therefore, be easily worked out under H_0. Since $(r_1, r_2, \ldots, r_n)$ is always a permutation of $(1, 2, \ldots, n)$, it is clear that the null distribution of W is the same for all possible realizations of $(r_1, r_2, \ldots, r_n)$; in other words, the Wilcoxon statistic W is stochastically independent of the rank vector if the null hypothesis is ture. Thus, the Wilcoxon test is not a conditional test in the sense the Fisher randomization test is. The sign test and the Wilcoxon test are typical examples of nonparametric, distribution-free, marginal tests.

Commenting on the three tests, Kempthorne and Folks (1971, p. 344) wrote: "Since the sign test uses only the signs of x_i, the Wilcoxon test uses only the signs and the ranks of $|x_i|$, and the Fisher test uses the x_i without condensation, the Fisher test is superior as a significance test." Thus, it seems that Kempthorne and Folks are giving poorer ratings to the sign test and the Wilcoxon test on the score that they violate the insufficiency principle to a greater extent than does the Fisher test. But how to measure the extent of such violations ! Do any of these tests violate the sufficiency principle ? Let us examine the question.

In the context of our nonparametric statistical model, the set of order statistics $x_{(1)}, x_{(2)}, \ldots, x_{(n)}$ is minimal sufficient. Each of the three test criteria T, S, and W can be written as a function of the order statistics, for example

$$T = \Sigma\, x_{(i)} = \Sigma |x_{(i)}|\, \delta_{(i)} ,$$

$$S = \frac{1}{2}(\Sigma\, \delta_{(i)} + n) ,$$

$$W = \Sigma\, r_{(i)}\, \delta_{(i)} ,$$

where $\delta_{(i)} = \text{sgn}\, x_{(i)}$ and $r_{(i)}$ is the rank of $|x_{(i)}|$ among $|x_{(1)}|, |x_{(2)}|, \ldots, |x_{(n)}|$. Since the sign and the Wilcoxon tests are based on the marginal distributions of S and W, respectively (and, of course, on their observed values), there is no violation of the sufficiency principle in these cases - only the insufficiency principle is at stake.

In the case of the Fisher test, it may appear on the surface that the sufficiency principle has been violated in view of the fact that the conditioning statistic $(|x_1|, |x_2|, \ldots, |x_n|)$ is not a function of the minimal sufficient statistic $(x_{(1)}, x_{(2)}, \ldots, x_{(n)})$. If we carefully examine the conditional distribution of T given $(|x_1|, |x_2|, \ldots, |x_n|)$, then it will be clear that the conditional distribution depends on the sample $(x_1, x_2, \ldots, x_n)$ only through the order statistics. The sufficiency principle is sometimes interpreted as a requirement that the data ought to be first reduced to the minimal sufficient statistic (thus sieving out all the postrandomization impurities) and then the reduced data interpreted in terms of the marginal distribution of the minimal sufficient statistic. To satisfy the statistical intuition of such a purist we have only to point out that the Fisher test will remain unaltered if the conditioning statistic is chosen to be the ordered rearrangement of $|x_{(1)}|, |x_{(2)}|, \ldots, |x_{(n)}|$. The Fisher test does not violate the sufficiency principle.

The choice of the test criterion $T = \Sigma\, x_i$ and the choice of the conditioning statistic $(|x_1|, |x_2|, \ldots, |x_n|)$ are arbitrary elements in the Fisher test. For instance, we may want to condition T with respect to the statistic $(|x_1 + x_2 + \ldots + x_k|, |x_{k+1}|, \ldots, |x_n|)$ for some chosen k $(1 \leq k \leq n)$. It is easily seen that any such conditioning will make the null distribution of T distribution free. For instance, if $k = n$, then the conditional null distribution of T, given $|x_1 + \ldots + x_n| = d$, is uniform over the two-point set $\{-d, d\}$. With such a conditioning the one-sided test of the null hypothesis will result in a significance level of $\frac{1}{2}$ whenever the observed value of T is positive ! Suppose we somehow convince ourselves that the only reasonable choice of a conditioning statistic is the one that corresponds to $k = 1$, the Fisher choice. (If $1 < k < n$, then the test procedure violates the sufficiency principle. The case $k = n$ is too ridiculous to deserve any serious consideration. And so on for any other conditioning statistic that one can think of.) Even then the question about the choice of the test criterion remains. Instead of $T = n\bar{x}$, why not choose the sample median $\tilde{x}$ as the test criterion? With the Fisher conditioning with respect to $(|x_1|, |x_2|, \ldots, |x_n|)$, the null distribution of $\tilde{x}$ is also distribution free. In our nonparametric setup, the sample median $\tilde{x}$ seems to be as reasonable a choice (as a test criterion) as the sample mean. Now, let us try to evaluate the significance level attained by the sample

$$(x_1, x_2, x_3, x_4, x_5) = (4, 7, 2, 3, 1).$$

The Fisher reference set consists of the 32 points

$$(\pm 4, \pm 7, \pm 2, \pm 3, \pm 1).$$

The observed sample mean is 3.4 and the median is 3. In the reference set there is only one point (viz., the sample itself) whose mean is at least as high as 3.4. In the same reference set, however, there are four points, namely, $(4, 7, \pm 2, 3, \pm 1)$ with median as high as 3. Therefore, with $\bar{x}$ as the test criterion the significance level (SL) of the data will be evaluated as 1/32, whereas with $\tilde{x}$ as the test criterion the data will be deemed to have attained $SL = \frac{1}{8}$. Note that every sample of five positive observations, irrespective of how far out or how scattered they are on the positive half-line, will be judged as significant (SL = 1/32) if $\bar{x}$ is the test criterion and not significant (SL = 1/8) if $\tilde{x}$ is the test criterion. With a sample of seven positive observations the SL will be 1/128 or 1/16 depending on whether $\bar{x}$ or $\tilde{x}$ is chosen as the test criterion. Consider the two samples (- 5, - 4, - 1, 6, 7, 8, 9) and (62, 63, 64, 65, 66, 67, 68). Does it make any sense to say that, with respect to the null hypothesis $\mu = 0$ with one-sided alternatives, the two samples are equally significant with SL = 1/16 ? But that is exactly what the Fisher test will do if $\tilde{x}$ is chosen to be the test criterion.

Let us take a short break from this ruthless cross-examination of the Fisher test with some speculation on Sir Ronald's later thoughts on the subject. The cross-examination will continue in Section 6.

5. Did Fisher Change His Mind?

In all fairness to Sir Ronald we have to admit that, apart from making a passing reference to the randomization test method in Chapter III of DE, Fisher did not have much else to do with this kind of test procedure. Twenty one years later, when Fisher came out with his last testament on statistics - Statistical

Methods and Scientific Inference (SI) - he had apparently forgotten all about his randomization test method. In the winter of 1954-55, Fisher visited the Indian Statistical Institute for a couple of months and gave an extensive series of lectures based on the manuscript of SI. Those lectures profoundly influenced my own thinking on statistics. In SI, Fisher discussed the logic of inductive inference, his new outlook on significance testing, fiducial inference methods, likelihood methods of inference, and conditioning and recovery of ancillary information, but nowhere do we find any mention of randomization analysis of data. Randomization as an ingredient of statistical designs was mentioned only once, and that appeared in the following passage (Fisher 1956, p. 98):

> . . . whereas in the Theory of Games a deliberately randomized decision (1934) may often be useful to give an unpredictable element to the strategy of play; and whereas planned randomization (1935-53) is widely recognized as essential in the selection and allocation of experimental material, it has no useful part to play in the formation of opinion, and consequently in the tests of significance designed to aid the formation of opinion in Natural Sciences. [Note : The year 1934 refers to a Fisher article on randomization in card play and 1935-53 refers to DE.]

On the suggestion of an associate editor of JASA this passage is quoted in full so that the readers of the article can make up their own minds on the following.

Questions : Isn't it surprising that Fisher had no more to say about randomization in 1956? Was Fisher disassociating himself from the randomization test by not mentioning the method in SI ? Does the remark about "formation of opinion" refer to postrandomization only ?

It should be recognized that Fisher's views on significance testing underwent a major change during the period 1935-1956. On p. 77 of SI he made a clear distinction between tests of significance as used in natural sciences with tests for acceptance as in quality-control theory. According to him the dissimilarities (between the two methods) lie in the population, or reference set, available for making statements of probability. Let us quote Fisher (SI, p. 77) on this point :

> Confusion under this head has on several occasions led to erroneous numerical values; for, where acceptance procedures are appropriate the population of lots of one or more items, which could be chosen for examination, is unequivocally defined. The source of supply has an objective empirical reality. Whereas, the only populations that can

> be referred to in a test of significance have no objective reality, being exclusively the product of the statistician's imagination .... The demand was first made, I believe, in connection with Behrens' test of . . . . significance . . . ., that the level of significance should be determined by repeated sampling from the same population, evidently with no clear realization that the population in question is hypothetical, that it could be defined in many ways . . . ., or, that an understanding, of what the information is which the test is to supply, is needed before an appropriate population, if indeed we must express ourselves in this way, can be specified.

Again, on p. 91 of SI, Fisher quoted himself (from a 1945 Sankhyā article dealing with the fiducial argument) as follows :

> In recent times one often repeated exposition of the tests of significance, ... , seems liable to lead mathematical readers astray, through laying down axiomatically, what is not agreed or generally true, that the level of significance must be equal to the frequency with which the hypothesis is rejected in repeated sampling of any fixed population allowed by hypothesis. This intrusive axiom, which is foreign to the reasoning on which tests of significance were in fact based seems to be a real bar to progress . . . .

It seems clear to me that, in 1956, Fisher's views on significance testing were somewhat close to the Bayesian position that the evidential content of data cannot be judged in sample space terms. Indeed, the 1945 quotation from Fisher might very well have been written by De Finetti himself. I am, therefore, not surprised at all that in SI Fisher mentioned neither the randomization test nor the lady-tasting-tea-type data analysis. For these are very extreme types of non-parametric data analysis in which the evidential meaning of the data is sought to be evaluated by referring it to a sample space that is formed by the statistician in his or her mind by imagining all the possible outcomes of the planned randomization input of the experiment. This will be made clearer in the next section.

Many of our contemporary statisticians are unaware of the fact that in the seventh edition of DE (1960), Fisher added what looks like a disclaimer in the form of a short section (See. 21.1; "Nonparametric" Tests) at the end of Chapter III. We quote this section in full.

In recent years, tests using the physical act of randomization to supply (on the Null Hypothesis) a frequency distribution, have been largely advocated under the name of Nonparametric tests. Somewhat extravagant claims have often been made on their behalf. The example of this section, published in 1935, was by many years the first of its class. The reader will realize that it was in no sense put forward to supersede the common and expeditious tests based on the Gaussian theory of errors. The utility of such nonparametric tests consists in their being able to supply confirmation whenever, rightly or, more often, wrongly it is suspected that the simpler tests have been appreciably injured by departures from normality.

They assume less knowledge, or more ignorance, of the experimental material than do the standard tests, and this has been an attraction to some mathematicians who often discuss experimentation without personal knowledge of the material. In inductive logic, however, an erroneous assumption of ignorance is not innocuous; it often leads to manifest absurdities. Experimenters should remember that they and their colleagues usually know more about the kind of material they are dealing with than do authors of textbooks written without such personal experience, and that a more complex, or less intelligible, test is not likely to serve their purpose better, in any sense, than those of proven value in their own subject.

Note Fisher's use of the phrase "physical act of randomization." The same phrase appears in the Fisher quotation in the opening paragraph of the previous section. Where is the physical act of randomization in the Fisher randomization test? The random entities $\delta_1, \delta_2, \ldots, \delta_n$ can hardly be called randomization variables. It is only under the null hypothesis that the δ_i's can be regarded as iid uniform ± 1 variables. The nonnull distribution of the δ_i's depends on the parameter of interest μ in a rather complex fashion. We should recognize the fact that in Section 21 of DE (1935) Fisher was not really concerned with the particular test situation that we have discussed in the previous section. He was talking about the problem of comparing two treatment effects under a wider hypothesis and was suggesting a (nonparametric) randomization analysis of data generated by paired comparisons on the basis of a physical act of randomization. In the next section we discuss this matter in some detail.

6. Randomization and Paired Comparisons

A scientist wants to test whether a so-called improved diet (treatment) is in effect superior to the standard diet (control). The scientist has 30 animals (subjects) with which to experiment. The scientist carefully pairs (blocks) the subjects into 15 homogeneous pairs. Let $\{(s_{1i}, s_{2i}) : i = 1,2 \ldots, 15\}$ be the set of 15 subject pairs. The subjects in each pair are of the same sex, come from the same litter, and so on. From each pair the scientist selects one subject for the treatment and the other one for control. The 30 responses (weight gain in so many weeks) are laid out as $\{(t_i , c_i) : i = 1,2, \ldots, 15\}$, where t_i and c_i are the responses of the treated subject and the control subject, respectively.

The scientist observes that

$$T = \Sigma t_i - \Sigma c_i$$

is a large positive number and also notes that

$$d_i = t_i - c_i > 0 \quad \text{for all} \quad i .$$

The scientist, therefore, concludes that he or she has obtained very strong evidence in favor of the hypothesis H_1 that the improved diet is really superior to the standard diet. For measuring the strength of the evidence the scientist consults a statistician.

The statistician decides to make a one-sided test of significance of the null hypothesis H_0 (that the two diets are the same in their short-term weight- gain effects) on the basis of the scientist's data. The statistician also thinks that $T = \Sigma d_i$ is an appropriate test criterion in this case. For finding the significance level of the observed value of T, the statistician has to find the null distribution of T. So what the statistician needs now is a nice reference set.

The response difference d_i between the ith pair of subjects can be explained in terms of a possible treatment difference and other possible nuisance factors like subject differences (which the scientist tried his or her best to control by blocking), virus infection, loss of appetite, and many such uncontrollable factors that may have acted differently on the two subjects in the ith pair. If hypothesis H_0 is true, then there is no treatment difference; so the response difference d_i must be presumed to be caused by the previously mentioned nuisance factors. As Fisher explained in DE, randomization enables the statistician to eliminate all these nuisance factors from the statistical argument. Let us see how this elimination is achieved in the present case.

Suppose the scientist had made 15 independent random decisions (on the basis of 15 tosses of a fair coin) as to which subject in the ith pair gets the improved diet (i = 1, 2, ..., 15). Having recorded the 15 response differences d_1, d_2, ..., d_{15} and having computed $T = \Sigma d_i$, the scientist can speculate about a hypothetical rerun of the experiment in which all but one of the experimental factors (controllable or uncontrollable) are supposedly held fixed at the level of the last experiment - the same 30 animals exactly as they were at the commencement of the last trial, paired the same way into 15 blocks, exactly the same set of animals coming down with the same kind of virus infections with the same effects on them, and so forth. The only thing that is allowed free play in the hypothetical rerun of the experiment is the random allocation of treatment - the fair coin has to be tossed again 15 times. If H_0 is true, then the response difference $d_i = t_i - c_i$ for the ith pair must have been caused by the nuisance factors (subject differences, virus infection, etc). In the hypothetical rerun of the experiment all such nuisance factors are supposedly held fixed at the past level. Therefore, in the new experiment the response difference for the ith pair can take only two values d_i or $-d_i$, depending on whether the treatment allocation for the ith pair is the same as in the past experiment or is different. If we denote the response differeces for the hypothetical experiment by $(d'_1, d'_2 \ldots, \; d'_{15})$ then it is clear that, under the null hypothesis H_0, the sample space (for the response differences) is the set R of 2^{15} points (vectors)

$$R = \{ (\pm d_1, \pm d_2, \ldots, \pm d_{15})\} ,$$

with all the points equally probable. This is the reference set that the statistician was looking for. Let $T' = \Sigma d'_i$. The significance level of the data

$$SL = Pr(T' \geq T | H_0)$$

is now computed as follows.

The statistician looks back on the data and notes that $d_i > 0$ for all i. Therefore, $T' \geq T$ if and only if $d'_i = d_i$ for all i. Hence, $LS = (\frac{1}{2})^{15}$. This is randomization analysis of data in its classical form.

The rest of this section is devoted to an evaluation of this particular data analysis. The evaluation is laid out in the form of a hypothetical sequence of remarks and counterremarks by the statistician, the scientist, and the author.

Statistician : Observe that the randomization test argument does not depend on any probabilistic assumptions. The randomization probabilities are fully understood and are completely under control. I do not have to assume that the treatment-allocation process was like a sequence of 15 Bernoulli trials with $p = \frac{1}{2}$. Surely, I can regard that as demonstrably correct. In this argument there is no mention of a population. The experimental animals do not have to be regarded as a random sample from a population of animals. This test is an ultimate nonparametric test. Not only do we not have to deal with model parameters, we do not have to contend with even a statistical model. There is no mention of a sample space X equipped with a σ-field A of events and a family P of probability measures, no measurement errors, no mention of a sequence of iid random variables with an unknown distribution function.

Scientist : I am greatly puzzled by your data analysis. Your analysis seems to depend only on the randomization probabilities and the observed fact that $d_i > 0$ for all i. The fact that the test criterion $T = \Sigma d_i$ attained a rather large value in this case does not seem to enter into the probability evaluation of $(\frac{1}{2})^{15}$.

Author : Suppose we choose the median of $d_1, d_2, \ldots, d_{15}$ to be the test criterion instead of $T = \Sigma d_i$. The significance level of the data will then be evaluated as $(\frac{1}{2})^8$. How can we explain the big difference between $(\frac{1}{2})^{15}$ and $(\frac{1}{2})^8$?

Scientist : I do not understand the relevance of the randomization probabilities. Why is it so crucial that the coin with which I made the treatment allocation be a fair coin? Suppose I had used a biased coin with $p = \frac{1}{4}$. Suppose for the ith pair (s_{1i}, s_{2i}) of experimental animals my treatment allocation was (t,c) or (c,t) depending on whether the ith toss of the biased coin resulted in a head or a tail. How significant would my present data have been then?

Author : Let me answer the question. The hypothetical rerun of the experiment will be defined as before, but this time the biased coin will define the randomization scheme. The reference set for $(d'_1, d'_2, \ldots, d'_{15})$ will still be the same set R. Note that in this case $\Pr(d'_i = d_i) = \frac{1}{4}$ or $\frac{3}{4}$ depending on whether, in the original experiment, the response difference d_i was associated with the (t, c) or the (c,t) treatment allocation. Therefore, the level of significance will be evaluated as

$$SL = \Pr(T' \geq T|H_0) = (\tfrac{1}{4})^m (\tfrac{3}{4})^{15-m},$$

where m is the number of (t,c) allocations in the original experiment. The larger the value of m is, the more significant are the data.

Scientist : This is patently absurd. How can the SL depend so largely on such an irrelevant data characteristic as m? It is relevant to know that the 30 animals have been paired into 15 homogeneous blocks. The manner of my labeling the two animals in the ith block as (s_{1i}, s_{2i}) does not seem to be of much relevance. The number m of treatment allocations of the type (t,c) seems to be of no consequence at all. I have not been asked about all the background information that I have on the problem. For instance, I happen to have made a nutrition analysis of the two diets. I know that the improved diet has a much higher protein content and is very rich in vitamins C and D. I know the results of several past experiments on the same set of animals when they were fed the standard diet. I know that six animals came down with virus infections during the experiment and that five of them were fed the improved diet. I am amazed to find that a statistical analysis of my data can be made without reference to these relevant bits of information.

Statistician : You are trying to make a joke out of an excellent statistical method of proven value, a method that originated in the mind of one of the two (Fisher and Einstein) really outstanding men of genius that the world has seen in this century. Your criticisms are based on an extreme example and then on a misunderstanding of the very nature of tests of significance. Tests of significance do not lead to probabilities of hypotheses. I do not believe in "belief probabilities." I do not believe that any useful purpose can be served by trying to quantify your knowledge in the form of a belief probability. Go to a Bayesian if you wish to make any input of your subjective beliefs in the data analysis process. It does not make much sense to set up a statistical model for the purpose of analyzing experimental data. The randomization analysis of data is so simple, so free of unnecessary assumptions that I fail to understand how anyone can raise any objection against the method. In the case of the present experiment you have in effect tossed a fair coin 15 times, have you not? So why confuse the issue by bringing in the case of an absurdly biased coin with $p = \frac{1}{4}$? Note that the probability of $(\frac{1}{2})^{15}$ that I have computed for you is a gambler's probability, a frequency probability, a propensity measure of a well-defined physical system. A belief probability it is not.

Scientist : Your probability of $(\frac{1}{2})^{15}$ is defined in terms of a hypothetical experiment, a rerun of the original experiment with everything (repeat everything) but the randomization part fixed at the level of the original experiment. But how can you even think of such an utterly impossible experiment? My experimental animals have changed - one of them died last week - the weather has changed, the virus epidemic is gone. I do not see how you can claim any objective reality for the randomization probability of $(\frac{1}{2})^{15}$. In any case, I knew all along that the null hypothesis could not possibly be true. So any probability computed under the supposition that the null hypothesis is true cannot have much of an objective reality.

Author : The computation $SL = (\frac{1}{2})^{15}$ was based on the supposition that in the hypothetical rerun of the experiment all the 2^{15} treatment-allocation patterns are equally probable. It is not clear from the argument that the scientist had to make all the 2^{15} possible allocations equally probable in the original experiment.

Scientist : This is a good time for me to confess that in fact I did not randomize over the full set of 2^{15} possible allocations. As a scientist I have been trained to put as much control into the experimental setup as I am capable of, to balance out the nuisance factors as far as possible. After carefully blocking the 30 subjects into 15 nearly homogeneous pairs, I could still detect differences within the subject pairs. There were differences in weight, height, some relevant blood characteristics, and a few other relevant features. I wanted the set of 15 treated subjects to be nearly equal to the set of 15 control subjects in some group characteristics like average weight, average height, and so on. I worked very hard on the project of striking a perfect balance between the treatment and the control groups. Finally, I found two such complementary groups and then decided on the treatment/control allocation to the two groups by a mental process that may be likened to the toss of a fair coin. I wonder what the significance level of my data is going to be in the light of this confession.

Statistician : Had I known about this before, I would not have touched your data with a long pole. Now the reference set for $(d'_1, d'_2, \ldots, d'_{15})$ consists of only the two points

$$(d_1, d_2, \ldots, d_{15}) \text{ and } (-d_1, -d_2, \ldots, -d_{15}),$$

and the significance level

$$SL = Pr(T' \geq T | H_0)$$

works out to be $\frac{1}{2}$ if $T > 0$ and 1 if $T \leq 0$. Your data is not significant at all.

Scientist (utterly flabbergasted) : But my experiment was better planned than a fully randomized experiment, was it not ? With my group control (in addition to the usual local control) I made it much harder for $T = \Sigma t_i - \Sigma c_i$ to be large in the absence of any treatment difference. In spite of this careful global control, I found that T is a large positive number and that every $t_i > c_i$. And you are telling me that, under the null hypothesis, it is as easy to get a result as significant as mine as it is to get a head from a single toss of a symmetric coin !

Statistician : My good man, you must realize that your experiment is no good. The prerandomization that you had carried out was not wide enough; the randomization sample space has only two points in it with a uniform probability distribution under the null hypothesis. Thus, the only attainable significance levels are $\frac{1}{2}$ and 1. Your experiment is not informative enough. I wish you had consulted me before planning your experiment. It appears that you do not have a clear understanding of the role of randomization in statistical experiments.

7. Concluding Remarks

So the randomization argument foundered on the rocks of restricted and unequal probability randomization. The statistician had the last word but lost the argument. The statistician was clearly wrong in characterizing the scientist's one-toss randomized experiment as uninformative. During the last 15 years, I have heard three very eminent statisticians characterizing the one-toss experiment as uninformative on the score that the sample space has only two points in it. That this cannot be so is easily seen as follows.

An urn contains two balls that are either both white or both black. The draw of a single ball from the urn is then fully informative, although the sample space has only two points in it. If this example seems to be too artificial, then consider the case of an urn in which the proportion of white balls is either $\frac{1}{4}$ or $\frac{3}{4}$. Consider the sequential sampling plan that requires drawing of balls one at a time and with replacements until the likelihood ratio either exceeds 100 or falls below 1/100. Suppose the outcome of this experiment is recorded as "below 1/100" or "above 100." This is a highly informative experiment with only two sample points in it.

It should be noted that the sample space of the experiment performed by the scientist had a huge number of points in it. The statistician took a thin cross-section of the sample space (after holding fixed all the relevant factors like subjects, treatment effects, recognizable nuisance factors, and error terms) and then found only two points in it. No wonder the scientist failed to understand the argument.

The scientist was correct in questioning the relevance of randomization at the data analysis stage. Prerandomization injects an element of uncertainty about the actual experimental layout. But that uncertainty is removed once the scientist goes through the randomization ritual early in the game. At the data analysis stage, why is it still necessary to find out about the details of the actual randomization process? The randomization exercise cannot generate any information on its own. The outcome of the exercise is an ancillary statistic. Fisher advised us to hold the ancillary statistic fixed, did he not?

Concluding Remarks

Our statistician is a most ardent admirer of R.A. Fisher. But he does not like the postfiducial (1936-62) Fisher. During the last 27 years of his astonishing career, we find Sir Ronald entertaining such counter-revolutionary thoughts as the conditionality and the likelihood principle and toying with the half-baked Bayesian idea of fiducial probability distribution.

We have noted earlier how the sufficiecy principle rejects postrandomization analysis of data. Similarly, the conditionality principle (see Basu 1973 for more on this) rejects prerandomization analysis of data. In view of Fisher's postfiducial rethinking on statistical inference, it was almost inevitable for him finally to insert that astonishing short section on nonparametric test in the seventh edition of The Design of Experiments.

CHAPTER XVI

THE FISHER RANDOMIZATION TEST : DISCUSSIONS

Chapter XVI was published in the Journal of the American Statistical Association in 1980 with discussions. The discussants were David V. Hinkley, Oscar Kempthorne, David A. Lane, D.V. Lindley, and Donald B. Rubin. These discussions and Basu's rejoinder are put together in this Chapter.

DAVID V. HINKLEY

Basu has provided us with an interesting and provocative critique of significance tests related to randomized experiments. It does seem to be true that there is not a unified Fisherian mathematical theory of significance tests. This should not be surprising, however, since Fisher was wont to warn of the dangers of routine, formal application of mathematical statistics without very careful regard for scientific context and operational meaning. Indeed, one might view Basu's paper as an illustration of Fisher's warning.

In terms of Fisher and randomization, the first four sections of the paper require little comment, since they deal with the separate topic of permutation and rank tests. Nevertheless, it is important to point out a fallacy in Basu's criticism of nonunique significance level (SL) in Section 4 : The data as such do not possess an SL, which instead attaches to a particular statistic. Moreover, it is important to recognize that in Section 4 there is only an abstract null probability model – not a general model – so that the statistician has no basis for choice of statistic : The scientist must specify the relevant statistical measure. The role of the statistician here is to ensure that a valid, operational interpretation of the chosen statistic can be, and is, made.

After confessing to a "ruthless cross-examination" of the wrong topic – the non-Fisherian nonparametric tests of Section 4 – Basu suggests that Fisher's silence in 1956 may be used to condemn the randomization test. This speculation seems unwarranted on two counts. First, I do not think that Fisher ever did recommend the randomization test for analysis of data, but rather that he introduced it as a device for demonstrating that randomization validates the usual normal-theory methods of analysis. This notion seems clear in Yates (1933), for example. Unfortunately, Basu has not chosen to discuss the connection between randomization and the validity of mathematical models. The second point is that Fisher's views on randomization would have been so widely known and accepted after 25 years that it would not have seemed necessary to repeat them in a book on statistical inference for parametric probability models. Fisher did repeatedly assert that randomization could guarantee the relevance of such abstract models-including the seemingly innocuous model in Section 4 - but he realized that randomization was "sufficient" (CP 204)[1] rather than necessary, since often "Nature has done the randomization for us" (CP 212). These last remarks should be borne in mind when reading the amusing developments in Section 6.

What are we to make of the Statistician and the Scientist? They are certainly an entertaining addition to the literature, but hardly enlightening or enlightened. The first serious issue seems to be that of the biased-coin design, for which the author provides the SL. Surely the SL given is an appropriately cautious evaluation in the worst case in which the experimenter knowingly takes advantage of the bias - cheats, that is. But apparently the Scientist did not cheat ("my labeling . . . does not seem to be of much relevance"), so that in effect the treatments were allocated at random within each pair : Nature has done the randomization for us. Thus the usual analysis is presumably valid, and if a randomization SL were computed it would still be $(\frac{1}{2})^{15}$. The Statistician has apparently mistaken Fisher's "sufficient" for "necessary."

What follows after the biased-coin episode is a series of irrational remarks and misunderstandings. The Statistician's dogmatic attitude is hardly characteristic of the statistician who inspired the Rothamsted song "Why ! Fisher can always allow for it" (Box 1978, pp. 138-139).

What was Fisher's position on randomization and the induced distribution of statistics? While this is not entirely clear down to the last detail, I think it clear enough to suggest that Basu has missed the point. For a brief introduction to the relevant parts of Fisher's work, see the lectures by Holschuh, Picard, and Wallace in Fienberg and Hinkley (1980). Highly informative and balanced accounts of the issues may be found in Yates (1970), particularly the 1965 Berkeley Symposium paper reviewing experimental design. As I see it, the purpose of randomization in the design of agricultural field experiments was to help ensure the validity of normal-theory analysis. Nature was not in the habit of doing the randomization. Studies by Tedin (1931) and others on uniformity trial data showed that for systematic designs (such as that finally described by Basu's Scientist) the usual properties of t and F tests did not hold in an operational sense. Thus standard significance tests were invalidated. "Student," among others, correctly pointed out that effects could be more precisely estimated from carefully chosen systematic designs. But, said Fisher, this was of no use if the estimated precision were too high, higher even than the valid estimates obtained from randomized experiments. Thus, for some systematic designs, the computed normal-theory SL corresponding to a theoretically precise effect was in fact appreciably larger than the "real" SL. This is

1 Fisher's papers are referred to by their numbering in the Collected Papers (Fisher 1974)

exactly what could happen in the case of the two-point design of Basu's Scientist, although it probably did not if nature has randomized. With the limited information given us by Basu, we cannot give a reliable standard error for the Scientist's accurate estimate $\bar{d}$, at least not one with a clear operational meaning. The apparently silly SL values ($\frac{1}{2}$ and 1) are a warning of possible difficulty, surely, nothing more.

The empirical evidence confronting Fisher certainly suggested the necessity of randomization in most field experiments, if the standard methods of analysis were to be used. In recent years it has become apparent that relatively simple spatial models can often account for some of the effects that randomization was designed to overcome; see Bartlett (1978), for example. Complete or partial failure to randomize can have adverse effects in other areas too, for example, in survey sampling in which systematic grids are randomly positioned on a sampling frame. In such a case systematic effects can accidentally (or purposely) change the variation, as I have seen myself. Cochran (1977, Ch. 8) discussed this problem in detail. For informative accounts of the importance of randomization in medical and public-policy studies, respectively, see Chapters 9, 10, and 11 of Bunker, Barnes, and Mosteller (1977) and Gilbert, Light, and Mosteller (1977). In all areas, the randomization distribution literally induced by the experimental randomization is of value in assessing the validity of a standard analysis. This, I think, is Fisher's message.

The final substantial issue of Basu's paper is that of the ancillarity of the design outcome. Technically Basu is quite correct, if the randomization has validated a parametric model - the design outcome is then ancillary by design ! It would, however, be as well not to forget the purpose of an ancillary statistic, since otherwise we are merely playing with abstract mathematical definitions. An ancillary statistic indicates the set of comparable cases against which to judge the observed sample and the statistical summary thereof. Usually "comparable cases" is taken to mean "equally informative samples" in some appropriate sense, as in Fisher's brief comments on the 2 × 2 table (CP 205). Admittedly this is not mathematically precise, but it seems to have the merit of common sense. It is often unnecessary, and sometimes plain foolish, to take an infinitesimal slice through an abstract space as the set of comparable cases, although the mathematical definition of ancillarity would require this. Lest Basu think that Fisher has been caught with his conditional pants down here, let me suggest that Fisher implicity invoked conditionality in his criticism of Knut-Vik Squares, which in Tedin's (1931) analyses correspond to an ancillary set of real import.

For a more constructive use of design ancillarity, consider a randomized block design (RBD) with four replicates of four treatments. Suppose that in the particular physical layout the selected design coincides with a 4 × 4 Latin Square and that the accidental block structure corresponds to a noticeable effect not due to the treatments. Here my qualitative notion of ancillarity would suggest that we analyze the experimental data as coming from a Latin Square design, that is, treat the 4 × 4 Latin Squares as that subset of RBD's that constitute the set of comparable cases. The Latin Square analysis would be exactly equivalent to using a covariate-adjusted RBD analysis with accidental block totals as covariates. (There is of course a question as to whether randomization validates the latter analysis.)

This short discussion has necessarily focused on my major misgivings with Basu's interesting paper. As to whether randomization tests are logically viable, I think Basu has not made a case. There may be no case in logic if, with John Clerk Maxwell, we believe that "the true logic for this world is the calculus of Probabilities." What we need to know is : Which probabilities?

OSCAR KEMPTHORNE

Basu states that we have no satisfactory answer to the question, "Why randomize?" Various workers have attempted to give "satisfying" partial answers to this question. Surely, Fisher (7th ed., 1960) did so, with extensive exposition in his The Design of Experiments. Then we can examine various writings of the 1930's.

We can cite Greenberg (1951) with the question as the title, as also was the title of the cited Kempthorne (1977). It would be useful to have an attempted exhaustive bibliography of the topic.

It is obvious that expositions that some regard as carrying some real force are not so regarded by Basu and, for example, by Harville (1975). Is there any possibility that discussion will resolve the disagreement? I believe not. But I do believe that discussion is useful. All of us surely subscribe to the absolute necessity of critical examination of our ideas.

My discussion consists of two parts : (a) reactions of Basu's essay and (b) a few comments on the nature and role of randomization.

Basu writes entertainingly, perhaps, but not informatively. He poses the question, "Can the Fisher randomization test pass the test of common sense?" We must, I suggest, force Basu to be explicit and clear. What is this "common sense" that Basu refers to? Presumably, it is Basu's "common sense." The philosophy of statistics is plagued with writers who talk about "the probability" only to tell us that they mean "my probability"; now we have "common sense," but it is "Basu's common sense."

Section 2 gives us prerandomization, postrandomization, and unrecorded randomization. This discussion is irrelevant. But it is useful, perhaps, to make a remark. It is ludicrous that Basu, a keen bridge player, does not, it seems, give a role to postrandomization. Let Basu play poker with significant (to him) payoff. Then if he does not use some sort of postrandomization, may be very informal, he will be "cleaned out." I speak from past personal experience of playing social, but nontrivial (it now bores me!) poker. I regret that I surmise that many of those who write about gambling do not practice it. Section 2 is a red herring.

Section 3 discusses the sufficiency principle. As he has written, and others too numerous to cite, this is a data-reduction principle. It was used by Fisher in his (inadequate) formulation of tests of significance and reached its summit for Fisher in his fiducial inversion (which I shall not discuss). Problems with both of these led to the also inadequate formulation of use of ancillaries. The Fisher prescription, "The (Basu-titled) Insufficiency Principle," reached its summit only with fiducial inference, which was, in fact, the only real inference that Fisher espoused. It is in these terms, I suggest, that the later Fisher must be examined.

Section 4 discusses the Fisher randomization test on the basis of the cited Kempthorne-Folks presentation. My only regret here is that Basu did not examine, it seems, an article by Kempthorne and Doerfler (1969). I found Basu's remarks on the test not violating the sufficiency principle interesting and possibly a justification of the quoted remark of Kempthorne and Folks about condensation. Also, it was comforting that Basu seems to conclude that $k = 1$ gives "the only reasonable choice of a conditioning statistic." On the choice of test criterion, the naive idea I had was that the alternative is a uniform shift so that the sample total contains, perhaps, the maximum total shift. Here, there surely is a question. To this I add that my own use of the randomization test is in the experimental setting in which the alternative hypothesis is that a treatment adds some quantity Δ to each and every unit to which it is assigned. Also on a technological level, a question of interest is whether the treatment gives a gain when applied to all the experimental units.

Section 5 discusses whether Fisher changed his mind. This has perplexed others, including me (Kempthorne 1966, 1974, 1975). I have commented (1975) on what appeared to me to be outright inconsistencies in Fisher's whole output. It serves no useful purpose to try to psychoanalyze this phenomenon, I suggest. I do suggest, however, that we frankly admit the occurrence of these inconsistencies. Clearly, Fisher (1956) was writing in part a polemic against acceptance procedures. In connection with one quotation, it is obvious that the population in a randomization test of a randomized experiment is "the product of the statistician's imagination."

(Why Fisher should include "exclusively," I do not know.) It is not clear to me that in 1956 Fisher had the position that "evidential content of data cannot be judged in sample space terms." On the matter of the "lady tasting tea" and randomized experiments, Fisher (1956) is, I judge, entirely silent, and that is surely a mystery. I have to state my opinion that I do not find Basu's psychoanalysis clear or convincing. On the Fisher (1960) quotation, Fisher is merely polemic, for some unknown reasons. In fact one can use Fisher's words against Fisher. One does not have knowledge of distribution. If one did, one would not be involved in transformation search, for instance. Any supposed statistician who believes he or she knows the model, for example, of normality and independence, is not a real statistician; that is surely obvious. The only interpretation of this is that Fisher was polemicizing, against what we can guess, but with no profit.

With respect to Basu's writing on "the physical act of randomization," I believe Basu is merely plain wrong. In a randomized experiment, the δ_i's have a known distribution whether or not the null distribution holds.

Section 6 gives in the first part what is, or should be, routinely taught on the randomized pair trial. This has been known for decades and an elementary substantively oriented exposition is that of Kempthorne (1961).

In the second part Basu gives a hypothetical interchange of a statistician and a scientist and the author. I suggest that this serves no useful purpose. Comments on sentences of this interchange follow.

1. Scientist : " The fact that . . . $T = \Sigma d_i$ attained a large value . . . does not seem to enter."

Comment : Precisely ! When is T large ? With reference to what is T large? If one has external information on the possible magnitude of T, then one will have an idea of what values of T are large. If one is a Bayesian, one claims to know the possible distribution of T. Clearly, if one is in this situation, one does not randomize. I believe Basu has no familiarity with the problems of evaluating drugs for human illnesses, of evaluating diets on humans or mice or whatever. He does not see the variability among humans "treated alike." Why, is a mystery to me. Basu seems to say : If you are an expert on cancer, then you know a probability model. Otherwise, you should withdraw from the field. I regret that I find the lack of knowledge that underlies Basu's thesis rather surprising, incongruous, and deplorable. I would like Basu to take up a "very small" branch of investigative science, learn all the available background, and then design and conduct, with aid, of course, his own research program. Because Basu is highly competent, I believe, in the game of bridge, I would like him to make a comparative trial of two bridge systems.

How would he do this? He surely has as good a background as any bridge player or writer. It is the absence of any effort on a problem outside the WFFing (constructing well-formed-formulas) of mathematical statistics that concerns me. As I have said before, we must be skeptical of individuals who write books on cooking but have never made a meal in a kitchen.

2. Scientist : "Why is it crucial that the coin . . . be a fair coin?"

Comment : From Basu's viewpoint, this is obviously irrelevant. From the viewpoint of the investigator who is not a Bayesian, the situation is different. If one does not regard experimentation as a process of investigation, with the value of the process being determined, partly at least, by its operating characteristics, the question is irrelevant. But a scientist who does not care about the operating characteristics of his or her observation procedure is a "pretty poor" scientist. This is my opinion, of course, but one that is shared by the great bulk of scientists, I am totally sure. Indeed, the question with respect to any substantive experimental outcome is whether other scientists can duplicate the results. This is all very elementary, and I will not waste valuable journal space discussing it. The question is irrelevant to Basu, because one should not use any coin. But for the person who accepts the idea that operating characteristics of a procedure are important and who regards significance tests as evidential, the answer is obvious. If one uses a collection of plans, of which one plan has probability .99, then the significicnce level regardless of outcome and regardless of whether there is a real treatment difference will be equal or exceed .99 with probability .99. One then has to discard the idea of significance tests - at least as they are used at present. If there are M plans, then using these with equal probability gives the possibility with huge treatment effects of obtaining a significance level of 1/M. So, for the significance tester, there is value to equal probabilities, or the "fair coin." From another point of view, the use of equal probabilities gives estimates with nice properties, an analysis of variance with nice properties, and, surely, a valid use of the central limit theorem for the distribution of the test criterion in comparative trials of reasonable size. The reply of the author does not consider, I think, operating characteristics.

3. Scientist : "This patently absurd . . . "

Comment : I can say, equally as Basu, that his writings on randomization are "patently absurd" - but, of course, this does not lead to improved understanding. Basu credits the scientist with all sorts of "background information." The scientist "knows etc." The scientist and Basu are entitled to "be amazed to find that a statistical analysis of my data can be made without reference to these relevant bits of information." Why? Because if the scientist really has these "bits of information" a decent statistician will attempt to take them into account. No one claims that the randomization test of significance is the beginning, middle, and end of statistical analysis. Finally, I must ask the question, "Has Basu worked intimately with any scientists with a real problem (as opposed to a circus trainer with 10 elephants)?"

4. Scientist : "But how can you even think of such an utterly impossible experiment?"

Comment : I, Kempthorne, can! A very simple answer! I follow this with a question to Basu, related to the very interesting arcane mathematics he sometimes does. "How can you, Basu, even think of observing a real number exactly?" Each of us has a mental problem. Let us rest the matter there.

5. Scientist : ". . . I did not randomize over the full set . . . "

Comment : For me as a randomizing significance tester, that presents no problems. Tell me what your randomization frame was, and I can proceed. I may well find that to interpret your results a repeated sampling principle is useless. I, or rather you, have to supply a prior and a probability distribution. This is, perhaps, no problem for you. But it is for me, because now I have to assess for myself how much belief to hold with respect to your opinion. That is the rub!

6. Statistician : "Your data are not significant at all."

Comment : Right on! Your data are not significant to me. They may be to you, of course.

7. Basu : "So the randomization argument founders on the rocks of restricted and unequal probability randomization."

Comment : I do not see the claimed foundering.

8. Basu : "The one toss experiment (is) uninformative . . . ."

Comment : It is uninformative to me in the absence of forcing relevant and supported prior or external information. With such information, the actual experiment is only a part of the total information. What is there to argue about?

9. Basu : "The outcome of the (randomization) exercise is an ancillary statistic."

Comment : Yes and no. This outcome does not depend on the probability model, but if one does not know the probability model, one cannot (or should not) characterize the randomization outcome as ancillary. Furthermore, Basu's own work (not cited, but very well known) shows that there are huge difficulties in strict formulation of ancillary statistics.

I have one final comment about Fisher (1956). It is clear from Chapter III that Fisher envisaged various forms of inference from tests of significance to distributions on unknown parameters. He did not, then, reject tests of significance in 1956. Furthermore, there is no evidence that he rejected his "lady tasting tea" example.

I close with the statement (which will be unknown to most readers of this journal) that Basu and I are very deep friends. The argumentation and the comments I make in this article must be interpreted with that background.

DAVID A. LANE

The scientist's experimental results contain evidence bearing on the superiority of the improved diet. He asks the statistician to evaluate this evidence. The statistician answers by computing a significance probability, $(\frac{1}{2})^{15}$, by means of Fisher's randomization test. The scientist is baffled :

How can the evidence in his results be measured by a computation that ignores so much relevant information : the magnitude of the difference in weight gain between the two groups of animals, the ingredients of the two diets, previous experience with the standard diet, knowledge of the experimental animals gathered before and during the experiment, the mechanisms of the growth process, and so forth?

The statistician's computation refers to a biologically irrelevant feature of the experiment, the physical properties of the device determining the assignment of animals to diet; how can such a computation connect to the biological problem of the superiority of the improved diet?

The answer to the first question is clear : The statistician's significance probability cannot summarize completely the evidence in the scientist's experiment. The scientist cannot get something for nothing. If the scientist wants to assess what his experimental results imply about the effects of his improved diet and the nature of the growth process, he must analyze them in terms of a statistical model that describes as much as possible of what he knows about the biology of the experiment. But there are rhetorical as well as inferential issues involved in discussing an experiment. One of the scientist's goals is to obtain public confirmation for the superiority of the improved diet. If this can be accomplished with a minimum of fuss and assumption, preliminary to the detailed, model-based analysis, and without contradicting explicitly or implicitly the results of that analysis, so much the better. Here, the randomization test may be of use.

The randomization test addresses the question : Might the two diets really be equally effective and the apparent superiority of the improved diet be attributed to chance variability? The success of the test depends on the relevance of the interpretation it requires for the notions of "equally effective diets" and "chance variability." According to the Fisherian foundation of the test, two treatments can be considered equally effective only if they would each elicit exactly the same response from each experimental unit. The experimenter may, however, be interested in a weaker notion of equality between two treatments : Their distributions for the responses over the experimental units (or over all potential recipients of the treatments) should coincide. For example, a physician may not believe that each cancer patient faces the same prospect for a cure from radiotherapy as from chemotherapy, but he or she still might want to entertain the hypothesis that the overall success rates of the two treatments might be the same. The randomization test would be of no help to the physician.

The way in which "chance variability" enters into his experiment should be carefully explicated by the scientist when he constructs the statistical model he will use for analyzing his results. The randomization test ignores this model and substitutes an alternative relation between chance and the experiment, based on a frequency distribution induced by the physical act that assigns animals to diets. The logical foundation of this relation is challenged by the scientist's second question.

Basu presses this challenge home and denies Fisher's dictum that the physical act of randomization validates the randomization test. I find his argument convincing, and yet it seems to me that the significance probability of $(\frac{1}{2})^{15}$ can possess a rhetorical force that tells for the superiority of the improved diet, without reference to the distribution induced by the physical randomization. To explain this, I need to describe certain thoughts that the scientist might have about his experiment.

For each of his 30 animals, the scientist has ideas at the beginning of the experiment about what the animal would weight at the end, were it fed the standard diet. Although it undoubtedly implies more precision than the scientist could readily supply, think of these ideas as generating 30 standard-diet predictive distributions. One hypothesis about the diets that the scientist might entertain - although we know he does not believe it ! -is H_0: Each animal would end up weighing the same under the standard diet as it would under the improved diet. In particular, if H_0 were true, the 30 standard-diet predictive distributions also describe the scientist's ideas about what the animals would weigh if they were fed the improved diet. Now suppose the scientist has paired his 30 animals so successfully that the predictive distributions for the two animals in each pair coincide. Moreover, suppose also that there are no patterns of covariation among his animals such that, if H_0 were true and he knew the outcome of the experiment for some group of pairs, the scientist's conditional predictive distributions for the two animals in each of the remaining pairs would differ. Call this state of knowledge - or lack of it!- about the experimental animals null neutrality.

Under H_0 and null neutrality, the scientist's predictive probability that the 15 animals on the improved diet all end up heavier than their partners is $(\frac{1}{2})^{15}$. In fact, the joint predictive distributions under H_0 and null neutrality induce the uniform distribution on the set $S = \{(\pm 1, \ldots, \pm 1)\}$, where the ith coordinate of a point in S is + 1 if d_i is positive and -1 otherwise. So the usual null distribution of the sign test derives from the scientist's predictive distributions under H_0 and null neutrality, without regard to the method of assignment of animal to diet. The null distribution for the Fisher randomization test can also be derived, with somewhat more tedious assumptions about conditional predictive distributions, in terms of the scientist's prior beliefs about the experimental outcome under H_0. Since the scientist's real beliefs about the superiority of the improved diet imply predictive distributions weighted toward large positive values for the d_i's, small values of the significance probability from the randomization test indicate small posterior probability near H_0, if the scientist had assessed null neutrality and fully probabilized the problem—hence the rhetorical if not inferential force of the $(\frac{1}{2})^{15}$.

What about the assessment of null neutrality? It is to be regarded as a rough approximation at best; if the scientist is willing to think hard enough, he can of course recognize differeces between any pair of animals. Still, if null neutrality holds approximately, so do the conclusions that follow from assuming it, which serve only as guidelines anyway. In this regard, it is not much different from assuming normality in measurement situations. Yet, just as with normality, it is an assessment not to make lightly—and,as I shall argue later, the physical act of randomization can play a role in deciding whether the assessment is appropriate.

To seem whether this or the Fisherian interpretation of the statistician's significance probability provides the sounder guidance for the scientist, it is useful to consider some extreme cases. The issue should not be whether these cases occur in practice, but whether the logic that you claim to follow in practice guides you rightly or wrongly when pressed into extremity.

Example 1 (a variant of Basu's biased coin) : The scientist achieves a successful pairing - null neutrality seems reasonable. He generates 15 random numbers on the university computer, associates each of these numbers with a distinct pair of animals, and assigns the first animal (first, relative to a list of the animals' cage addresses) in each pair to the improved diet, if the pair's random number is even. It turns out that, in each pair, the animal that received the improved diet ends up heavier.

Just as the scientist is about to write up his results, however, the computer center informs him that because of a faulty program, only about 40 percent of the random digits the generator produced during the experiment were even.

Example 2 : Same story, but the scientist knew about the generator's quirk before he chose the numbers.

Example 3 : The scientist is not so lucky as in example 1, or perhaps he knows more about his experimental animals : In each pair, he can identify one animal that seems to have more growth potential than the other. This time, the random-number generator is working fine. Surprisingly, all the animals that the scientist judged to have higher growth potential get assigned to the improved diet. And they all end up heavier.

Basu carefully - and, as far as I can see, successfully - argues that Fisher's logic leads to a significance probability different from $(\frac{1}{2})^{15}$ for example 2. The same argument must apply to example 1, since Fisher's logic allows the scientist's knowledge of the randomizing mechanism to enter into the analysis only when he writes down a probabilistic model for it - and since this model attempts to represent the mechanism's physical properties, he must use whatever he knows when he analyzes the experiment, not what he thought he knew when he generated the random numbers. The significance probability derived from the Fisherian logic, as discussed by Basu, is singularly unattractive as a measure of evidence, depending as it does on an artifact of the method of listing cages.

Fisher's logic, tied to the random-number generator and imaginary repetitions, cannot fault the calculation of a significance probability of $(\frac{1}{2})^{15}$ in example 3. The design of the experiment may be at fault here, and the experiment itself quite uninformative scientifically, but this does not seem to stand in the way - in the Fisherian framework - of analyzing its evidential content by the $(\frac{1}{2})^{15}$ significance probability.

Interpreting the statistician's significance probability in terms of the scientist's predictive probability distribution changes this analysis completely. In example 1, the significance probability of $(\frac{1}{2})^{15}$ is unchanged by the computer center's information, since the probability refers to the scientist's thoughts at the commencement of the experiment, to which the information is irrelevant. At first sight, the same holds in example 2, since the probability does not refer to the method of assignment of animal to diet. But the scientist is interested in sharing his assessment of null neutrality : He wants his readers to feel the force of his argument, and so his assessment must be theirs. From this point of view, using a biased or arbitrary mode of assignment is to invite suspicion of loading the experiment in the scientist's favor - perhaps unconsciously, as in the famous Lanarkshire milk experiment ("Student" 1931). Randomly assigning animals to diets with public probability $\frac{1}{2}$ is a way of guaranteeing the honest - to the public and the scientist himself - of his subjective assessment that both animals in a pair had the same standard-diet predictive distribution.

In example 3, the scientist cannot assess null neutrality, and so the significance probability of $(\frac{1}{2})^{15}$ does not apply. Here, he can block his pairs according to his predictive distributions for the d_i's, to ensure as informative an experiment as possible. Again, he can employ one or more physical acts of randomization as a check and guarantee of his subjective assessments of these distributions. The experiment can of course be analyzed, but the null distribution for the randomization test will no longer follow Fisher's frequency distribution and will necessarily be somewhat less open to general agreement.

D. V. LINDLEY

What is one to do with this paper but applaud it ? Another incoherent procedure has its nature clearly displayed. Here is an encore that I have used in class to suggest that the randomization test does not "pass the test of common sense."

The example is artificial in that the experiment is very small, but this has the virtue of simplifying the arithmetic. The same principle holds for a larger and more realistic experiment at the expense of computations that might obscure the essential ideas. Two scientists are to conduct an experiment to compare a treatment, T, thought to improve the yield, with a control C. Four units are to be used, two each for T and C. The six possible assignments of T and C to the units are listed in the first column of the tabulation appearing two paragraphs after this one. The first scientist, A, decides to select one of the six designs at random. The second scientist, B, feels that the first and last designs would be unsatisfactory, because all the treatments and all the controls come together, and therefore selects a design at random from the four remaining. (In practice, as mentioned before, larger sets of designs would be used). Both A and B carry out their respective randomizations and both come up with the design TCTC, in the second row of the tabulation. On implementing the design, both scientists obtain the results 5, 4, 3, 2 shown in the final row of the tabulation. The total for the treated units is 8, that for the control

6, and the effect is measured by the difference, 2. So far the scientists agree, but now see what happens if they use the randomization argument for analysis.

Had the observed values arisen from any other of the designs that might have been used, the differences would have been those listed in the second column of the tabulation. Consider scientist B first. Scientist B excluded the first and last designs, and so the possible differences are (2, 0, 0, -2), of which the first, the one actually obtained, is the largest. Hence the result is significant at 25 percent, because all designs had the same 25 percent chance of being used. Scientist A, however, included the first and last designs in the randomization so must include the differences 4 and -4 that could have arisen by use of them. Of all six differences, 4 is the largest and 2, the one actually observed, the next largest. Hence the chance of the observed difference, or more extreme differences, is 2 out of 6 and the result is significant at $33\frac{1}{3}$ percent.

There, then, are two scientists who have performed exactly the same experiment, TCTC, obtained exactly the same result, and yet one is quoting a significance level substantially in excess of the other. And the reason for this difference in level is that A contemplated doing experiments that B did not (viz., those in the first and last rows of the tabulation), although, in fact, A did not perform one of these experiments. Expressed slightly differently, the analysis of the results of the experiment depended on what might have been done, but in fact was not done. Certainly in this context, in which the only probability ideas leading to the level are the equal probabilities involved in the random assignment, the argument seems unsatisfactory.

		Designs	Differences
A		TTCC	4
A	B	TCTC	2
A	B	TCCT	0
A	B	CTTC	0
A	B	CTCT	- 2
A		CCTT	- 4
Results		5432	

The whole concept of A and B reaching substantially different conclusions seems so absurd that the randomization-analysis argument has to be dismissed. There are two defenses : first, that in practice substantial differences (like 25 and $33\frac{1}{3}$ percent) are not observed and that the results are typically the same as normal theory. In that case, why not use normal theory? The second defense is that A and B ought to argue differently because B thought that the first and last experiments might be unsatisfactory, whereas A did not. In other words, both scientists had different ideas before the experiment; is it not reasonable that the two scientists should have different ideas afterwards? This argument violates the claim often made for significance tests - that they allow the data to speak for themselves and are not affected by considerations outside the data - and if admitted plays straight into the Bayesian camp, where the ideas of prior information are considered explicitly.

A minor comment is that it is perhaps a little unfair to say that there is "no satisfactory answer to the question : Why randomize?" The work of Rubin (1978) has at least made a substantial contribution to the answer. The answer for me is tied up with what we mean by random. (Basu's definition of randomization, in the first section of Section 2, is in terms of randomness.) I suggest that X is random, given H, if X is independent of any A, given H; that is, if $p(A|X, H) = p(A|H)$. The idea is that the generation of X, whether by a random mechanism, or by pseudorandom numbers, is unconnected with anything else. It is thus a subjective notion, in that what you consider random, I might not; though, in practice, we observe a lot of agreement among people. The value of randomization in design may then be illustrated by an experiment to test the efficacy of treatment T in aiding the recovery R of a patient. We require the probability of a patient's recovery were the patient to be given a treatment, $p(R|T, D)$, using data D from a planned experiment. This may differ from $p(R|T, D,A)$, where A is some factor unrecognized by us. (Had it been recognized it could have been planned for in the acquisition of D.) In order to make reasonably sure that our design does not confound the effects of T and A, we may assign treatments at random, that is, independent of A. This does not ensure lack of confounding but reduces its possibility to an acceptable level. Thus prerandomization has a place in coherent analysis : Basu shows that postrandomization is incoherent.

DONALD B. RUBIN

Basu's article on Fisher's randomization test for experimental data (FRTED) is certainly entertaining. Although much of the paper is devoted to the thesis that Fisher changed his views on FRTED, apparently the primary point of the paper is to argue that FRTED is "not logically viable." Admittedly, FRTED is not the ultimate statistical weapon, even in randomized experiments, but calling it illogical is rather bizarre.

Basu criticizes FRTED through two primary arguments. His first line of criticism follows from his attack on a nonparametric test labeled in Section 4 as "Fisher's randomization test." But this test was not proposed by Fisher and is not a logical variant of FRTED; consequently, these criticisms are not of FRTED. I believe that Basu agrees with this contention because in concluding this first criticism he states, "Where is the physical act of randomization in the Fisher randomization test? ... We should recognize the fact that in Section 21 of Design of Experiments (1935a) Fisher was not really concerned with the particular test situation that we have discussed in the previous section." Basu's second line of criticism of FRTED takes the form of a discussion between a statistician and a scientist; I find this discussion so confused that it is easier for me to challenge the argument indirectly by clearly describing FRTED than directly by correcting particular misconceptions.

In the paired comparison experiment, let Y_{ij} be the response of the ith unit ($i = 1, \ldots, 2n$) if exposed to treatment $j (j = 1,2)$, where $Y = \{Y_{ij}\}$ is the $2n \times 2$ matrix of values of Y_{ij}. The assumption that such a representation is adequate may be called the stable unit treatment value assumption : If unit i is exposed to treatment j, the observed value of Y will be Y_{ij}, that is, there is no interference between units (Cox 1958b, p. 19) leading to different outcomes depending on the treatments other units received and there are no versions of treatments leading to "technical errors" (Neyman 1935b). If Y were entirely observed, we could simply calculate the effect of the treatments for these 2n units; for example, $Y_{i1} - Y_{i2}$ would be an obvious measure of the effect of treatment 1 versus treatment 2 for the ith unit, and the average value of $Y_{i1} - Y_{i2}$ would be a common measure of the typical effect of treatment 1 versus treatment 2 for these 2n units. Because each unit can be exposed to only one treatment, we cannot observe both Y_{i1} and Y_{i2}, and so we will have to draw inferences about the unknown values of Y from observed values of Y.

Let $T = (T_1, \ldots, T_{2n})$ be the indicator for treatment received : $T_i = 1$ if the ith unit received treatment 1 and $T_i = 2$ if the ith unit received treatment 2; if $T_i = 1$, Y_{i1} is observed and Y_{i2} is missing, whereas if $T_i = 2$, Y_{i2} is observed and Y_{i1} is missing. In order to avoid confusion about the inferential content of indices, suppose that the unit indices i are simply a random permutation of (1, ..., 2n). The pairing of the units in the paired comparison experiment will be represented by X, where $X_i = 1$ for the two units in the first pair, ..., and $X_i = n$ for the two units in the nth pair. Other characteristics of units can be coded in other variables, but for simplicity assume for now that only values of Y, X, and T will be used for drawing inferences, where Y is partially observed and both X and T are fully observed.

Fisher Randomization Test

Both randomization and Bayesian inferences for unobserved Y values require a specification for the conditional distribution of T given (Y, X), say Pr(T | Y, X). The physical act of randomization in the experiment (e.g., the physical act of haphazardly pointing to a starting place in a table of random numbers) is designed to ensure that all scientists will accept the specification Pr(T | Y, X) = Pr(T | X). In the paired comparison experiment,

$$\Pr(T|X) = \begin{cases} 0 & \text{if } T_i = T_j \text{ for any } i \neq j \text{ s.t. } X_i = X_j \\ 2^{-n} & \text{otherwise.} \end{cases} \qquad (1)$$

If treatments are assigned using characteristics Z of the units that are correlated with Y (the scientist's confessed experiment at the end of Sec.5), then Pr(T | Y,X) = Pr(T | X) would generally not be acceptable. For example, if treatment assignments are determined by tossing biased coins where the bias favors the first unit in each pair receiving treatment 1 (Z = order of unit in pair), then whether Pr(T | Y,X) = Pr(T | X) is generally acceptable depends on the scientific view of the partial correlation between Z and Y given X; if the order "does not seem to have much relevance," then Pr(T | X,Y) = Pr(T | X) may be plausible with (1) as the accepted specification for Pr(T | Y, X). Of course, even if unit order is randomly assigned within pairs, one could decide to record its values and use Pr(T | X,Z) to draw inferences; this is analogous to recording the random numbers used to assign treatments and observing that given them no randomization took place (i.e., Pr(T | X, Z) = 1 for one value of T and 0 for all other values of T). In order to make sensible use of FRTED, we cannot condition on numbers accepted a priori to be unrelated to Y.

Suppose that we wish to consider the hypothesis H_0 that $Y_{i1} = Y_{i2}$ for all i,or any other sharp null hypothesis such that given H_0 and the observed values in Y, all values of Y are known. Under H_0 and accepting specification (1), the difference in observed averages $\bar{y}_d = \Sigma\, Y_{i1}(2 - T_i)/n - \Sigma\, Y_{i2}(T_i - 1)/n$, or any other statistic, has a conditional distribution given Y and X consisting of 2^n equally likely known values. Because the expectation of $\bar{y}_d$ over this distribution is zero, values of $\bar{y}_d$ far from zero are a priori considered to be more extreme than values near zero. The proportion of possible values as extreme or more extreme than the observed value of $\bar{y}_d$, that is, the significance level of FRTED is not a property solely of the data and the null hypothesis but also of the statistic and the definition of extremeness of the statistic. If the observed value of $\bar{y}_d$ is extreme (e.g., if the significance level is less than 1 in 20), then we must believe that

1. H_0 is false with the result that the treatments have an effect; or

2. Pr(T | Y, X) = Pr(T | X) is false with the result that the 2^n values of $\bar{y}_d$ are not a priori equally likely; or

3. An a priori unusual (extreme) event took place.

The physical act of randomization is designed to rule out option 2 and consequently leave us believing either that an a priori unusual event has taken place or that H_0 is false.

I see nothing illogical about the FRTED; it is relevant for those rare situations when a purely confirmatory test of an a priori sharp hypothesis is to be made using an a priori defined statistic having an associated a priori definition of extremeness. On this point, I find myself in total agreement with the following statement of Brillinger, Jones, and Tukey (1978, p. F-1):

> If we are content to ask about the simplest null hypothesis, that our treatment ("seeding") has absolutely no effect in any instance, then the randomization, that must form part of our design, provides the justification for a randomization analysis of our observed result. We need only choose a measure of extremeness of result, and learn enough about the distribution of this result
>
> - for the observed results held fixed
> - for re-randomizations varying as is permitted by the specification of the designed process of randomization.
>
> If p% of the values obtained by calculating as if a random re-randomization had been made are more extreme than (or equally extreme as) the value associated with the actual randomization, then p% is an appropriate measure of the unlikeliness of the actual result.
>
> Under this very tight hypothesis, this calculation is obviously logically sound.

Of course, there are limitations of FRTED of which Fisher was well aware. For example, the null hypothesis that $Y_{i1} = Y_{i2}$ for all i may not be very realistic; when Neyman (1935b) criticized the FRTED for Latin Squares, Fisher (1935a) replied:

> [The null hypothesis that "the treatments were wholly without effect"] may be foolish, but that is what the Z-test [FRTED] was designed for, and the only purpose for which it has been used ... Dr. Neyman thinks that another test would be more important [one for the average treatment effect being zero]. I am not going to argue that point. It may be that the question which Dr. Neyman thinks should be answered is more important than the one I have proposed and attempted to answer ... I hope he will invent a test of significance, and a method of experimentation, which will be as accurate for questions he considers to be important as the Latin Square is for the purpose for which it was designed.

More complicated questions, such as those arising from the need to adjust for covariates brought to attention after the conduct of the experiment, simultaneously estimate many effects, or generalize results to other units, require statistical tools more flexible than FRTED. Such tools are essentially based on a specification for Pr(Y|Z), where now Y refers to outcome variables in general, X refers to blocking and design variables, and Z refers to covariates. Fisher (1935a) was certainly willing to specify particular distributional forms for data in experiments, and I believe that he was simply advocating such an attack whenever justified in his "astonishing short section on nonparametric tests in the seventh edition of DE." This desire to condition on all relevant information is obviously very Bayesian.

I believe (Rubin 1978) that Bayesian thinking, which requires specifications for both Pr(T |Y, X,Z) and Pr(Y |X, Z) and draws inferences conditional on all observed values, provides, in principle, the most effective framework for inference about causal effects. Other statisticians view the specification Pr(Y|X, Z) as something to be avoided in principle: "For crucial comparisons ... the appropriate role for the classical kind of parametric analysis would seem to be confined to assistance in the selection of the test statistics to be used ... in a randomization analysis" (Brillinger, Jones, and Tukey 1978, p. F-5). Using the test statistic (in conjunction with the null hypothesis and definition of extremeness) to summarise all scientific knowledge relevant for data analysis seems to be unduly restrictive. Although much care is needed in applying Bayesian principles because of the sensitivity of inference to the specification Pr(Y |X, Z), the increased flexibility and directness of the resulting inferences make the Bayesian approach scientifically more satisfying.

On this point, perhaps Basu and I are actually in substantial agreement. FRTED cannot adequately handle the full variety of real data problems that practicing statisticians face when drawing causal inferences, and for this reason it might be illogical to try to rely solely on it in practice.

Rejoinder

D. BASU

Let me begin by thanking Hinkley, Lane, Lindley, Rubin, and my good friend Kemp for their many interesting comments. I also offer my apologies to them for my inability, because of an eye condition needing surgical treatment, to read the discussions for myself. They were read out to me, and so I may have missed out on some of the many issues raised. I thank Carlos Pereira for his help in putting together this reply.

Rubin wonders about the relevance of the material discussed in Section 4 . Let me explain why I challenged the Fisher nonparametric test - the first nonparametric test by many years, as Fisher (DE 1960) put it. The logic of the test is essentially the same as that of the paired-comparison test discussed in Section 6. Both are conditional tests of a very extreme kind. In the nonparametric test, the statistic $(|x_1|, |x_2|, \ldots, |x_n|)$ is held fixed; the δ_i's define the reference set. In the randomization test of Section 6, everything but the design outcome is held fixed. Kempthorne and Folks (1971) labeled the nonparametric test as the Fisher randomization test even though, as I explained at the end of Section 5, the δ_i's cannot really be likened to a set of randomization variables. (Kemp disputes this, but then he disputes almost everything I said.) Each of my difficulties with the nonparametric test also persists with the randomization test. For instance, why must we choose $\bar{x}$ (in Sec. 6, d) as the test criterion and not the median $\tilde{x}$? With $n = 7$ and each $x_i > 0$, the significance level (SL) works out as 1/128 with $\bar{x}$ as the criterion and as 1/16 with $\tilde{x}$ as the criterion. Neither Kemp or Hinkley answers my question. At one place Kemp mumbles about the central limit theorem, but that is hardly relevant for my sample size. Hinkley makes the curious suggestion that the choice of the test criterion is not a statistical problem. How to justify holding $|x_1|, |x_2|, \ldots, |x_n|$ fixed in the nonparametric test ? Why not hold $|\bar{x}|$ fixed instead ? In the latter case, the SL is either $\frac{1}{2}$ or 1. In Section 6, when the scientist admitted that he had made a one-toss restricted randomization, the statistician declared the experiment to be uniformative because, for every possible outcome of the experiment, the SL is either $\frac{1}{2}$ or 1. Kemp agrees with the statistician. But Kemp, why? Should we not treat such value-loaded terms like significant or informative with greater respect?

When I said that the Fisher randomization test is not logically viable - Rubin calls the characterization "bizarre" and Kemp, in classical debating style, queries my system of logic - I only meant that the logic of the test procedure is not viable. How else can you characterize a test procedure that falls to pieces when confronted with the slightly altered circumstances of a restricted or unequal probability randomization? I am happy to note that Lane and Lindley agree with me on this point.

My working definition of a Bayesian fellow traveler is one who has trouble in understanding a P value as the level of significance attained by the particular data. Rubin, who claims to be a Bayesian, seems to be quite at home with significance testing. George Box is another notable exception to my working definition.

Let us try to make some sense - please Kemp, do not ask me to define sense - of the P value of 2^{-15} in Section 6. Suppose each of the 15 subject pairs is indistinguishable to the scientist. Also suppose that the scientist believes that there is no treatment difference. No doubt then the scientist will be surprised if, at the end of the experiment, he finds that each of the 15 treated subjects gains more weight than the corresponding control subjects. The SL of 2^{-15} may be regarded as a measure of this element of surprise. It is a probability (measure of doubt) that existed in the mind of the scientist before the experiment and under the assumed circumstances. As Lane observes, this probability does not depend on the nature of prerandomization. But Kemp, the frequentist, refuses to interpret the SL in terms of such nonexistent belief probabilities.

If the scientist cannot truly distinguish between the subjects in each block, then "Nature has done the randomization for us," says Hinkley, and so he cannot understand the point in all the fuss that I am making. But our scientist, like most scientists, can distinguish between the subjects in each block - one subject is heavier, the other one is older and so on. Mother Nature is asking for a helping hand, and so the scientist must randomize! But the scientist can still distinguish between the subjects in each pair. How can we evaluate his surprise index? So we very sternly tell the scientist, "Randomize and close your eyes!" The scientist randomizes, closes his eyes, but still refuses to be greatly surprised in the end. Because, he says, he knew all along that the improved diet is superior to the standard diet. At this point Kemp will perhaps say, "I am surprised that you can write so much on surprise without even defining the term."

Many of my esteemed colleagues believe that postrandomization is a useful statistical device. I know my friend Kemp well enough to say that he is not one among them. He agrees with Fisher, Lindley, and me that postrandomization has no place in scientific thinking. But, today, fighting for every inch of the ground, Kemp is trying to prove me wrong even on this issue. Perhaps one can play better poker by wearing a mask, making hand signals instead of using one's vocal chords, and carrying a randomizer hidden in one's pocket. But does Kemp really think that our scientist is engaged in something like a poker game against Mother Nature? Why does he not advise the scientist also to wear a mask?!

I have no objection to prerandomization as such. Indeed, I think that the scientist ought to prerandomize and have the physical act of randomization properly witnessed and notarized. In this crooked world, how else can he avoid the charge of doctoring his own data? In order to make the device a superior cosmetic agent it may be necessary to make the extent of prerandomization sufficiently wide. In Basu(1978b) I have mentioned a few noncosmetic uses of the prerandomization device.

Lindley agrees wholeheartedly with my criticisms of the Fisher randomization test. But, disagreeing with me on what he calls a "minor point," he suggests that there may be a place for randomization in a subjective Bayesian theory of statistics. All I know is that L.J. Savage had similar thoughts but he never spelt them out for us. I may have something to say on the Rubin (1978) thesis on another occasion.

Hinkley and Rubin quote from the prefiducial Fisher to dispute me on the randomization test. In the thirties, Fisher knew that the unrestricted, equal probability randomization test closely parallels the traditional test based on the Gaussian law. So Lindley is asking, "Why not use normal theory?" I remember having seen a Fisher quotation (from the prefiducial time) saying that the randomization test provides a logical justification for the parametric tests based on the normal theory. So Kemp is asking us to discard the normal theory and use the randomization logic instead. In Section 21(a) of DE (1960), we find Fisher summarily discarding the Kempthorne thesis on experimental designs. Kemp says that no useful purpose can be served by trying to "psychoanalyze" the mind of Fisher. But what purpose does it serve to dismiss much of Fisher's later writings as mere polemics?

I cannot understand what Hinkley is trying to communicate with his comments on the ancillarity of the design outcome. Is it "plain foolish" to regard the design outcome as an experimental constant? Since there are only a finite number of design outcomes, how can one get an "infinitesimal slice" of the sample space by holding the ancillary statistic fixed? As I pointed out, it is the randomization-test argument that rests on an infinitesimal slice of the sample space by holding fixed everything but the design outcome. The Bayesian recommendation is to hold the data fixed and to speculate about the still-variable parameters. When you push the Fisher conditionality argument to the limit, you become a Bayesian.

On the ancillarity issue, Kemp adopts the proverbial Chinese philosophy of seeing no evil. He is in effect saying, "How can there be an ancillary statistic when there is no probabilistic statistical model and, therefore, no parameters?" I have no difficulty in recognizing the 60 parameters $\omega = \{(x_i, y_i) : i = 1,2 \ldots, 30\}$ in the scientist's diet problem - x_i and y_i are, respectively, the would-be treatment and control responses of subject i at the planning stage of the experiment. Let us suppose that the scientist's parameter of interest is $\theta = \bar{x} - \bar{y}$. Consistent with his prior opinion ξ on ω, the scientist has a prior opinion η on θ. After the experiment, the scientist, having observed 15 of the x_i's, and the complementary set of 15 y_i's, must have drastically revised his prior opinion ξ to a new opinion ξ^*. Consistent with ξ^*, the scientist has then an opinion η^* on the parameter of interest θ.

According to DeFinetti, probability, like beauty, exists only in the mind; it is a formal representation of opinion on parameters. The subjective Bayesian thesis on statistics deals with the process of opinion changes in the very limited context of what we may call statistical parameters. The Bayesian thesis appears to me to be coherent and pertinent to the real issues of scientific inference. That the Bayesian paradigm is useful is slowly gaining recognition. Fuller recognition will take time. But by then it will perhaps be time for us to move on to a more useful paradigm.

When it comes to changing one's opinion on a scientific paradigm, the mind of a stubborn scientist - for that matter, the minds of a whole community of trained scientists - certainly does not, perhaps cannot, follow any logic. In his Scientific Autobiography and Other Papers (1949, pp. 33-34). Max Planck wrote, "A new scientific truth does not triumph by convincing the opponents and making them see the light, but rather because its opponents eventually die, and a new generation grows up that is familiar with it." It rarely happens that Saul becomes Paul.

PART III

MISCELLANEOUS NOTES AND DISCUSSIONS

CHAPTER XVII

LIKELIHOOD AND PARTIAL LIKELIHOOD

1. Introduction

During the fifty years (1912-1962) that R.A. Fisher dominated the field of statistical research, he came out with many innovative ideas like likelihood, sufficiency, ancillarity, asymptotic efficiency, information and intrinsic accuracy, pivotal quantities and fiducial distribution, conditionality argument and recovery of ancillary information, analysis of variance and covariance, randomization analysis of experimental data, etc. Of these new concepts, likelihood is certainly the first and the foremost and perhaps the only one that is likely to endure the severe test of time. Even though likelihood plays a central role in current statistical theory, a great deal of controversy and confusion surround the usage of the notion. This article, a natural sequel to an earlier long essay (Basu, 1973) on the subject, is an elaboration of some aspects of the controversy.

Let θ be the parameter of interest and x the observed data (sample). In the absence of troublesome nuisance parameters, the likelihood function $L(\theta) = p(x|\theta)$ is the conditional probability of the particular data x, given θ . [In the nondiscrete case, p is the conditional probability density with respect to a σ-finite dominating measure.] In 1912, when Fisher rediscovered the Gaussian notion of likelihood and the maximum likelihood (ML) estimate $\hat{\theta}$, he had in mind the one and only observed data x and, therefore, only one likelihood L, and the corresponding ML estimate $\hat{\theta}$. He looked upon L as a scale of comparative support lent by the data x to various possible values of the unknown θ. A constant (θ-free) factor could therefore be freely discarded from the likelihood function or added on to it. The ML estimate $\hat{\theta}$ is the value of θ that is best supported by the data. In the twenties, however, Fisher (1922, 1925) took the now familiar viewpoint that the sample x is a random variable and began to investigate the sampling properties of $\hat{\theta} = \hat{\theta}(x)$. The likelihood L became a quantity — a function of both θ and x. The celebrated Fisher quantity $\frac{\partial}{\partial\theta}\log L$ was introduced and information was defined as

$$I(\theta) = V_\theta(\frac{\partial}{\partial\theta}\log L) = E_\theta(-\frac{\partial^2}{\partial\theta^2}\log L).$$

It was demonstrated that when x consists of a large number of i.i.d. components, the ML estimator $\hat{\theta}$ is distributed (conditionally, given θ) as $N(\theta, 1/I(\theta))$. This celebrated

Editor's Note : This chapter is based on an invited article presented by Basu at an International Symposium on Probability and Statistics at the University of Sao Paulo, Sao Paulo, Brazil in the summer of 1982 and published in the proceedings of the Symposium as "A Note on Likelihood".

Fisher theorem is one of the four cornerstones of the sample space theory (also called Neyman-Pearson theory) of statistics; the other three cornerstones being Karl Pearson's Chi-squared goodness of fit test, Student's t-test, and Neyman-Pearson's confidence statements. Since the quality of $\hat{\theta}$ as an estimator of θ depends on $I(\theta)$, which is an average characteristic of $L(\theta)$ for varying x, it became axiomatic to many of our contemporary statisticians that the unique likelihood scale determined by the observed sample x cannot be the only basis for judgement making about the unknown parameter θ. The Neyman-Pearson thesis on inference (decision) making may be briefly stated as : "The data x can be evaluated only in the context of a well defined sample space $\mathcal{X}$ and an agreed to model (a family of probability distributions on the sample space) involving the parameter θ. A decision rule with acceptable average performance characteristics needs to be worked out relative to the model and then applied to the particular data x". An antithesis to the Neyman-Pearson thesis is what George Barnard, influenced by the later writings of R.A. Fisher, called the Likelihood Principle. The principle asserts : "If the data analysis objective is not to check on the correctness of the assumed model but to make judgements (relative to the model) about the parameter θ, then all the information contained in the data is fully summarized in the particular likelihood function $L(\theta)$". The principle has been discussed in great length in Basu (1973) and will be the basis of the subsequent discussion in this article. We first consider the case where θ is one-dimensional and no nuisance parameters are present.

2. Likelihood and Information

Information is what information does. It changes opinion. Only a Bayesian knows how to characterize his/her prior opinion on θ as a prior distribution $q(\theta)$. This prior opinion is changed, by the data x, to the posterior opinion

$$q^*(\theta) = q(\theta)L(\theta)/\Sigma q(\theta)L(\theta)$$

Since the likelihood changes the prior to the posterior, it is all the information that the data supply. The likelihood principle is therefore, a self-evident principle from the Bayesian viewpoint.

It $L(\theta)$ is flat over the whole parameter range, that is, if the conditional probability of the particular sample does not depend on θ, then $q^* = q$ irrespective of what q is. The data, unable to change any opinion on θ, must be regarded as fully noninformative in this case. On the other hand, if the likelihood curve is unimodal and the total mass (area) under it is tightly packed around the ML estimate $\hat{\theta}$, then the data is highly informative in a sense. This is the kind of situation that we usually come across when the data x consist of a large number of i.i.d. components.

With a likelihood $L(\theta)$ that is highly informative in the above sense, write $\Lambda(\theta) = \log L(\theta)$ and consider the Taylor expansion

$$\Lambda(\theta) = \Lambda(\hat{\theta}) + (\theta-\hat{\theta})\Lambda'(\hat{\theta}) + \frac{1}{2}(\theta-\hat{\theta})^2 \Lambda''(\hat{\theta}) + \ldots$$

Ignoring terms of higher order than the second in the above expansion and noting that $\Lambda'(\hat{\theta}) = 0$, we have

$$\Lambda(\theta) \doteq \Lambda(\hat{\theta}) - \frac{1}{2}(\theta-\hat{\theta})^2 J(\hat{\theta}),$$

where $J(\theta) = -\Lambda''(\theta)$ and $\doteq$ means approximate equality. That is,

$$L(\theta) \doteq L(\hat{\theta})\, e^{-\frac{1}{2}(\theta-\hat{\theta})^2 J(\hat{\theta})}.$$

The factor $L(\hat{\theta})$, being θ-free, may be ignored. Clearly, it is $J(\hat{\theta})$ that determines the approximate shape and the quality of the likelihood function. If $J(\hat{\theta})$ is large and the prior opinion $q(\theta)$ relatively flat around $\hat{\theta}$, then the posterior distribution $q^*(\theta)$ will (approximately) be $N(\hat{\theta}, I/J(\hat{\theta}))$. In this situation, the two statistics $\hat{\theta}$ and $J(\hat{\theta})$ may be regarded as (approximately) summarizing all the information supplied by the data. Furthermore, the statistic $J(\hat{\theta})$ may be regarded as a numerical measure of the quality of information supplied by the particular data.

It is not clear what steps Fisher actually took to arrive at his information function $I(\theta)$. Perhaps he tried to find the average value $E_\theta[J(\hat{\theta})]$ of $J(\hat{\theta})$ over all possible samples and then simplified the problem to

$$I(\theta) = E_\theta[J(\theta)].$$

Perhaps he then noted the alternative representation of $I(\theta)$ as the variance of the score statistic (quantity) $\Lambda'(\theta)$,

$$I(\theta) = V_\theta[\Lambda'(\theta)] = E_\theta[\Lambda'(\theta)]^2,$$

which is how the Fisher information is usually defined in textbooks.

In the i.i.d. case, with $x = (x_1, x_2, \ldots, x_n)$, the sample size enters as a factor of $J(\theta)$ and, therefore, of $I(\theta)$. Under suitable regularity conditions, Fisher's asymptotic sampling distribution result

$$\sqrt{I(\theta)}\,(\hat{\theta}-\theta) \stackrel{a}{\sim} N(0, 1) \tag{2.1}$$

is true but cannot be used directly because $I(\theta)$, being a function of the parameter θ, is unknown. With $I(\theta)$ estimated by $I(\hat{\theta})$ the asymptotic result

$$\sqrt{I(\hat{\theta})}\,(\hat{\theta}-\theta) \overset{\rightarrow}{\sim} N(0,\ 1) \tag{2.2}$$

is also true. It is (2.2), rather than (2.1), that is the basic asymptotic theorem that justifies the method of maximum likelihood.

Since $I(\theta)$ and, therefore, $I(\hat{\theta})$ cannot be obtained from the particular likelihood function $L(\theta)$, any inference making usage of the asymptotic result (2.2) will violate the likelihood principle. As explained earlier, it is the asymptotic result

$$\sqrt{J(\hat{\theta})}\,(\theta-\hat{\theta}) \overset{\rightarrow}{\sim} N(0,\ 1)\ , \tag{2.3}$$

where θ is regarded as (approximately) normally distributed with mean $\hat{\theta}$ and variance $1/J(\hat{\theta})$, that has a natural Bayesian interpretation under suitable conditions. If the two statistics $I(\hat{\theta})$ and $J(\hat{\theta})$ happen to be equal then the two asymptotic results (2.2) and (2.3), though quite different in theory, may be regarded as indistinguishable in their common usage. Strange as it may seem, $I(\hat{\theta}) = J(\hat{\theta})$ in many familiar situations. The following result explains why.

Proposition : If the statistical model belongs to an exponential family of order one, then $I(\hat{\theta}) = J(\hat{\theta})$ for all x.

Proof: Since $L(\theta) = \exp[a(x) + b(\theta) + c(\theta)T(x)]$, we have $\Lambda'(\theta) = b'(\theta)+c'(\theta)T$, $J(\theta) = -\Lambda''(\theta) = -b''(\theta) - c''(\theta)T$ and $I(\theta) = E_\theta[J(\theta)] = -b''(\theta) - c''(\theta)\tau(\theta)$, where $\tau(\theta) = E_\theta(T)$. Since $\hat{\theta}$ is the ML estimate, we have

$$b'(\hat{\theta}) + c'(\hat{\theta})T = 0 \quad \text{for all } x \tag{2.4}$$

Since $E_\theta[\Lambda'(\theta)] = 0$ for all θ we have

$$b'(\hat{\theta}) + c'(\hat{\theta})\,\tau(\hat{\theta}) = 0 \quad \text{for all } x. \tag{2.5}$$

Since $c(\theta)$ has to be strictly monotone in θ, $c'(\hat{\theta}) \neq 0$ and, therefore, $T(x) = \tau(\hat{\theta})$ for all x. Hence, $I(\hat{\theta}) = J(\hat{\theta})$ for all x.

That $I(\hat{\theta})$ can be quite different from $J(\hat{\theta})$, is seen from the following example where the model belongs to an exponential family of order two.

Example : Let $x = (x_1, x_2, \ldots, x_n)$ be n i.i.d. variables with common distribution $N(\theta, \theta^2)$, where $\theta > 0$. In this case it can be easily verified that

$\Lambda'(\theta) = n[-\frac{1}{\theta} - \frac{\bar{x}}{\theta^2} + \frac{S^2}{\theta^3}]$, where $\bar{x}$ is the sample mean and $S^2 = \Sigma x_i^2/n = \bar{x}^2 + s^2$. The ML estimate is $\hat{\theta} = (\sqrt{5\bar{x}^2 + 4s^2} - \bar{x})/2$.

It is easy to verify that

$$J(\theta) = -\Lambda''(\theta) = n\left[-\frac{1}{\theta^2} - \frac{2\bar{x}}{\theta^3} + \frac{3s^2}{\theta^4} \right],$$

$$J(\hat{\theta}) = \frac{n}{\hat{\theta}^2}\left[\frac{\bar{x}}{\hat{\theta}} + 2 \right]$$

and

$$I(\theta) = E_\theta[J(\theta)] = \frac{3n}{\theta^2} .$$

Therefore, $I(\hat{\theta}) = J(\hat{\theta})$ if and only if $\bar{x} = \hat{\theta}$. Note that $\bar{x} = \hat{\theta}$ if and only if $\bar{x} = s$.

The Fisher information $I(\theta)$ and its estimate $I(\hat{\theta})$ are of no relevance at the data interpretation stage. Even the role of $J(\hat{\theta})$ is questionable in these days of high speed computers when we can plot the likelihood curve even for very complex non i.i.d. models.

3. Partial Likelihood

Consider now the common situation where the model and, therefore, the likelihood is polluted by a set of nuisance or incidental parameters $\phi = (\phi_1, \phi_2, \dots, \phi_k)$. A vast literature has grown around the problem of eliminating the nuisance parameters. For one who is accustomed to analyzing the data in some sample space terms, the multiplicity of methods proposed and notions discussed are indeed very perplexing. In contrast, the Bayesian recipe is pretty well defined and straightforward at least in theory. A Bayesian, concerned only with the task of analyzing the information supplied by the particular likelihood function $L(\theta,\phi)$ that corresponds to the observed data, has only to get into a spell of introspection to arrive at his/her prior distribution

$$q(\theta, \phi) = q(\theta)q(\phi|\theta)$$

of (θ, ϕ) and then to compute the posterior distribution

$$q^*(\theta, \phi) = c\, L(\theta, \phi)\, q(\theta, \phi),$$

where c is the normalizing constant. The marginal posterior distribution $q^*(\theta)$ of θ is then obtained by integrating out ϕ from $q^*(\theta, \phi)$. Clearly, $q^*(\theta)$ is the normalized version of $L_e(\theta)\, q(\theta)$, where

$$L_e(\theta) = \sum_\phi L(\theta, \phi)\, q(\phi|\theta).$$

The subscript e in $L_e(\theta)$ denotes the fact that L_e is obtained from the full likelihood L after elimination of the nuisance parameter.

The task of figuring out one's conditional prior opinion $q(\phi \mid \theta)$ on ϕ for each given value of θ is not an easy one. With a nuisance parameter ϕ of high dimension, the Bayesian has to look for a simpler way to arrive at an eliminated likelihood $L_e(\theta)$.

Two traditional elimination methods that originated in the writings of R.A. Fisher are those of marginal and conditional likelihoods. Let $y = y(x)$ be a statistic, a part of the full data x, and let z be the remainder part. The full likelihood $L(\theta, \phi)$ is the (conditional) probability $p(x|\theta,\phi)$ of the total experience x. Suppose it so happens that when the likelihood is factored as

$$p(x|\theta, \phi) = p(y|\theta, \phi)\, p(z|y, \theta, \phi) \tag{3.1}$$

the first factor $p(y \mid \theta, \phi)$ is found to be ϕ-free. If the scientist suppresses z from the data and reports only y as the sample experience, then the likelihood is said to be marginalized to $L_e(\theta) = p(y \mid \theta, \phi)$. On the other hand, if it is the second factor in (3.1) that is ϕ-free, then the scientist may adopt the conditioning method of eliminating the nuisance parameter by simply regarding y as just another of the many experimental constants and putting forward z as the only sample experience. The likelihood then comes in the ϕ-free form $L_e(\theta) = p(z|y, \theta, \phi)$.

Before we dismiss the marginal and conditional likelihood methods as statistical heresies, let us ponder for a moment on what constitutes scientific data. A scientist, interested in a parameter θ, is reporting on some experience that he/she gained through observations and experimentations related to the parameter. Of course the scientist has to discard a substantial portion of his/her total experience as irrelevant to (non-evidential on) the problem at hand. Suppose $z_1, z_2, \ldots, z_m$ are the different components of the scientist's relevant data presented in their chronological order of observation. The total experience of the scientist may be represented as

$$x = (y_1, z_1, y_2, z_2, \ldots, y_m, z_m, y_{m+1}),$$

where, y_1 is all the experience that the scientist had just prior to the experience z_1, and similarly for y_2, y_3, etc. Were it possible to set up a statistical model for x in terms of the parameter of interest θ and a set of nuisance parameters, then the full likelihood would have been

$$p(x|\theta, \phi) = p(y_1|\theta,\phi)\, p(z_1|y_1, \theta,\phi)\, p(y_2|y_1, z_1, \theta,\phi) \tag{3.2}$$

When the scientist is presenting $z_1, z_2, \ldots, z_m$ as the experimental or observational data, it is clear that he is regarding $y_1, y_2, \ldots, y_m$ as experimental constants and y_{m+1} (the experience after z_m) as irrelevant. The likelihood that the scientist reports is

$$L(\theta,\phi)= p(z_1|y_1, \theta,\phi)\ p(z_2|y_1,z_1,y_2, \theta,\phi)\ p(z_3| \ldots) \tag{3.3}$$

The likelihood (3.3) is obtained by omitting from (3.2) all the factors of the form $p(y_i | \ldots)$. If it so happens that the factors of the type $p(z_i | \ldots)$ are all ϕ-free, then the likelihood (3.3) involves only the parameter of interest θ.

D.R. Cox (1975) called a likelihood of the type (3.3) a partial likelihood. Clearly, marginal and conditional likelihoods are particular variants of partial likelihood. What we are saying here is that, in a sense, all likelihoods are partial likelihoods. When a scientist reports the outcome of a sequence of Bernoulli trials as a sequence of successes and failures and then calculates the likelihood as $\theta^r(1-\theta)^{n-r}$, we know for sure that the scientist has omitted from the report many ancillary experiences that he/she gathered before, after and in between the trials.

The partial likelihood method for elimination of nuisance parameters may be formally stated as follows. Represent the sample x as a sequence y_1, z_1, y_2, $\ldots$, y_m, z_m, y_{m+1} of 2m+1 statistics — the number m may depend on x — in such a manner that, when the full likelihood $p(x | \theta, \phi)$ is factored in the manner of (3.2), all factors of the form $p(z_i | \ldots)$ are ϕ-free and none of the other factors are. The product of the ϕ-free factors of the full likelihood is the partial likelihood $L_e(\theta)$ for the parameter of interest θ.

In the absence of a clearcut prescription for decomposing the sample into the y and the z bits, the above can hardly be called a definition of partial likelihood. One sample may be decomposed in more than one way and another may not be decomposible. Since the number m of z-bits varies from sample to sample, it is hard to visualize a particular z_i as a measurable function on a sample space. For one who is used to interpreting the likelihood function in some sample space terms, there is the added discomfiture that, in general, the partial likelihood cannot be interpreted as a probability function on a sample space. As a rule, the partial likelihood $L_e(\theta)$ is different from the marginal and the conditional probability function of $(z_1, z_2, \ldots, z_m)$.

It is possible to derive a score function $\frac{\partial}{\partial\theta} \log L_e$ from the partial likelihood and then indulge in some Fisher asymptotics, but the relevance of such mathematics is hard to fathom. As we have said before, the question, "How informative is the partial likelihood?", cannot be answered by deriving some average sample space characteristic of the likelihood. It is the shape of the particular likelihood curve that ought to provide an answer to the question.

Just as in the case of the prior, the only thing that is objective about the likelihood is that it is a tool fashioned by the mind of a scientist that is focused on the object of enquiry, the parameter. Like any other tool, a likelihood can be faulted and can also be misused.

CHAPTER XVIII

A DISCUSSION ON THE FISHER EXACT TEST

I begin with a note of dissent. The Fisher-Yates conditional test for the 2×2 categorical data is called "exact" not because the test is "based on the theories of R.A. Fisher" but because the computation of the attained level of significance (the P-value) requires no mathematical approximation beyond what is already involved in the choice of the statistical model. The normal test is inexact because the null distribution of the test statistic T_N is N(0,1) only as an approximation.

Let us limit the discussion to the following data (the outcomes of two independent binomial experiments) as reported in Fisher (1956 p. 86):

	Died	Survived	Total
Treated	3	0	3
Control	0	3	3
Total	3	3	6

In Professor Berkson's notations, we have $n_1 = n_2 = 3$ and $(a,c) = (3,0)$. The marginal totals n_1, n_2 are fixed by the experimental layout. The sample vector (a,c) varies over a 16 point sample space, the observed sample being (3,0). It takes a lot of courage to recommend a normal or a chi-squared test of significance in this case.

The statistical problem is to test the null hypothesis $P_1 = P_2$ against the alternatives $P_1 > P_2$. The data $(a,c) = (3,0)$ is the most extreme one in the 16 point data set. What is a measure of the strength of evidence provided by this particular data against the null hypothesis?

Professor Berkson is talking about three different types of significance levels— the nominal, the effective and the attained. He calls a test exact if the nominal and the effective levels are equal. I find all these very curious. Does anyone ever end a statistical report with a remark like: "The null hypothesis is rejected at the nominal level of 0.05, the P-value is 0.04," and then hasten to report the effective level of significance of his test procedure? If significance level is to be understood as a measure of the strength with which observed data tend to corroborate the null

---

Editor's Note : This chapter is based on a discussion by Basu on Joseph Berkson's paper : In dispraise of the exact test, Journal of Statistical Planning and Inference 3, (1979), 189-192.

hypothesis, then, among the three notions of significance level, the P-value is the only one that appears to make some sense. However, I regard the P-value also as a very misleading concept, but that isn't the topic under discussion.

My mind goes back about 22 years when I heard George Barnard making a spirited defense of the conditional test in terms of the above very example. At the end of the talk I simply asked: "Why did you hold the marginals fixed?" A little flustered, George answered my question with the question: "What else can you do?" The rest of this discussion is addressed to these two questions.

Holding a+c fixed at its observed value 3, Fisher reduced the 16 point sample space to the 4 point reference set {(0,3), (1,2), (2,1), (3,0)} . Under the null hypothesis the relative probability of the observed sample (3,0) in the above reference set is $\frac{1}{20}$. Recognizing (3,0) as the most extreme sample point, Fisher evaluated the one-sided tail area probability (P-value) as $\frac{1}{20}$. Professor Berkson is rebelling against this kind of data analysis. Although my sympathies lie with Professor Berkson, I cannot help asking him: "What took you so long?"

Why not hold both a+c and $|a - c|$ fixed at their observed values? We then have the 2 point reference set {(0,3),(3,0)} and so the P-value climbs to $\frac{1}{2}$. On the other hand, if we take the full 16 point sample space as the reference set, then the tail area probability is $p^3(1-p)^3$, where p is the nuisance parameter (the common value of P_1, P_2) in the composite null hypothesis. Since $p^3(1-p)^3 \leqslant \frac{1}{64}$, why not declare the data to be highly significant at the level of $\frac{1}{64}$? Fisher's (1956, pp. 87-88) response to the question is a typical example of the kind of obscurantism with which Sir Ronald used to mystify his critics and charm his admirers.

It is not quite correct to identify the notion of ancillarity with that of noninformativeness. A statistic is ancillary if its sampling distribution does not involve the model parameters. If we condition the data with respect to an ancillary statistic then the likelihood function is not altered. Fisher often regarded the likelihood function as the sole carrier of all the information in the data about the model parameters, and that is probably why he saw nothing wrong in conditioning with respect to an ancillary statistic. Although the statistic a+c is not ancillary in the technical sense of the term, Fisher regarded it as ancillary in the extended (and somewhat dubious) sense that there is no recoverable bit of information in a+c about the parameter of interest (that is, the question of equality or otherwise of P_1,P_2) in the absence of any prior information about the parameters. Much has been written [Cox (1958), Barnard (1963), Sprott (1975), Barndorff-Nielson (1975)] about this particularly nebulous concept of no information or nonformation. But nothing tangible has emerged from these investigations.

Shorn of a data dependent multiplicative constant, the likelihood function is

$$L(P_1, P_2) = (\frac{P_1}{1-P_1})^a (\frac{P_2}{1-P_2})^c (1 - P_1)^{n_1} (1 - P_2)^{n_2}.$$

Reparametrizing to the cross-ratio $\Psi = P_1(1 - P_2)/(1 - P_1) P_2$ as the parameter of interest and to $\phi = P_2/(1 - P_2)$ as the nuisance parameter, the likelihood function takes on the new look

$$L(\Psi, \phi)= \Psi^a \phi^{a+c} (1 + \Psi\phi)^{-n_1} (1 + \phi)^{-n_2}.$$

The two parameters Ψ, ϕ are variation independent, with $1 \leqslant \Psi < \infty$(since $P_1 \geqslant P_2$) and $0 < \phi < \infty$. For every fixed value of Ψ, the statistic $a + c$ is clearly sufficient for ϕ. That is, $a + c$ is specific sufficient for ϕ [refer to Basu (1977) for more on this] and so the conditional distribution of the variable a, given $a + c$, depends only on the parameter of interest Ψ. This fact was recognized by Fisher (1935, p. 50). Conditioning with respect to $a + c$ does a neat job of eliminating the nuisance parameter ϕ, but so also does conditioning with respect to the wider specific sufficient statistic $(a + c, |a - c|)$. But see what happens when we choose to reparametrize the model in a different fashion.

Let $\theta = (P_1 - P_2)/(1 - P_2)$ be the parameter of interest and $\pi = P_2$ the nuisance parameter. Observe that $0 \leqslant \theta < 1$ and $0 < \pi < 1$ and that they are variation independent. The likelihood function is

$$L(\theta,\pi) = (\theta + \pi - \theta\pi)^a [(1 - \theta)(1 - \pi)]^{n_1-a} \pi^c (1 - \pi)^{n_2-c}.$$

Now, for each fixed value of $\theta > 0$, the minimum sufficient statistic for π is the full data (a, c) and so there is no way we can eliminate π from the likelihood function by invoking the Fisher conditionality argument. Is there a logically compelling reason why we should reparametrize the model in terms of (Ψ, ϕ) and not in terms of (θ, π)? I cannot regard the Fisher conditionality argument as anything but an adhoc method that appears to succeed once in a while but fails completely when the same problem is restated in a slightly different form.

Since π ($= P_2$) refers to the control population, the scientist may have a lot of prior opinion about it. About the parameter of interest $\theta = (P_1 - P_2)/(1-P_2)$ he may not have any well formulated opinion other than $0 \leqslant \theta < 1$. After a careful probing of the mind of the scientist it may seem reasonable to represent his prior opinion about θ, π in terms of a density function like

$$q(\theta, \pi) = B(m, n)\, \pi^{m-1} (1 - \pi)^{n-1}, \quad 0 \leqslant \theta < 1, \; 0 < \pi < 1.$$

It should be clear by now how I intend to answer George Barnard's counter question at this time. My answer is: "Act like a Bayesian. Integrate out the nuisance parameter from the likelihood function after weighting it with a function of θ, π that is a rough representation of the scientist's prior opinion about the parameters."

Data interpretation is not an objective scientific method. There cannot be a mindless weighing of evidence. Can I be truly objective unless I am completely ignorant of the subject?!

CHAPTER XIX

A DISCUSSION ON SURVEY THEORY

I did not intend to participate in today's discussion on fundamental questions. But, Mr. Chairman, you have put me on the spot with the request that I throw some light on some of the hotly debated issues. If I succeed only in further muddying the water of clear thinking, please remember that you asked for it.

Yesterday, with an electrifying last-minute remark on Carl Sarndal's joint paper, Richard Royall ushered my herd of circus elephants (Basu, 1971, p. 212) into the arena. Since Royall is not here today, let me begin with an amplification of that remark.

Let $P = (1, 2, \ldots, N)$ be the population and let $\phi = (y_1, y_2, \ldots, y_N)$ be the universal parameter, the parameter of interest being $Y = \Sigma y_k$. On the basis of some known auxiliary characters $x_1, x_2, \ldots, x_N$ of the population units, a survey design is chosen and then set in motion, resulting in the sample $\mathbf{x} = (s,r,\mathbf{y})$, where $s \subset P$ is the sample label set, $r \subset s$ is the set of respondents, and $\mathbf{y} = \{y_i : i \in r\}$ is the observation vector. For each $k \in P$, let $\pi_k = \text{Prob}(k \in s)$ be the known inclusion probability and $q_k = \text{Prob}(k \in r \mid k \in s)$ be the usually unknown conditional response probability for unit k given that unit k is selected. For the sake of the present argument, let us suppose that $q_k = q(y_k)$ is a known function of y_k and that $\pi_k q_k > 0$ for all $k \in P$.

Now, $\pi_k q_k$ is the probability that unit k is selected and then responds, that is, the probability that $k \in r$. Consider the Horvitz-Thompson estimate $\hat{Y} = \Sigma y_i/\pi_i q_i$, where $i \in r$. That $\hat{Y}$ is a design unbiased estimator of Y is immediately apparent if we rewrite $\hat{Y}$ in the equivalent form $\hat{Y} = \Sigma y_k J_k/\pi_k q_k$, where $k \in P$ and J_k is the indicator of the event $k \in r$. Despite its design unbiasedness, Royall is concerned about the face validity of $\hat{Y}$ as an estimate of Y. What is the rationale for attaching a large weight, namely $(\pi_i q_i)^{-1}$, to an observed y that is associated with a small selection response probability $\pi_i q_i$? If $\hat{Y}$ is a reasonable estimate of Y, then $\hat{Y} - \Sigma y_i$, $i \in r$, ought to be regarded as a reasonable estimate of the total y value of the set $P - r$ of unobserved units, How does one convince oneself and the client about this reasonableness? Is $\hat{Y}$ a consistent estimate in some sense? For what superpopulation model ζ is the estimate ζ-unbiased?

Editor's Note : This discussion took place at a Symposium on Incomplete Data held in Washington, D.C. in the Summer of 1979. Professor William Madow was in the chair at the particular session. This article is listed as Basu (1979b) in the references.

If the ratios $\{y_k/\pi_k q_k : k \varepsilon P\}$ are nearly equal to each other and if the effective sample size $n_1 = \#(r)$ is equal to its expected value $\Sigma \pi_k q_k$, $k \varepsilon P$, then $\hat{Y}$ is a reasonable estimate of $Y = \Sigma y_k$, $k \varepsilon P$. Since neither of the two conditions are likely to be fulfilled in practice, $\hat{Y}$ as an estimate of Y will be dimensionally wrong in most instances. I find that Carl Sarndal is quite aware of this difficulty. He proposes to solve the problem, like Hajek (1971b, p. 236) did, in the following manner. Suppose we know of numbers $\{a_k : k \varepsilon P\}$ such that the ratios $\{y_k/a_k : k \varepsilon P\}$ are nearly the same. We then modify the Horvitz-Thompson estimator $\hat{Y}$ to the ratio type estimator

$$\hat{Y}_1 = \Sigma\, a_k\, \Sigma(y_i/\pi_i q_i)/\Sigma(a_i/\pi_i q_i), \quad k \varepsilon P,\ i \varepsilon r\ .$$

The modified estimate $\hat{Y}_1$ is dimensionally correct but is not design unbiased. So we are prepared to sacrifice design unbiasedness for the sake of accuracy. But why are we still clinging on to the $\pi_i q_i$'s? What is the logical necessity for weighting y_i and a_i with $(\pi_i q_i)^{-1}$? Why not use the classical ratio estimate $\hat{Y}_R = \Sigma a_k\ [\ \Sigma y_i / \Sigma\ a_i]$, $k \varepsilon P$, $i \varepsilon r$, instead? Can the ratio estimate $\hat{Y}_R$ be justified in the present case of nonresponse? What if the ratios y_k/a_k are larger for the respondents than for the nonrespondents? In that case $\hat{Y}_R$ will overestimate Y.

Please permit me a small digression at this stage. Yesterday, I was taken aback by a forthright assertion by Ken Brewer that the ratio estimate makes sense only in the case of equal probability sampling. Brewer has no objection to model-based estimates so long as the sampling plan is chosen so that the estimates have good sampling properties. In a typical multipurpose survey project, the model-based thinking of a Bayesian or a prediction theorist will generate a large number of estimates of a great many parameters of interest. What choice of sampling plan will make all these different estimates look good in a repeated sampling sense? I do not really understand how V.P. Godambe and others aim to achieve robustness through careful randomization. Surely they do not advise us to abandon the model-based approach to survey sampling. But I hear them make alarming noises about the risk of model failure and they marvel about the naive courage of the model builders. For myself, I often wonder at the courage of a randomizer who pins all his faith on a table of random numbers and refuses to analyze in depth the unique sample that he has actually obtained.

The inclusion probabilities $\{\pi_k : k \varepsilon P\}$ play a major role in V.P. Godambe's survey theory. In the full response situation, that is, when $q_k = 1$ for all $k \varepsilon P$, the likelihood principle tells us (Basu, 1969b) that the inclusion and other design probabilities are irrelevant at the data analysis stage. I find it a little curious that Godambe invokes his discarded likelihood principle to question the relevance of the response probabilities $\{q_k : k \in P\}$ at the data analysis stage. In effect, he is saying that a Bayesian cannot possibly have any use for knowing how the q_k's are related to the corresponding y_k's, because, when a unit k is observed (that is, when $k \in r$), the Bayesian knows y_k itself and can therefore dispense with the knowledge of $q_k = q(y_k)$. Godambe is in error because he forgets the fact that the sample **x** contains the set $s - r$ of sampled units that did not respond. Knowing the function q, we may be able to put restrictions on the y values for the units in $s - r$. Let me quote yet another of my unrealistic examples to elucidate this point.

Suppose each y_k is either 0 or 1 and that q_k is 1 or α when y_k is 0 or 1, respectively. It does not matter whether α is known or is estimated from the sample. With a sample $\mathbf{x} = (s, r, \mathbf{y})$ the problem is to estimate the population total $Y = \Sigma y_k$. Let n, n_1 and $n_2 = n - n_1$ be, respectively, the number of elements in the sets s, r and $s - r$; that is, n_1 is the number of respondents and n_2 the number of nonrespondents. Now, the parameter of interest Y can be broken down into three parts as follows :

$$Y = Y_1 + Y_2 + Y_3 = \Sigma y_i + \Sigma y_j + \Sigma y_k ,$$

where $i \in r$, $j \in s - r$ and $k \in P - s$. With the sample $\mathbf{x} = (s, r, \mathbf{y})$ before us, we know for sure that $Y_1 = n_{11}$, the number of respondents i with $y_i = 1$. In view of our knowledge that $q_k = 1$ whenever $y_k = 0$, we know that $Y_2 = n_2$. So the only parameter that remains to be estimated is the Y_3 part of Y. The sampling design that selected the label set s has not entered into the discussion yet. Can we forget about it? To a Bayesian, or semi-Bayesian modelist like Richard Royall, the answer is clearly yes. A lazy Bayesian approach to the problem of estimating Y_3 would be to check whether the label set s can be regarded as a representative sample from P and, if so, to assume that the proportion of ones is about the same for the two sets s and $P - s$. In other words, the lazy Bayesian estimates Y_3 by $(N - n)(Y_1 + Y_2)/n = (N - n)(n_{11} + n_2)/n$ and, therefore, estimates Y by $N(n_{11}+n_2)/n$. Observe that the specific knowledge about the q_k's have entered into the estimation procedure but the π_k's have not. Also note that the estimate $N(n_{11} + n_2)/n$ is not a particular case of the ratio estimate $\hat{Y}_R$ considered earlier.

I am intrigued by Brewer and Sarndal's sixfold classification of survey theoreticians. But I am really astonished by Don Rubin's assertion that at the grass root level of survey practice there is hardly any differece between the Bayesian and other approaches to survey sampling. If by this Rubin means that survey practitioners usually go about their individual prodding ways, generally oblivious to all the storms raised by the theoreticians, then he is probably right. However, if Bayesian methods really take hold of survey practice and if active (not lazy, that is) Bayesians begin to come out of their exchangeable prior shells, then, I am sure, survey practice will be completely revolutionized. But of course these are very big ifs.

Dennis Lindley (1965) used to say that most of the traditional statistical methods are right, but they are believed to be right for the wrong reasons. I believe Lindley has now changed his mind about that. It is about 12 years now since I finally came to the sad conclusion that most of the statistical methods that I had learned from pioneers like Karl Pearson, Ronald Fisher and Jerzy Neyman and survey practitioners like Morris Hansen, P. C. Mahalanobis and Frank Yates are logically untenable. It is my interest in survey theory that finally forced me to this unhappy conclusion.

CHAPTER XX

A NOTE ON UNBIASED ESTIMATION

1. Summary

It is shown that even in very simple situations (like estimating the mean of a normal population) where a uniformly minimum variance unbiased estimator of the unknown population characteristic is known to exist, no best (even locally) unbiased estimator exists as soon as we alter slightly the definition of variance.

2. Introduction

Let $(\mathscr{X}, \mathscr{A})$ be an arbitrary measurable space (the "sample space") and let $\{P_\theta\}$, $\theta \in \Omega$, be a family of probability measures on $\mathscr{A}$. A real-valued function $\mu = \mu_\theta$ of θ is "estimable" if it has an "unbiased estimator". An unbiased estimator of μ is a mapping $\eta = \eta_x$ of the "sample space" $\mathscr{X}$ onto the space of all probability measures over the σ-field of all the Borel sets on the real line such that

(i) $T_x = \int_{-\infty}^{\infty} t \, d\eta_x$ is an $\mathscr{A}$-measurable function of x,

(ii) $\mu_\theta \equiv \int_{\mathscr{X}} T_x \, dP_\theta$ for all $\theta \in \Omega$.

If, for every $x \in \mathscr{X}$, the whole probability mass of η_x is concentrated at one point, say T_x, then η_x (or equivalently T_x) is called a nonrandomized estimator. With reference to a given loss or weight function $w(t, \theta)$, which is a Borel-measurable function of the real variable t for every fixed $\theta \in \Omega$, an unbiased estimator η_x of μ_θ is better than an alternative unbiased estimator η'_x at the point $\theta = \theta_0$ if

$$\int_{\mathscr{X}} dP_{\theta_0} \int_{-\infty}^{\infty} w(t, \theta_0) \, d\eta_x < \int_{\mathscr{X}} dP_{\theta_0} \int_{-\infty}^{\infty} w(t, \theta_0) \, d\eta'_x.$$

We consider only such estimators η_x for which $\int_{-\infty}^{\infty} w(t, \theta) \, d\eta_x$ is an $\mathscr{A}$-measurable function of x for all $\theta \in \Omega$.

Hodges and Lehmann (1950) noted that if, for every $\theta \in \Omega$, the loss function $w(t, \theta)$ is a convex (downwards) function of the variable t, then the class of nonrandomized estimators of μ is essentially complete. Barankin (1949) and Stein (1950) considered the particular case where $w = | t - \mu_\theta |^s$ for $s \geqq 1$ and proved, under a few regularity assumptions, that there always exists an unbiased estimator which is locally the best at a given value of $\theta = \theta_0$. Simple examples may be given to show that there need not exist a uniformly best unbiased estimator even in the simplest case of

Editor's Note : This chapter is based on Basu (1955b).

$s = 2$. If, however, there exists a complete sufficient statistic for θ and if w is convex (downwards) for every fixed $\theta \in \Omega$, then there exists an essentially unique uniformly best unbiased estimator for every estimable parametric function μ_θ. The convexity of the loss function is essential in the proofs of the above results. We demonstrate in the next section how a slight departure from the convexity of the loss function might destroy all these results.

3. Nonexistence of a best unbiased estimator

Let us assume that $w(t, \theta) \geqq w(\mu_\theta, \theta) = 0$ for all t and θ. That is, we assume that the loss function is nonnegative and that there is no loss when the estimate hits the mark. Let U be the class of all unbiased estimators η_x of μ_θ for which the risk function

$$r(\theta|\eta) = E[w(t, \theta)|\eta, \theta] = \int_{\mathscr{X}} dP_\theta \int_{-\infty}^{\infty} w(t, \theta)\, d\eta_x$$

is defined for all θ. We prove the following

Theorem. If for every fixed $\theta\epsilon\Omega$ the loss function $w(t,\theta)$ is bounded in every finite interval $|t - \mu_\theta| \leqq A$, and is $o(|t - \mu_\theta|)$ as $|t - \mu_\theta| \to \infty$, then

$$\inf_{\eta\, \epsilon U} r(\theta|\eta) \underset{\theta}{\equiv} 0 .$$

Proof. Let $T = T_x$ be a nonrandomized unbiased estimator of μ_θ. The existence of an unbiased estimator clearly implies the existence of such a T_x. Consider now the randomized estimator $\eta^{(\delta)} = \eta_x^{(\delta)}$ which, for any $x \in \mathscr{X}$, has its entire probability mass concentrated at the two points μ_{θ_0} and $(T_x - \mu_{\theta_0})/\delta + \mu_{\theta_0}$ on the real line in the ratio $1 - \delta$ to δ, with $0 < \delta < 1$. It is easily verified that $\eta^{(\delta)}$ is an unbiased estimator of μ_θ and that

$$r(\theta_0|\eta^{(\delta)}) = E[\omega(t, \theta_0)|\eta^{(\delta)}, \theta_0]$$

$$= E[\delta w(H/\delta + \mu_{\theta_0}, \theta_0)|\theta_0], \quad \text{where } H = T_x - \mu_{\theta_0} .$$

Since $w = o(|t - \mu_{\theta_0}|)$ as $|t - \mu_{\theta_0}| \to \infty$, given $\varepsilon > 0$ we can determine A so large that

$$w(t, \theta_0) \leqq \varepsilon|t - \mu_{\theta_0}| \quad \text{whenever } |t - \mu_{\theta_0}| \geqq A .$$

Let $B = \sup_{|t-\mu_{\theta_0}|<A} w(t, \theta_0) < \infty$. Then

$$r(\theta_0 | \eta^{(\delta)}) = \{\int_{|H| < \delta \cdot A} + \int_{|H| \geqq \delta A}\} \delta\omega(H/\delta + \mu_{\theta_0}, \theta_0)\, dP_{\theta_0}$$

$$\leqq \delta B + \varepsilon E(|H| | \theta_0).$$

Since ε and δ are arbitrary and B depends only on ε, it follows that $\inf_{\eta \varepsilon U} r(\theta_0|\eta) = 0$. Since θ_0 is arbitrary, the theorem is proved.

Now, if $w(t, \theta_0) > 0$ for $t \neq \mu_{\theta_0}$, then $r(\theta_0 | \eta)$ can be zero only if η_x gives probability one to μ_{θ_0} for almost all x with respect to the measure P_{θ_0}. In the usual circumstances, η_x then would not be an unbiased estimator of μ_θ.

Thus, this theorem shows that if we work with a loss function satisfying the conditions of the theorem, even locally best unbiased estimators would not exist in all the familiar situations in which we are interested. In particular estimation problems, it will be easy to see that the theorem holds even in the restricted class U* of all nonrandomized estimators of μ. In the next section we consider the classical problem of estimating the mean of a normal population, but with a slightly altered definition of variance.

4. The Case of the Normal Mean

Let $x = (x_1, x_2, \ldots, x_n)$ be a random sample from $N(\theta, 1)$. The problem is to get a good unbiased estimator of θ with the loss function

$$w(t, \theta) = \begin{cases} (t - \theta)^2, & |t - \theta| \leqq a, \\ a^{3/2}|t - \theta|^{1/2}, & |t - \theta| > a, \end{cases}$$

where a is an arbitrarily large constant.

Let $\bar{x}$ and s^2 be the sample mean and variance, respectively, and let c_δ be the upper 100δ per cent point of the probability distribution of s^2, where $0 > \delta > 1$. Consider the nonrandomized estimator

$$T^{(\delta)} = T_x^{(\delta)} = \begin{cases} \theta_0 & \text{when } s^2 \leqq c_\delta \\ (\bar{x} - \theta_0)/\delta + \theta_0 & \text{when } s^2 > c_\delta. \end{cases}$$

Since the distribution of s^2 is independent of θ and $\bar{x}$, it follows that $T^{(\delta)}$ is a function of x and δ alone and that $T^{(\delta)}$, for every fixed δ with $0 < \delta < 1$, is an unbiased estimator of θ. Also

$$r(\theta_0 | T^{(\delta)}) = E[\delta w\{(\bar{x} - \theta_0)/\delta + \theta_0, \theta_0\} | \theta_0]$$

$$= \int_{|\bar{x}-\theta_0| \leqq a\delta} \delta\left(\frac{\bar{x} - \theta_0}{\delta}\right)^2 \phi(\bar{x})\, d\bar{x} + \int_{|\bar{x}-\theta_0| > a\delta} \delta a^{3/2} \left|\frac{\bar{x} - \theta_0}{\delta}\right|^{1/2} \phi(\bar{x}) d\bar{x}$$

$$< \delta a^2 + \delta^{1/2} a^{3/2} E(|\bar{x} - \theta_0|^{1/2} | \theta_0),$$

where $\phi(\bar{x})$ is the frequency function of $\bar{x}$ when $\theta = \theta_0$. Thus $r(\theta_0 | T^{(\delta)}) \to 0$ as $\delta \to 0$. Therefore

$$\inf_{T \in U^*} r(\theta | T^{(\delta)}) \equiv 0, \qquad -\infty < \theta < \infty,$$

where U^* is the class of all nonrandomized unbiased estimators of θ.

When the constant a is very large, the modification to the usual definition of variance apparently is very negligible, yet this slight change of variance completely wrecks the theory of unbiased estimation. Not even locally best unbiased estimators exist, let alone a uniformly best one.

In the construction of $T^{(\delta)'}$, the independence of s^2 and $\bar{x}$ is not essential. As a matter of fact, we can replace s^2 by any real-valued statistic Y whose conditional distribution, given $\bar{x}$, is continuous. We then replace c_δ by $c'_\delta(\bar{x})$, where $c'_\delta(\bar{x})$ is, say, the upper 100δ per cent point of the conditional distribution of Y given $\bar{x}$. From the sufficiency of $\bar{x}$ it follows that $c'_\delta(\bar{x})$ is independent of δ, and the rest of the proof follows through. Under similar circumstances the general theorem proved earlier will remain true in the restricted class U^* of all nonrandomized unbiased estimators of μ_θ.

CHAPTER XXI

THE CONCEPT OF ASYMPTOTIC EFFICIENCY

1. Summary

Partly of an expository nature this note brings out the fact that an estimator, though asymptotically much less efficient (in the classical sense) than another, may yet have much greater probability concentration (as defined in this article) than the latter.

2. Definitions

Let $\{X_i\}$, $i = 1, 2, \ldots$ be an infinite sequence of independent and identically distributed random variables whose common distribution function F is known to belong to a family Ω of one dimensional distribution functions. Let $\mu = \mu(F)$ be a real valued functional defined on Ω. By an estimator $T = \{t_n\}$ of μ we mean a sequence of real valued measurable functions of $\{X_i\}$, where t_n is a function of $X_1, X_2, \ldots, X_n$ only ($n = 1, 2, \ldots$). The estimator T is said to be an asymptotically normal estimator of μ if there exists a sequence $\{\sigma_n(F)\}$ of positive numbers such that as $n \to \infty$

$$\{t_n - \mu(F)/\sigma_n(F)\} \Longrightarrow N(0, 1) \quad \text{for all } F \in \Omega$$

where $\Longrightarrow$ stands for convergence in law and N(0, 1) for the standard normal variable. The sequence $\{\sigma_n(F)\}$ is called the asymptotic standard deviation of T. A necessary and sufficient condition in order that $\{\sigma_n(F)\}$ and $\{\sigma_n'(F)\}$ may both be called the asymptotic standard deviation of T is

$$\lim_{n \to \infty} \{\sigma_n(F)/\sigma_n'(F)\} \equiv 1 \quad \text{for all } F \in \Omega.$$

A necessary and sufficient condition in order that the asymptotically normal estimator T is also consistent is

$$\lim_{n \to \infty} \sigma_n(F) \equiv 0 \quad \text{for all } F \in \Omega.$$

Let $\mathcal{T}$ be the family of all consistent asymptotically normal estimators of μ. We consider only the space $\mathcal{T}$.

Editor's Note : This chapter is based on Basu (1956a)

3. The Partial Order of Efficiency

Two elements T and T' of $\mathcal{J}$ are said to be equally efficient (or equivalent) if they have the same asymptotic s.d.s, i.e. if

$$\lim_{n \to \infty} \{\sigma_n(F)/\sigma'_n(F)\} \equiv 1 \quad \text{for all } F \in \Omega \tag{3.1}$$

where $\{\sigma_n(F)\}$ and $\{\sigma'_n(F)\}$ are the corresponding asymptotic s. d.'s.

It is easily verified that the above equivalence relation is reflexive, symmetric, and transitive.

If $$\limsup_{n \to \infty} \{\sigma_n(F)/\sigma'_n(F)\} \leqslant 1 \quad \text{for all } F \in \Omega$$

and $$\liminf_{n \to \infty} \{\sigma_n(F)/\sigma'_n(F)\} < 1 \quad \text{for some } F \in \Omega$$

then we say that T is more efficient than T' and write $T \supset T'$. It is easily seen that the relation $\supset$ induces a partial order on $\mathcal{J}$.

It is known that there do not exist a maximal element in $\mathcal{J}$ with respect to the partial order $\supset$, i.e. there do not exist any element $T \in \mathcal{J}$ which is either equivalent to or more efficient than any alternative $T' \in \mathcal{J}$. As a matter of fact it has been demonstrated (LeCam, 1953) how given any $T \in \mathcal{J}$ we can always find a $T' \in \mathcal{J}$ such that $T' \supset T$.

4. The Partial Order of Concentration

The estimator $T = \{t_n\}$ of μ is consistent if for all $\varepsilon > 0$ and $F \in \Omega$

$$p_n(\varepsilon, F) = P\{|t_n - \mu| > \varepsilon | F\} \to 0 \text{ as } n \to \infty .$$

If we work with the simple loss function that is zero or one according as the error in the estimate is $\leqslant \varepsilon$ or $> \varepsilon$ then $p_n(\varepsilon, F)$ is the risk (or expected loss) when the estimator is used with observations on $X_1, X_2, \ldots, X_n$ only.

The rapidity with which $p_n(\varepsilon, F) \to 0$ may be considered to be a measure of the asymptotic accuracy or concentration of T. This motivates the following definition of a partial order on $\mathcal{J}$ (and as a matter of fact on the wider family of all consistent estimators of μ).

Definition : The estimator T with the associated sequences of risk functions $p_n(\varepsilon, F)$ is said to have greater concentration than T' with the associated sequences $p'_n(\varepsilon, F)$ if, for all $\varepsilon > 0$ and $F \in \Omega$,

$$\limsup_{n \to \infty} \{p_n(\varepsilon,F)/p'_n(\varepsilon, F)\} \leq 1$$

with the limit inferior being < 1 for some $\varepsilon > 0$ and some $F \in \Omega$. We then write $T > T'$.

Intuitively it may seem reasonable to expect that $T \supset T'$ implies $T > T'$. That this is not so is demonstrated in the next section. An example is given where

$$\lim_{n \to \infty} \frac{\sigma_n(F)}{\sigma'_n(F)} \equiv 0 \qquad \text{for all } F \in \Omega, \tag{4.1}$$

whereas

$$\lim_{n \to \infty} \frac{p_n(\varepsilon, F)}{p'_n(\varepsilon,F)} \equiv \infty \qquad \text{for all } \varepsilon > 0 \text{ and } F \in \Omega. \tag{4.2}$$

5. An Example

Let each of the X_i's be $N(\mu , 1)$, the problem being to estimate μ.

Let
$$\overline{X}_n = \sum_1^n X_i/n \text{ and } S_n = \sum_1^n (X_i - \overline{X}_n)^2 .$$

Then $\overline{X}_n$ and S_n are mutually independent random variables and the distribution of S_n is independent of μ. Let a_n be the upper 100/n % point of S_n and let

$$H_n = \begin{cases} 0 & \text{if } S_n \leq a_n \\ 1 & \text{if } S_n > a_n \end{cases}$$

Now let
$$T = \{t_n\}$$

where
$$t_n = (1 - H_n) \overline{X}_n + n H_n$$

and
$$T' = \{t'_n\}$$

where
$$t'_n = \overline{X}_{[\sqrt{n}]}.$$

(By [x] we mean the largest integer not exceeding x .)

Since
$$P(H_n = 0) = 1 - \frac{1}{n} \to 1,$$

it follows (vide Cramer, p. 254) that $\sqrt{n}(t_n - \mu) = \sqrt{n}(\bar{X}_n - \mu) + \sqrt{n}\, H_n(n - \bar{X}_n)$

$$\Rightarrow N(0, 1)$$

when μ is the true mean.

Hence, $T \in \mathcal{T}$ with asymptotic s.d. $= \{n^{-1/2}\}$. Also $T' \in \mathcal{T}$ with asymptotic s.d. $= \{n^{-1/4}\}$.

Therefore (4.1) is satisfied. Again, since $\bar{X}_n$ is independent of H_n it follows that, for every $n > \mu + \varepsilon$,

$$P(|t_n - \mu| > \varepsilon|\mu) = P(H_n = 0)\, P(|\bar{X}_n - \mu| > \varepsilon|\mu) + P(H_n = 1)$$

$$= \frac{1}{n} + o\left(\frac{1}{n}\right)$$

because $\qquad P(|\bar{X}_n - \mu| > \varepsilon|\mu) = o\left(\frac{1}{n}\right)$, as may be easily verified.

Whereas $\qquad P(|t'_n - \mu| > \varepsilon|\mu) = o\left(\frac{1}{n}\right)$

Therefore (4.2) also is satisfied.

It may be noted that in the example given the s.d. of t_n is not asymptotically equal to the asymptotic s.d. of T. But this can be easily arranged to be true by, say, taking a_n for the upper $100/n^4$ % point of S_n.

CHAPTER XXII

STATISTICS INDEPENDENT OF A COMPLETE SUFFICIENT STATISTIC

1. Introduction

If $\{P_\theta\}$, $\theta \varepsilon \Omega$, be a family of probability measures on an abstract sample space $\mathcal{S}$ and T be a sufficient statistic for θ then for a statistic T_1 to be stochastically independent of T it is necessary that the probability distribution of T_1 be independent of θ . The condition is also sufficient if T be a boundedly complete sufficient statistic. Certain well-known results of distribution theory follow immediately from the above considerations. For instance, if $x_1, x_2, \ldots, x_n$, are independent N(μ , σ)'s then the sample mean $\bar{x}$ and the sample variance s^2 are mutually independent and are jointly independent of any statistic f (real or vector valued) that is independent of change of scale and origin. It is also deduced that if x_1, $x_2, \ldots, x_n$ are independent random variables such that their joint distribution involves an unknown location parameter θ then there can exist a linear boundedly complete sufficient statistic for θ only if the x's are all normal.Similar characterizations for the Gamma distribution also are indicated.

2. Definitions

Let ($\mathcal{S}, \mathcal{A}$) be an arbitrary measurable space (the sample space) and let $\{P_\theta\}$,$\theta \varepsilon \Omega$, be a family of probability measures on $\mathcal{A}$.

Definition 1 : Any measurable transformation T of the sample space ($\mathcal{S}$, $\mathcal{A}$) onto a measurable space ($\mathcal{T}, \mathcal{B}$) is called a statistic. The probability measures on $\mathcal{B}$ induced by the statistic T are denoted by $\{P^T_\theta\}$,$\theta \varepsilon \Omega$.

For every $\theta \varepsilon \Omega$ and $A \varepsilon \mathcal{A}$ there exists an essentially unique real valued $\mathcal{B}$-measurable function $f_\theta(A|t)$ on $\mathcal{T}$ such that the equation

$$P_\theta(A \cap T^{-1}B) \int_B f_\theta (A|t)\, dP^T_\theta \quad (1)$$

holds for every $B \varepsilon \mathcal{B}$. The set of points t for which $f_\theta(A|t)$ falls outside the closed interval (0, 1) is of P^T_θ-measure zero for every $\theta \varepsilon \Omega$. We call $f_\theta(A|t)$ the conditional probability of A given that T = 1 and that θ is the true parameter point.

Definition 2 : A statistic T is said to be independent of the parameter θ if, for every $B \varepsilon \mathcal{B}$, $P^T_\theta(B)$ is the same for all $\theta \varepsilon \Omega$.

Editor's Note : This chapter is based on Basu (1955c)

Definition 3 : The two statistic and T_1, with associated measurable spaces $(\mathcal{T}, \mathcal{B})$ and $(\mathcal{T}_1, \mathcal{B}_1)$ respectively, are said to be stochastically independent of each other if, for every $B \varepsilon \mathcal{B}$ and $B_1 \varepsilon \mathcal{B}_1$

$$P_\theta(T^{-1}B \cap T_1^{-1} B_1) \equiv P_\theta(T^{-1}B)P_\theta(T_1^{-1} B_1)$$

for all $\theta \varepsilon \Omega$.

Now,

$$P_\theta(T^{-1}B \cap T_1^{-1} B_1) \equiv \int_B f_\theta(T_1^{-1} B_1|t)dP_\theta^T .$$

It follows, therefore, that a necessary and sufficient condition in order that T and T_1 are stochastically independent is that the integrand above is essentially independent of t, i.e.

$$f_\theta(T_1^{-1} B_1|\, t) = P_\theta(T_1^{-1} B_1) = P_\theta^{T_1}(B_1)$$

for all $t \varepsilon T$ excepting possibly for a set of P_θ^T -measure zero.

Definition 4 : The statistic T is called a sufficient statistic (Halmos and Savage, 1949) if for every $A \varepsilon \mathcal{A}$ there exists a function $f(A|t)$ which is independent of θ and which satisfies equation (1) for every $\theta \varepsilon \Omega$.

Let G be the class of all real valued, essentially bounded, and $\mathcal{B}$ -measurable functions on $\mathcal{T}$.

Definition 5 : The family of probability measures $\{P_\theta^T\}$ is said to be boundedly complete (Lehmann and Scheffe, 1950) if for any $g \varepsilon G$ the identity

$$\int_T g(t)dP_\theta^T \equiv 0 \quad \text{for all } \theta \varepsilon \Omega \tag{2}$$

implies that $g(t) = 0$ excepting possibly for a set of P_θ^T -measure zero for all θ. $\{P_\theta^T\}$ is called complete if the condition of essential boundedness is not imposed on the integrand in (2). The statistic T is called complete (boundedly complete) if the corresponding family of measures $\{P_\theta^T\}$ is so.

3. Sufficiency and Independence

For any two statistics T_1 and T we have for any $B_1 \varepsilon \mathcal{B}_1$

$$P_\theta^{T_1}(B_1) = P_\theta(T_1^{-1} B_1) = \int_{\mathcal{T}} f_\theta(T_1^{-1} B_1|\, t)dP_\theta^T . \tag{3}$$

Now if T be a sufficient statistic then the integrand is independent of θ and if, moreover, T_1 is stochastically independent of T then the integrand is essentially independent of t also. Thus, the right hand side of (3) is independent of θ and so we have

Theorem 1 : Any statistic T_1 stochastically independent of a sufficient statistic T is independent of the parameter θ.

That the direct converse of the above result is not true will be immediately apparent if we take for the sufficient statistic T the identity mapping of $(\mathcal{S}, \mathcal{A})$ into itself. No statistic T_1 independent of θ will then be stochastically independent of T excepting in the trivial situation where T_1 is essentially equal to a constant. We, however, have the following weaker but important converse.

Theorem 2 : If T be a boundedly complete sufficient statistic then any statistic T_1 which is independent of θ is stochastically independent of T.

Proof : Since T is sufficient the integrand in (3) is independent of θ. It is also essentially bounded. Now the left hand side of (3) is independent of θ since T_1 is independent of θ. Hence, from bounded completeness of $\{P_\theta^T\}$ it follows that the integrand in (3) is essentially independent of t as well. That is, T_1 is stochastically independent of T.

In the next section we demonstrate how the above theorem may be used to get a few interesting results in distribution theory.

4. Some Characterizations of Distributions with Location and Scale Parameters

Let $\underset{\sim}{x} = (x_1, x_2, \ldots, x_n)$ be a random variable in an n-dimensional Euclidean space whose probability distribution involves an unknown location parameter μ and a scale parameter $\sigma > 0$. Then any measurable function $f(x_1, x_2, \ldots, x_n)$ which is independent of change of origin and scale, i.e.

$$f\left(\frac{x_1 - a}{b}, \ldots, \frac{x_n - a}{b}\right) \equiv f(x_1, \ldots, x_n)$$

for all a and $b > 0$ is independent of the unknown parameter (μ, σ). Now, if there exists a boundedly complete sufficient statistic T for (μ, σ) then f must be stochastically independent of T. For example, if $x_1, x_2, \ldots, x_n$ are independent observations on a normal variable with mean μ and s.d. σ then it is well known that $T = (\bar{x}, s)$ is a sufficient statistic ($\bar{x}$ is the sample mean and s the sample s.d.). The completeness of T follows from the unicity property of the bivariate Laplace transform. It then follows from Theorem 2 that any measurable function $g(\bar{x}, s)$ of $\bar{x}$ and s is stochastically independent of any measurable function $f(x_1, x_2, \ldots, x_n)$ of the observations that is independent of change of origin and scale. The functions g and f need not be real valued. For instance, we may have

$$g = (\Sigma x_i^2, \sum_{i \neq j} x_i x_j)$$

and
$$f = \left(\frac{\Sigma(x_i - \bar{x})^3}{s^3}, \frac{\Sigma(x_i - \bar{x})^4}{s^4}, \ldots\right).$$

Again the stochastic independence of $\bar{x}$ and s follows from the fact that, for any fixed σ, the statistic $\bar{x}$ is a complete sufficient statistic for μ and that s, by virtue of its being independent of change of origin, is independent of the location parameter μ.

Now let $x_1, x_2, \ldots, x_n$ be independent random variables with joint d.f. $F_1(x_1 - \theta), F_2(x_2 - \theta), \ldots, F_n(x_n - \theta)$. Since θ is a location parameter it follows that any linear function $\Sigma a_i x_i$ with $\Sigma a_i = 0$ is independent of θ. If $\Sigma b_i x_i$ is a boundedly complete sufficient statistic for θ then from Theorem 2 it follows that $\Sigma a_i x_i$ is independent of $\Sigma b_i x_i$.

Now, since $\Sigma b_i x_i$ is a sufficient statistic it follows that every $b_i \neq 0$. For, if possible, let $b_j = 0$. Then x_j is stochastically independent of $\Sigma b_i x_i$ and so from Theorem 1 x_j is independent of the parameter θ which contradicts the assumption that the d.f. of x_j is $F_j(x_j - \theta)$. Again, we can take all the a_i's different from zeros. Thus, the two linear functions $\Sigma a_i x_i$ and $\Sigma b_i x_i$ (with non-zero coefficients) of the independent random variables $x_1, x_2, \ldots, x_n$ are stochastically independent. Therefore, all the x_i's must be normal variables. We thus have the following:

Theorem 3 : If $x_1, x_2, \ldots, x_n$, are independent random variables such that their joint d.f. involves an unknown location parameter θ then a necessary and sufficient condition in order that $\Sigma b_i x_i$ is a boundedly complete sufficient statistic for θ is that $b_i > 0$ and that x_i is a normal variable with mean θ and variance b_i^{-1} ($i = 1, 2, \ldots, n$).

Let us now turn to the case of the Gamma variables. Let $x_1, x_2, \ldots x_n$ be independent Gamma variables with the same scale parameter $\theta > 0$, i.e., the density function of x_i is

$$f_i(x)dx = \frac{1}{\Gamma(m_i)\theta^{m_i}} x^{m_i - 1} e^{-x/\theta} dx \qquad (x \geq 0, \theta > 0, m_i > 0).$$

It is clear then that Σx_i is a sufficient statistic for θ and its completeness follows from the unicity property of the Laplace transform. Thus, we at once have the well known result that Σx_i is stochastically independent of any function $f(x_1, x_2, \ldots, x_n)$ that is independent of change of scale (i.e. independent of θ).

Recently it has been proved by R. G. Laha that if $x_1, \ldots, x_n$ are independent and identically distributed chance variables and if Σx_i is independent of $\Sigma a_{ij} x_i x_j / (\Sigma x_i)^2$ then (under some further assumptions) all the x_i's must be Gamma variables. Using this result we can immediately get a characterization of the Gamma distribution analogous to Theorem 3.

CHAPTER XXIII

STATISTICS INDEPENDENT OF A SUFFICIENT STATISTIC

1. Introduction

Let X be a random variable (sample) taking values in an arbitrary sample space $\mathcal{X}$ with the associated σ-field of measurable sets $\mathcal{A}$ and the family of probability measures $\{P_\theta\}$, $\theta \in \Omega$. By a statistic $T = T(X)$ we mean a measurable characteristic of the sample X, i.e., T is an $\mathcal{A}$-$\mathcal{B}$ measurable transformation of the measurable space $(\mathcal{X}, \mathcal{A})$ into some measurable space $(\mathcal{T}, \mathcal{B})$. The family of induced (by the mapping T) probability measures on $\mathcal{T}$ is denoted by $\{P_\theta T^{-1}\}$, $\theta \in \Omega$.

If $P_\theta T^{-1}$ is the same for all $\theta \in \Omega$ then it is clear that an observation on the random variable T will be of no use for making any inference about the parameter θ. In this case we may say that the statistic T contains no information about the parameter θ. On the other hand if T be a sufficient statistic then we may say that T contains the whole of the information about θ that is contained in the sample X. Barring these two extreme situations it is not possible to make a general assessment of how much (or what proportion) of information is contained in a particular statistic. The author feels that the question 'How much information is contained in T?' should be rephrased as 'How effective an observation on T is for making a particular inference about θ?' Clearly the answer will depend on the kind of inference (tests of hypotheses, point or interval estimation etc.) that we wish to make and also on our idea (or criterion) of effectivity. An element of arbitrariness is bound to enter into any attempted definition of the amount of information in a statistic.

One interesting feature of Fisher's definition (1921) of the amount of information is that it is additive for independent statistics. That is, if T_1 and T_2 are any two statistics that are independent for every $\theta \in \Omega$ then the amount of information in (T_1, T_2) is equal to the sum of the informations contained in T_1 and T_2 separately. This, however, does not appear to the author to be a necessary requirement for a satisfactory definition of information. It is possible to think of situations where T_1 and T_2 are equally informative (identically distributed for example) and are independent of one another but still, when an observation on T_1 is given, very little extra information will be supplied by an observation on T_2. For example, suppose we have a population whose distribution we know to be either N(0, 1) or N(5, 1). A single observation from the population will identify the true distribution with a great measure of certainty. Given one observation from the population very little

---

Editor's Note : This chapter is based on Basu (1958a)

extra information will be obtained from a second observation from the population. Surely the total information contained in two independent observations from the population is much less than twice that contained in a single observation. The following is a more extreme example.

Suppose it is known that a bag contains 10 identical balls numbered $\theta+1$, $\theta+2, \ldots, \theta+10$ where the unknown parameter θ takes any one of the values 0, 10, 20, 30, ... Suppose two balls are drawn one by one with replacement and let T_i be the number on the i-th ball drawn (i = 1, 2). Here T_1 and T_2 are identically distributed independent statistics and each is sufficient for θ. Given an observation on T_1 the distribution of T_2 gets completely specified and hence T_1 contains as much information as is contained in (T_1, T_2).

Independence of statistics is sometimes loosely interpreted as follows : – 'If the statistic T_2 is independent of T_1 then knowing what the realization of T_1 has been in a particular trial gives us no information about the possible realization of T_2 in the same trial.' When the probability measure on the sample space is only partially known the above interpretation of independence is no longer true. The example in the previous paragraph very forcefully brings this point out.

2. Statistics Independent of a Sufficient Statistic

In the previous section we have given an example to show that a statistic can be independent of a sufficient statistic and still contain a great deal of information about the parameter. The example demonstrates that Theorem 1 in Basu (1955c) is not true in the generality stated there. Under some mild restrictions, however, the theorem remains true.

Let T be a sufficient statistic and let $A \in \mathcal{A}$ be any fixed event that is independent of T. From the sufficiency of T it follows that there exists a $\mathcal{B}$-measurable real valued function $f(A|t)$ on $\mathcal{T}$ (called the conditional probability of A given T = t) such that for any $B \in \mathcal{B}$

$$P(A \cap T^{-1}B) = \int_B f(A|t) d P_\theta T^{-1} \quad \text{for all } \theta \in \Omega .$$

From the independence of the event A and the statistic T it follows that for any $\theta \in \Omega$,

$$f(A|t) = P_\theta(A) \tag{2.1}$$

almost everywhere $[P_\theta T^{-1}]$ in t.

From (2.1), we cannot conclude that $P_\theta(A)$ is the same for all $\theta \varepsilon \Omega$.

If the two measures $P_{\theta_1} T^{-1}$ and $P_{\theta_2} T^{-1}$ on $(\mathcal{T}, \mathcal{B})$ overlap, [i.e., for any set $B \varepsilon \mathcal{B}$, $P_{\theta_1} T^{-1}(B) = 1$ implies that $P_{\theta_2} T^{-1}(B)$ is positive], then it is very easy to see that (2.1) implies the equality of $P_{\theta_1}(A)$ and $P_{\theta_2}(A)$.

Let us write $\theta_1 \Leftrightarrow \theta_2$ if $P_{\theta_1} T^{-1}$ and $P_{\theta_2} T^{-1}$ overlap. The equality of $P_\theta(A)$ and $P_{\theta'}(A)$ can be deduced if there exists a finite number of parameter points θ_1, $\theta_2, \ldots, \theta_k$ such that

$$\theta \Leftrightarrow \theta_1 \Leftrightarrow \theta_2 \ldots \Leftrightarrow \theta_k \Leftrightarrow \theta' . \tag{2.2}$$

We say θ and θ' are connected (by the statistic T) if there exists $\theta_1, \theta_2, \ldots, \theta_k$ satisfying (2.2).

Thus, we have the

Theorem : If T be a sufficient statistic and if every pair of θ's in Ω are connected (by T), then any event A independent of T has the same probability for all $\theta \varepsilon \Omega$.

As a corollary we at once have that under the conditions of the above theorem any statistic T_1 independent of the sufficient statistic T contains no information about the parameter.

CHAPTER XXIV

THE BASU THEOREMS

The theorems are related to the notions of sufficiency, ancillarity and conditional independence. Let X denote the sample and θ the parameter that completely specifies the sampling distribution P_θ of X. An event E is ancillary if $P_\theta(E)$ is θ-free, i.e., $P_\theta(E) = P_{\theta'}(E)$ for all θ, $\theta' \varepsilon \Theta$, the parameter space. A statistic $Y = Y(X)$ is ancillary if every Y-event (i.e., a measurable set defined in terms of Y) is ancillary (see ANCILLARY STATISTICS). A statistic T is sufficient if, for every event E, there exists a θ-free version of the conditional probability function $P_\theta(E|T)$ (see SUFFICIENCY). The event E is (conditionally) independent of T if, for each $\theta \varepsilon \Theta$, the conditional probability function $P_\theta(E|T)$ is P_θ-essentially equal to the constant $P_\theta(E)$. The statistic Y is independent of T if every Y-event is independent of T. (Independence is a symmetric relationship between two statistics.)

The theorems originated in the following query. Let X_1, X_2 be independent, identically distributed (i.i.d.) $N(\theta,1)$ and let $Y = X_1 - X_2$. Clearly, Y is ancillary and, therefore, so also is every measurable function h(Y) of Y. The statistic Y is shift invariant in the sense that $Y(X_1 + a, X_2 + a) = Y(X_1, X_2)$ for all X_1, X_2 and a. It is easy to see that every shift-invariant statistic is a function of Y and vice versa. Therefore, every shift invariant statistic is ancillary. Is the converse true?

That the answer has to be in the negative is seen as follows. The statistic $T = X_1 + X_2$ is sufficient and is independent of the ancillary statistic $Y = X_1 - X_2$. Let A be an arbitrary T-event and B_1, B_2 be two distinct Y-events such that $P_\theta(B_1) = P_\theta(B_2) = \alpha$, where $0 < \alpha < 1$ is a constant chosen and fixed. Consider the event $E = AB_1 \cup A^cB_2$. The T-events, A, A^c are independent of the Y-events B_1,B_2; therefore,

$$P_\theta(E) = P_\theta(AB_1) + P_\theta(A^cB_2)$$

$$= P_\theta(A)P_\theta(B_1) + P_\theta(A^c)\, P_\theta(B_2)$$

$$= \alpha[P_\theta(A) + P_\theta(A^c)]$$

$$= \alpha \qquad \text{for all } \theta.$$

Editor's Note : This chapter is based on Basu (1982a) which is an insertion in Kotz-Johnson : Encyclopedia of Statistical Sciences, Vol. 1, pp 193-196.

Thus, E is ancillary even though it is not shift-invariant (not an Y-event). How do we characterize the class of ancillary events in this case?

Consider an arbitrary ancillary event E with $P_\theta(E) \equiv \alpha$. Since $T = X_1 + X_2$ is sufficient, there exists a θ-free version $f(T)$ of the conditional probability function $P_\theta(E|T)$ (see SUFFICIENCY). Now, $E_\theta f(T) = P_\theta(E) \equiv \alpha$, so $f(T) \alpha$ is a bounded function of T that has zero mean for each $\theta \varepsilon \Theta$. But the statistic T is complete in the sense that no nontrivial function of T can have identically zero mean. Therefore, the event $f(T) \neq \alpha$ is P_θ-null for each $\theta \varepsilon \Theta$. In other words, $P_\theta(E|T) = P_\theta(E)$ a.s., $[P_\theta]$ for each θ. That is, every ancillary E is independent of $T = X_1 + X_2$. Is the converse true?

Let T be an arbitrary sufficient statistic and let E be independent of T. Let $f(T)$ be a θ-free version of $P_\theta(E|T)$. Then, for each $\theta \varepsilon \Theta$, $f(T) = P_\theta(E)$ a.s. $[P_\theta]$. If $\mathscr{X}_\theta$ is the set of all sample points for which $f(T) = P_\theta(E)$, then $P_\theta(\mathscr{X}_\theta) = 1$ for all $\theta \varepsilon \Theta$. If $P_{\theta_1}(E) \neq P_{\theta_2}(E)$, then the two sets $\mathscr{X}_{\theta_1}$ and $\mathscr{X}_{\theta_2}$ are disjoint and so P_{θ_1} and P_{θ_2} have disjoint supports, which is a contradiction in the present case. Thus the class of ancillary events may be characterized as the class of events that are independent of $X_1 + X_2$.

The "Basu theorems" are direct generalizations of the foregoing results and may be stated as follows.

Theorem 1. Let T be sufficient and boundedly complete Then a statistic Y is ancillary only if it is (conditionally) independent of T for each θ.

The measures P_θ and $P_{\theta'}$ are said to overlap if they do not have disjoint supports. The family $\mathscr{P} = \{P_\theta\}$ of measures on a space $\mathscr{X}$ is said to be connected if for all θ, θ' there exists a finite sequence $\theta_1, \theta_2, \ldots, \theta_k$ such that every two consecutive members of the sequence $P_\theta, P_{\theta_1}, P_{\theta_2}, \ldots, P_{\theta_k}, P_{\theta'}$ overlap. For example, if under P_θ, $-\infty<\theta<\infty$, the random variables $X_1, X_2, \ldots, X_n$ are i.i.d. with a common uniform distribution concentrated on the interval $(\theta, \theta+1)$, then the family $\mathscr{P} = \{P_\theta\}$ is connected even though P_θ and $P_{\theta'}$ do not overlap whenever $|\theta - \theta'| \geqslant 1$.

Theorem 2. Let $\mathscr{P} = \{P_\theta\}$ be connected and T be sufficient. Then Y is ancillary if it is (conditionally) independent of T for each θ.

Neither the condition of bounded completeness (in Theorem 1) nor that of connectedness (in Theorem 2) can be entirely dispensed with. If $T = X$, the whole sample, then it is sufficient. Consider, therefore, a case where a nontrivial ancillary statistic $Y = Y(x)$ exists. Such an Y cannot be independent of $T = X$, because, if it were,then Y has to be independent of itself, which it cannot be unless it is a constant (a trivial ancillary). On the other hand, if $\mathscr{P}$ is not connected, then it is typically true that there exists nonempty proper subsets $\mathcal{X}_0 \subset \mathcal{X}$ and $\Theta_0 \subset \Theta$ of the sample space $\mathcal{X}$ and the parameter space Θ, respectively, such that

$$P_\theta(\mathcal{X}_0) = \begin{cases} 1 & \text{for all } \theta \in \Theta_0 \\ 0 & \text{for all } \theta \in \Theta - \Theta_0 . \end{cases}$$

Koehn and Thomas (1975) called as a set $\mathcal{X}_0$ a splitting set. A splitting set (event) $\mathcal{X}_0$ is clearly not ancillary; however, for every $\theta \in \Theta$, it is P_θ-equivalent either to the whole space $\mathcal{X}$ (the sure event) or to the empty set ϕ (the impossible event). Therefore, the non-ancillary event $\mathcal{X}_0$ is independent of every other event E; that is, $\mathcal{X}_0$ is independent of the sufficient statistic T = X. Basu (1958a) gave a pathological example of a statistical model, with a disconnected $\mathcal{P}$, where we have two independent sufficient statistics.

Consider the following three propositions :

(a) T is sufficient.

(b) Y is ancillary.

(c) T and Y are (conditionally) independent for each $\theta \in \Theta$.

Under suitable conditions (a) and (b) together imply (c) (Theorem 1) and (a) and (c) together imply (b) (Theorem 2). The following theorem completes the set.

Theorem 3. If (T, Y) is jointly sufficient, then (b) and (c) imply (a).

Theorem 1 is often used to solve diverse problems involving sampling distributions. The following example illustrates this.

Example. Let $X_1, X_2, \ldots$ be a sequence of mutually independent gamma variables with shape parameters $\alpha_1, \alpha_2, \ldots$ (i.e., the PDF of X_n is $Cx^{\alpha_n - 1} e^{-x}$, $x > 0$, $n = 1, 2, \ldots$.) Let $T_n = X_1 + X_2 + \ldots + X_n$ and $Y_n = T_n/T_{n+1}$, $n = 1, 2, \ldots$. It is easy to show that T_n has a gamma distribution with shape parameter $\alpha_1 + \alpha_2 + \ldots + \alpha_n$ and that Y_n has a beta distribution with parameters $\alpha_1 + \alpha_2 + \ldots + \alpha_n$ and α_{n+1} . But it is not clear why $Y_1, Y_2, \ldots, Y_{n-1}, T_n$ have to be mutually independent. This is seen as follows.

Introduce a scale parameter θ into the joint distribution of $X_1, X_2, \ldots$. That is, suppose that the X_n's are mutually independent and that the PDF of X_n is $c(\theta)\, x^{\alpha_n - 1} e^{-x/\theta}$, $x > 0$, $\theta > 0$, $n = 1, 2, \ldots$. Regard the α_n's as known positive constants and θ as the unknown parameter. With $\mathbf{X}^{(n)} = (X_1, X_2, \ldots, X_n)$ as the sample, $T_n = X_1 + X_2 + \ldots + X_n$ is a complete sufficient statistic The vector-valued statistic $\mathbf{Y}^{(n-1)} = (Y_1, Y_2, \ldots, Y_{n-1})$ is scale-invariant. Since θ is a scale parameter, it follows that $\mathbf{Y}^{(n-1)}$ is an ancillary statistic. From Theorem 1 it then follows that, for each n, $\mathbf{Y}^{(n-1)}$ is independent of T_n. Since $(\mathbf{Y}^{(n-1)}, T_n)$ is a function of the first n X_i's, the pair $(\mathbf{Y}^{(n-1)}, T_n)$ is independent of X_{n+1}. Thus $\mathbf{Y}^{(n-1)}$, T_n and X_{n+1} are mutually independent. It follows at once that $\mathbf{Y}^{(n-1)}$ is independent of $Y_n = T_n/(T_n + X_{n+1})$. Therefore, for each $n \geqslant 2$, the vector $(Y_1, Y_2, \ldots, Y_{n-1})$ is independent of Y_n and this means that the Y_i's are mutually independent.

(Refer to Hogg and Craig (1956) for several interesting uses of Theorem 1 in proving results in distribution theory.)

Basu (1955c) stated and proved Theorem 1 in the generality stated here. At about the same time, Hogg and Craig proved a particular case of the theorem. Basu(1955c) stated Theorem 2 without the condition of connectedness for $\mathcal{P}$. This has resulted in the theorem being incorrectly stated in several statistical texts. Basu (1958a)stated and proved Theorem 2 in the general form stated here. Koehn and Thomas (1975)noted that the proposition "every" Y that is conditionally independent of a sufficient T is ancillary" is true if and only if there do not exist a splitting set as specified above. It turns out, however, that the two notions of connectedness and nonexistence of splitting sets coincide for all typical statistical models.

The Basu theorems are of historical interest because they established a connection between the three apparently unrelated notions of sufficiency, ancillarity and independence. That the three notions really hang together is not easy to see through if we adopt an orthodox Neyman-Pearson point of view. However, if we take a Bayesian view of the matter and regard θ as a random variable with a (prior) probability distribution ξ and the model $\mathcal{P}$ as a specification of the set of conditional distributions of X given θ, then the notions of ancillarity and sufficiency will appear to be manifestations of the notion of conditional independence.

For a model $\mathcal{P}$ and for each prior ξ, consider the joint probability distribution Q_ξ of the pair (θ, X). A statistic $Y = Y(X)$ is ancillary if its conditional distribution, given θ, is θ-free. In other words, Y is ancillary if, for each joint distribution Q_ξ of (θ, X), the two random variables Y and θ are stochastically independent. A statistic $T = T(X)$ is sufficient if the conditional distribution of X, given θ and T, depends only on T (i.e., the conditional distribution is θ-free). Sufficiency of T may, therefore, be characterized as follows:

Definition. The statistic T is sufficient if, for each Q_ξ, X and θ are conditionally independent given T.

Thus a neo-Bayesian version of Theorem 1 may be stated as :

Theorem 1(a). Suppose that, for each Q_ξ, the variables Y and θ are stochastically independent and also X and θ are conditionally independent given T. Then Y and T are conditionally independent given θ provided that the statistic T is boundedly complete in the sense described earlier.

Refer to Florens and Mouchart (1977) for more on the Bayesian insight on the theorems and also to Dawid (1979) for a clear exposition on conditional independence as a language of statistics.

REFERENCES

Ajgaonkar, S.G. Prabhu (1965). On a class of linear estimates in sampling with varying probabilities without replacements. J. Amer. Statist. Assoc. 60, 637-642.

Akaike, H. (1982). On the fallacy of the likelihood principle. Statistics and Probability Letters 1, 75-78.

Amari, S. (1982). Geometrical theory of asymptotic ancillarity and conditional inference. Biometrika 69, 1-18.

Andersen, E. B. (1967). On partial sufficiency and partial ancillarity. Skandinavist Aktuarietidskrift 50, 137-52.

Andersen, E.B. (1970). Asymptotic properties of conditional maximum likelihood estimators. J. Roy. Statist. Soc. B 32, 283-301.

Andersen, E. B. (1971). A strictly conditional approach in estimation theory. Skand. Aktuarietidskr. 54, 39-49.

Anscombe, F.J. (1957). Dependence of the fiducial argument on the sample rule. Biometrika 33, 464- 469.

Bahadur, R.R. (1954). Sufficiency and statistical decision functions. Ann. Math. Statist. 25, 423-462.

Barankin, E.W. (1949). Locally best unbiased estimates. Ann. Math. Statist. 20, 477-501.

Barnard, G.A. (1947a). A review of 'Sequential Analysis' by Abraham Wald. J. Amer. Statist. Assoc. 42, 658-669.

Barnard, G.A. (1947b). The meaning of significance level. Biometrika 34, 179-182.

Barnard, G.A. (1949). Statistical inference (with Discussion). J. Roy. Statist. Soc. B 11, 115-149.

Barnard, G.A. (1962). Comments on Stein's 'A remark on the likelihood principle'. J. Roy. Statist. Soc. A 125, 569-573.

Barnard, G.A. (1963). Some logical aspects of the fiducial argument. J. Roy. Statist. Soc. B 125, 111-114.

Barnard, G.A. (1967). The use of the likelihood function in statistical practice. Proc. Fifth Berkeley Symp., University of California Press, 1, 27-40.

Barnard, G.A. (1971). Scientific inferences and day-to-day decisions. In Foundations of Statistical Inference, V.P. Godambe and D.A. Sprott (eds.). Holt, Rinehart and Winston, Toronto.

Barnard, G.A. (1974). On likelihood. In the Proceedings of the Conference on Foundational Questions in Statistical Inference, O. Barndorff-Nielsen, P. Blaesild, and G. Schou (eds.). Department of Theoretical Statistics, University of Aarhus.

Barnard, G.A. (1975). Conditional inference is not inefficient. Scandinavian J. of Statist. 3, 132-134.

REFERENCES

Barnard, G.A. (1980). Pivotal inference and the Bayesian controversy (with Discussion). In Bayesian Statistics, J. M. Bernardo, M. H. DeGroot, D.V. Lindley, and A.F.M. Smith (eds.). University Press, Valencia.

Barnard, G.A. and Godambe, V.P. (1982). Allan Birnbaum, A memorial article. Ann. Statist. 10, 1033-1039.

Barnard, G.A., Jenkins, G.M., and Winsten, C. B. (1962). Likelihood inference and time series. J. Roy. Statist. Soc. A 125, 321-372.

Barnard, G.A.and Sprott,D.A.(1971). A note on Basu's examples of anomalous ancillary statistics (with Discussions). In Foundations of Statistical Inference, V.P. Godambe and D.A. Sprott (Eds.). Holt, Rinehart and Winston, 163-176.

Barnard, G.A. and Sprott, D.A. (1983). The generalized problem of the Nile: robust confidence sets for parametric functions. Ann. Statist. 11, 104-113.

Barndorff-Nielsen, O. (1971). On Conditional Statistical Inference. Matematisk Institute, Aarhus University.

Barndorff-Nielsen, O. (1973). Exponential Families and Conditioning. Sc. D. thesis, Department of Mathematics, University of Copenhagen, Denmark.

Barndorff-Nielsen, O. (1975). Nonformation. Biometrika 63, 567-571.

Barndorff-Nielsen, O. (1978). Information and Exponential Families in Statistical Theory. Wiley, New York.

Barndorff-Nielsen, O. (1980). Conditionality resolutions. Biometrika 67, 293-310.

Barnett, V. (1982). Comparative Statistical Inference (2nd Edition). John Wiley and Sons, New York.

Bartlett, M.S. (1936). Statistical information and properties of sufficiency. Proc. Royal Soc. A 154, 124.

Bartlett, M.S. (1978). Nearest neighbour models in the analysis of field experiments (with discussion). J. Roy. Statist. Soc. B 40, 147-174.

Basu, D. (1951). On the independence of linear functions of independent chance variables. Bull. Int. Statist. Inst. 33, pt. 2, 83-96.

Basu, D. (1952a). On a class of admissible estimators of the normal variance. Sankhyā, 12, Pts. 1 & 2, 57-62.

Basu, D. (1952b). On symmetric estimators in point estimation with convex loss functions. Sankhyā, 12, Pts. 1 & 2, 45-52.

Basu, D. (1952c). On the minimax approach to the problem of point estimation. Proc. Nat. Inst. of Sc., India 18, No. 4, 287-299.

Basu, D. (1953a). Choosing between two simple hypotheses and the criterion of consistency. Proc. Nat. Inst. of Sc., India 19, No. 6, 841-849.

Basu, D. (1953b). Some contributions to the Theory of Statistical Inference. D. Phil. thesis submitted to the University of Calcutta.

Basu, D. (1954). On the optimum character of some estimators used in multistage sampling problems. Sankhyā 13, Pt. 4, 363-368.

REFERENCES

Basu, D. (1955a). An inconsistency of the method of maximum likelihood. Ann. Math. Statist. 26, No. 1, 144-145.

Basu, D. (1955b). A note on the Theory of Unbiased Estimation. Ann. Math. Statist. 26, No. 2, 345-348. [Chapter XX]

Basu, D. (1955c). On statistics independent of a complete sufficient statistic. Sankhyā 15, Pt. 4, 377-380. [Chapter XXII]

Basu, D. (1956a). The concept of asymptotic efficiency. Sankhyā 17, Pt. 2, 193-196. [Chapter XXI]

Basu, D. (1956b). A note on the multivariate extension of some theorems related to the univariatre normal distribution. Sankhyā 17, Pt. 3, 221-224.

Basu, D. (1958a). On statistics independent of a sufficient statistic. Sankhyā 20, Pts. 3 & 4, 223-226. [Chapter XXIII]

Basu, D. (1958b). On sampling with and without replacements. Sankhyā 20, Pts. 3 & 4, 287-294.

Basu, D. (1959). The family of ancillary statistics. Sankhyā 21, Pts. 3 & 4, 247-256.

Basu, D. (1964). Recovery of ancillary information. Contributions to Statistics, Pergamon Press, Oxford, pp 7-20. Republished in Sankhyā 26, (1964), 3-16). [Chapter I].

Basu, D. (1965a). Sufficiency and model-preserving transformations. University of North Carolina. Inst. of Statistics, Mimeo Series No. 420.

Basu, D. (1965b). Problems related to the existence of minimal and maximal elements in some families of statistics (subfields). Proc. Fifth Berkeley Symp. on Maths. and Probability, 1, 41-50. University of California Press.

Basu, D. (1969a). On sufficiency and invariance. In Essays in Prob. and Statistics, R.C. Bose et al (eds.), University of North Carolina, 61-84. [Chapter VIII].

Basu, D. (1969b). Role of the sufficiency and likelihood principles in survey theory. Sankhyā A 31, Pt. 4, 441-453. [Chapter XI].

Basu, D. (1971). An essay on the logical foundations of survey sampling, part one (with discussions). In Foundations of Statistical Inference, V.P. Godambe and D.A. Sprott (eds.). Holt, Rinehart and Winston, Toronto. [Chapters XII and XIII].

Basu, D. (1973). Statistical information and likelihood (with discussions). In Proceedings of Conference on Foundational Questions of Statistical Inference, O. Barndorff-Nielsen, P. Blaesild and G. Schou (eds.). Dept. of Theoretical Statistics, University of Aarhus, Denmark. Republished in Sankhyā A 37 (1975), 1-71. [Chapters II, III, IV & V].

Basu, D. (1977). On the elimination of nuisance parameters. J. Amer. Statist. Assoc. 72, No. 358, 355-366. [Chapter VII]

Basu, D. (1978a). On partial sufficiency: a review. J. Statistical Planning and Inference 2, 1-13. [Chapter VI]

REFERENCES

Basu, D. (1978b). On the relevance of randomization in data analysis (with discussions). In Survey Sampling and Measurement, N. K. Namboodiri (ed.), Academic Press : New York, 267-339. [Chapter XIV]

Basu, D. (1979a). Discussion on Joseph Berkron's paper "In dispraise of the Exact Test". J. Statistical Planning and Inference 3, 189-192. [Chapter XVIII]

Basu, D. (1979b). A discussion on survey theory. Incomplete Data in Sample Surveys, 3, Sec 8, 1983, 407-410, Academic Press: New York. [Chapter XIX]

Basu, D. (1980). Randomization analysis of experimental data : the Fisher Randomization Test. (with discussions). J. Amer. Statist. Assoc. 75, No. 371, 575-595. [Chapters XV and XVI]

Basu, D. (1981). On ancillary statistics, pivotal quantities and confidence statements. In Topics in Applied Statistics, Y.P. Chaubey and T.D. Dwivedi (eds.), Concordia University, Montreal, 1-29. [Chapter IX]

Basu, D. (1982a). Basu Theorems. Kotz-Johnson: Encyclopedia of Statistical Sciences, 1, 193-196, John Wiley : New York. [Chapter XXIV]

Basu, D. (1982b). A note on likelihood. Proc. National Symp. on Prob. and Statistics, University of Sao Paulo, Brazil. [Chapter XVII]

Basu, D. and Cheng, S.C. (1981). A note on sufficiency in coherent models. Internat. Jl. Math. & Math. Sci. 4, No. 3, 571-582.

Basu, D. and Ghosh, J. K. (1967). Sufficient statistics in sampling from a finite universe. Bull. Int. Statist. Inst. 42,BK.2, 850-859. [Chapter X]

Basu, D. and Ghosh, J. K. (1969). Invariant sets for translation-parameter families of measures. Ann. Math. Statist. 40, No. 1, 162-174.

Basu, D. and Pereira, Carlos A. de B. (1982). On the Bayesian analysis of categorical data : the problem of nonresponse. Jl. Statistical Planning and Inference 6, 345-362.

Berger, J. (1984a). In defense of the likelihood principle: axiomatics and coherency. In Bayesian Statistics II, J. M. Bernardo, M. H. DeGroot, D. Lindley and A. Smith (eds.).

Berger, J. (1984b). Bayesian salesmanship. In Bayesian Inference and Decision Techniques with Applications: Essays in Honor of Bruno deFinetti, P.K. Goel and A. Zellner (eds.). North-Holland, Amsterdam.

Berger, J. (1984c). The frequentist viewpoint and conditioning. To appear in the Proceedings of the Berkeley Conference in Honor of J. Kiefer and J.Neyman, L. LeCam and R. Olshen (eds.). Wadsworth, Belmont California.

Berger, J. (1984d). A review of J. Kiefer's work on conditional frequentist statistics. To appear in The Collected Works of Jack Kiefer (L. Brown, I. Olkin, J. Sacks, H. Wynn, eds.).

Berger, J. and Wolpert, R.L. (1984). The Likelihood Principle. Institute of Mathematical Statistics. Lecture Notes-Monograph Series, Vol. 6.

REFERENCES

Birnbaum, A. (1961). On the foundations of statistical inference: binary experiments. Ann. Math. Statist. 32, 414-435.

Birnbaum, A. (1962). On the foundations of statistical inference (with discussion). J. Amer. Statist. Assoc. 57, 269-326.

Birnbaum, A. (1968). Likelihood. In International Encyclopedia of the Social Sciences, Vol. 9.

Birnbaum, A. (1969). Concepts of statistical evidence. In Philosophy, Science, and Method: Essays in Honor of Ernest Nagel, S. Morgenbesser, P. Suppes and M. White (eds.). St. Martin's Press, New York.

Birnbaum, A. (1970). Statistical methods in scientific inference. Nature 225, 1033.

Birnbaum, A. (1972). More on concepts of statistical evidence. J. Amer. Statist. Assoc. 67, 858-861.

Birnbaum, A. (1977). The Neyman-Pearson theory as decision theory and as inference theory: with a criticism of the Lindley-Savage argument for Bayesian theory. Synthese 36, 19-49.

Blackwell, D. (1951). Comparison of experiments. Proc. second Berkeley Symp., University of California Press, 93-102.

Bondar, J. V. (1977). On a conditional confidence principle. Ann. Statist. 5, 881-891.

Box, G.E.P. (1980). Sampling and Bayes' inference in scientific modelling and robustness (with discussion). J. Roy. Statist. Soc. B 143, 383-430.

Box, J.F. (1978). R. A. Fisher. The Life of a Scientist, New York. John Wiley and Sons.

Brillinger, D.R., Jones, L.V., and Tukey, J. W. (1978). "The Role of Statistics in Weather Resources Management." Report of the Statistical Task Force to the Weather Modification Advisory Board.

Brown, L. D. (1967). The conditional level of Student's t-Test. Ann. Math. Statist. 38, 1068-1071.

Brown, L.D. (1978). A contribution to Kiefer's theory of conditional confidence procedures. Ann. Statist. 6, 59-71.

Buehler, R. J. (1971). Measuring information and uncertainty. In Foundations of Statistical Inference, V.P. Godambe and D.A. Sprott (eds.). Holt, Rinehart and Winston, Toronto.

Buehler, R. J. (1982). Some ancillary statistics and their properties (with discussion). J. Amer. Statist. Assoc. 77, 581-594.

Buehler, R. J. and Fedderson, A.P. (1963). Note on a conditional property of Student's t. Ann. Math. Statist. 34, 1098-1100.

Bunker, J. P., Barnes, B.A., and Mosteller, F. (1977). Costs, Risks, and Benefits of Surgery. Oxford University Press, New York.

Burkholder, D. L. (1961). Sufficiency in the undominated case. Ann. Math. Statist. 32, 1191-1200.

REFERENCES

Cassel, C.M., Sarndal, C.E. and Wretman, J.H. (1977). Foundations of Inference in Survey Sampling. Wiley, New York.

Cheng, S.C. (1978). A mathematical study of sufficiency and adequacy. Ph.D. thesis submitted to the Florida State University.

Cochran, W.G. (1977). Sampling Techniques (3rd edn.). John Wiley and Sons, New York.

Cornfield, J. (1966). Sequential trials, sequential analysis and the likelihood principle. The American Statist. 20, No. 2, 18-23.

Cornfield, J. (1969). The Bayesian outlook and its application (with discussion). Biometrics 25, 617-657.

Cox, D. R. (1958a). Some problems connected with statistical inference. Ann. Math. Statist. 29, 357-372.

Cox, D. R. (1958b). Planning of Experiments. John. Wiley & Sons, New York.

Cox, D. R. (1971). The choice between alternative ancillary statistics. J. Roy. Statist. Soc. B. 33, 251-255.

Cox, D. R. (1975). Partial likelihood. Biometrika 62, 269-276.

Cramer, H. (1946). Mathematical Methods of Statistics. Princeton University Press.

Darling, D.A. and Robbins, Herbert (1967). Series of Notes published in the Proceedings of the U.S. National Academy of Sciences, beginning will be Vol.57, 1188-92.

Darmois, G. (1951). Sur diverses properties characteristiques de la loi de probabilite de Laplace-Gauss, Bull. Int. Stat. Inst. 33 pt. 2, 79-82.

Darmois, G. (1953). Analyse Generele des liaisons stochastiques - Etude particuliere de l'analyse factorielle lineaire. Rev. Inst. Internat. Stochastiques, 21, 2-8.

Dawid, A. P. (1975). On the concepts of sufficiency and ancillarity in the presence on nuisance parameters. J. Roy. Statist. Soc. B 37, 248-258.

Dawid, A. P. (1979). Conditional independence in statistical theory. J. Roy. Statist. Soc. B, 41, 1-31.

Dawid, A. P. (1980). A Bayesian look at nuisance parameters. In Bayesian Statistics, J. M. Bernardo, M. H. DeGroot, D. V. Lindley and A. F. M. Smith (eds.). University Press, Velencia.

Dawid, A. P. (1981). Statistical inference. In Encyclopedia of Statistical Sciences, S. Kotz and N. L. Johnson (eds.). Wiley, New York.

De Finetti, B. (1972). Probability, Induction, and Statistics. Wiley, New York.

De Finetti, B. (1974). Theory of Probability, Volumes 1 and 2. Wiley, New York.

De Groot, M. H. (1973). Doing what comes naturally: interpreting a tail area as a posterior probability or as a likelihood ratio. J. Amer. Statist. Assoc. 68, 966-969.

REFERENCES

Dempster, A. P. (1974). The direct use of likelihood for significance testing. In the Proceedings of the Conference on Foundational Questions in Statistical Inference, O. Barndorff-Nielsen, P. Blaesild and G. Schou (eds.). Department of Theoretical Statistics, University of Aarhus.

Desraj (1968). Sampling Theory, McGraw-Hill.

Durbin, J. (1961). Some methods of constructing exact tests. Biometrika 48, 41-55.

Durbin, J. (1970). On Birnbaum's theorem on the relation between sufficiency, conditionality and likelihood. J. Amer. Statist. Assoc. 65, 395-398.

Dynkin, E. B. (1965). Markov Processes, Vol. II. Springer-Verlag, Berlin.

Edwards, A. W. F. (1972). Likelihood. Cambridge University Press, Cambridge.

Edwards, A.W.F. (1974). The history of likelihood. Int. Statist. Rev. 42, 9-15.

Efron, B. and Hinkley, D. V. (1978). Assessing the accuracy of the maximum likelihood estimator: observed versus expected Fisher information. Biometrika 65, 457-482.

Ericson, W. A. (1969). Subjective Bayesian models in sampling finite populations. J. Roy. Statist. Soc. B 31, 195-233.

Farrell, R. H. (1962). Representation of invariant mesures. Illinois J. of Math. 6, 447-467.

Fienberg, S. E. and Hinkley, D.V. (1980). R. A. Fisher : An Appreciation, Lecture Notes in Statistics. Springer-Verlag, New York.

Fisher, R. A. (1912). On an absolute criterion for fitting frequency curves, Messeng. Math. 41, 155-160.

Fisher, R. A. (1920). A mathematical examination of the methods of determining the accuracy of an observation by the mean error and the mean square error. Monthly Notices of the Royal Astronomical Society, 80, 758-770. (Also reproduced in Fisher 1950).

Fisher, R. A. (1922). The mathematical foundations of theoretical statistics, Philosophical Transactions of the Roy. Soc. London, Ser. A 222, 309-368. (Also reproduced in Fisher 1950).

Fisher, R. A. (1925). Theory of statistical estimation, Proceedings of the Cambridge Society, 22, 700-725.

Fisher, R. A. (1930). Inverse probability. Proceedings of Cambridge Philosophical Society, 26, 528-35. (Also reproduced in Fisher 1950).

Fisher, R. A. (1934a). Two new properties of mathematical likelihood, Proceedings of the Royal Society, Ser. A, 144, 285.

Fisher, R. A. (1934b). Statistical Methods for Research Workers, 5th edn. Oliver & Boyd, Edinburgh.

Fisher, R. A. (1935a). (7th Edn, 1960). The Design of Experiments. Oliver & Boyd, Edinburgh.

Fisher, R. A. (1935b). Discussion of "Statistical problems in agricultural experimentation" by J. Neyman, J. Roy. Statist. Soc. II, 2, 154-180.

Fisher, R. A. (1935c). The logic of inductive inference. J. Roy. Statist. Soc. 98, 39-54.

Fisher, R. A. (1936). Uncertain inference, Proceedings of the American Academy of Arts and Sciences, 71, 245-258.

Fisher, R. A. (1950). Contributions to Mathematical Statistics. John Wiley and Sons, London.

Fisher, R. A. (1956). Statistical Methods and Scientific Inference. Oliver & Boyd, London.

Fisher, R. A. (1974). Collected Papers of R. A. Fisher. University of Adelaide, Adelaide, Australia. (Papers are referred to by number).

Florens, J. P. and Mouchart, M. (1977). Reduction of Bayesian experiments. CORE Discuss. Paper 1737.

Fraser, D. A. S. (1956). Sufficient statistics with nuisance parameters. Ann. Math. Statist. 27, 838-842.

Fraser, D. A. S. (1963). On the sufficiency and likelihood principles. J. Amer. Statist. Assoc. 58, 641-647.

Fraser, D. A. S. (1968). The Structure of Inference. Wiley, New York.

Fraser, D. A. S. (1977). Confidence, posterior probability and the Buehler example. Ann. Statist. 5, 892-898.

Gilbert, J. P., Light, R. J. and Mosteller, F. (1977). "Assessing Social Innovations: An Empirical Base for Policy", in Statistics and Public Policy, W. B. Fairley and F. Mosteller (eds.). Addison-Wesley, Menlo Park, Calif.

Godambe, V. P. (1955). A unified theory of sampling from finite populations. J. Roy. Statist. Soc. B 17, 269-278.

Godambe, V. P. (1960). An admissible estimate for any sampling design. Sankhyā 22, 285-288.

Godambe, V. P. (1965). Contributions to the unified theory of sampling, Review of the International Statistical Institute, 33, 242-258.

Godambe, V. P. (1966a). A new approach to sampling from finite population, I : sufficiency and linear estimation. J. Roy. Statist. Soc. B 28, 310-319.

Godambe, V. P. (1966b). A new approach to sampling from finite population II: distribution-free sufficiency, J. Roy. Statist. Soc. B 28, 320-328.

Godambe, V. P.(1968a). Bayesian sufficiency in survey sampling. Ann. Inst. Stat. Math. (Japan), 20, 363-373.

Godambe, V. P.(1968b). Some aspects of the theoretical developments in survey sampling; In New Developments in Survey Sampling, Johnson, N. L. and Smith, H. (eds.). Wiley Interscience, (1969), 27-58.

REFERENCES

Godambe, V. P. (1975). A reply to my critics, Sankhyā, Series C, 37, 53-76.

Godambe, V. P. (1979a). Comments on Dawid's paper: Conditional independence in statistical theory, J. Roy. Statist. Soc. 41, 1-31.

Godambe, V. P. (1979b). On Birnbaum's mathematically equivalent experiments. J. Roy. Statist. Soc. B 41, 107-110.

Godambe, V. P. (1980). Unpublished research report.

Godambe, V. P. (1982). Likelihood principle and randomization. In Statistics and Probability: Essays in Honour of C. R. Rao. G. Kallianpur, P. R. Krishnaiah and J. K. Ghosh (eds.), North-Holland, Amsterdam.

Godambe, V. P. and Joshi, V. M. (1965). Admissibility and Bayes estimation in sampling finite populations - Parts I, II and III. Ann. Math. Statist. 36, 1707-1742.

Godambe, V. P. and Thompson, M. E. (1971). Bayes, fiducial and frequency aspects of Statistical inference in regression analysis in survey sampling. J. Roy. Statist. Soc. Series B 33, 361-390.

Good, I. J. (1950). Probability and the Weighing of Evidence. Griffin, London.

Good, I. J. (1965). The Estimation of Probabilities. M. I. T. Press, Cambridge.

Good, I. J. (1976). The Bayesian influence, or how to sweep subjectivism under the carpet. In Foundations of Probability Theory, Statistical Inference, and Statistical Theories of Science, Vol. II, W. L. Harper and C.A. Hooker (eds.). Reidel, Dordrecht.

Greenberg, B. G. (1951). Why randomize? Biometrika 7, 309-322.

Hacking, I. (1965). Logic of Statistical Inference. Cambridge University Press, Cambridge.

Hájek, J. (1959). Optimum strategy and other problems in probability sampling, Casopis Pest. Math., 84, 387-423.

Hájek, J. (1965). On basic concepts of statistics, Proceedings of the Fifth Berkeley Symposium, University of California Press, California, 1, 139-162.

Hájek, J. (1967). On basic concepts of statistics. In Proceedings of the Fifth Berkeley Symposium on Mathematical Statistics and Probability, LeCam and J. Neyman (eds.). University of California Press, Berkeley.

Hájek, J. (1971a). Limiting properties of likelihoods and inference. In Foundations of Statistical Inference, V. P. Godambe and D. A. Sprott (eds.). Holt, Rinehart and Winston, Toronto.

Hájek, J. (1971b). Discussion on D. Basu : an essay on logical foundations of survey sampling, Part I. In Foundations of Statistical Inference, V. P. Godambe and D. A. Sprott (eds.). Holt-Rinehart, New York.

Hall, W. J., Wijsman, R.A. and Ghosh, J. K. (1965). The relationship between sufficiency and invariance. Ann. Math. Statist. 36, 575-614.

REFERENCES

Halmos, P. R. and Savage, L. J. (1949). Application of the Randon-Nykodym theorem to the theory of sufficient statistic. Ann. Math. Statist. 20, 225-241.

Hanurav, T. V. (1962). On Horvitz-Thompson estimator, Sankhyā, A 24, 429-436.

Hanurav, T. V. (1968). Hyper-admissibility and optimum estimators for sampling finite populations, Ann. Math. Statist. 39, 621-642.

Harville, D. A. (1975). "Experimental Randomization : Who Needs It?" The American Statist. 29, 27-31.

Hege, V. S. (1967). An optimum property of the Horvitz-Thompson estimate, J. Amer. Statist. Assoc. 2, 1013-1017.

Hill, B. (1973). Review of "Likelihood" by A. W. F. Edwards. J. Amer. Statist. Assoc. 68, 487-488.

Hill, B. (1974). Review of "Bayesian Inference in Statistical Analysis" by G. E. P. Box and G. Tiao. Technometrics 16, 478-479.

Hill, B. (1975). Abberant behavior of the likelihood function in discrete cases. J. Amer. Statist. Assoc. 70, 717-719.

Hill, B. (1981). On some statistical paradoxes and non-conglomerability. In Bayesian Statistics, J. M. Bernardo, M. H. DeGroot, D. V. Lindley and A. F. M. Smith (eds.). University Press, Valencia.

Hinkley, D. V. (1978). Likelihood inference about location and scale parameters. Biometrika 65, 253-262.

Hinkley, D. V. (1979). Predictive likelihood. Ann. Statist. 7, 718-728.

Hinkley, D. V. (1980a). Fisher's development of conditional inference. In R. A. Fisher: An Appreciation, S. E. Fienberg and D. V. Hinkley (eds.). Springer-Verlag, New York.

Hinkley, D. V. (1980b). Likelihood as approximate pivotal distribution. Biometrika 67, 287-292.

Hinkley, D. V. (1983). Can frequentist inferences be very wrong? A conditional 'yes'. In Scientific Inference, Data Analysis, and Robustness, G. E. P. Box, T. Leonard and C. F. Wu (eds.). Academic Press, New York.

Hodges, J. L. and Lehmann, E. L. (1950). Some problems of minimax point estimation. Ann. Math. Statist. 21, 182-197.

Hodges, J. L. and Lehmann, E. L. (1973). Wilcoxon and t-test for matched pairs of typed subjects. J. Amer. Statist. Assoc. 68, 151-158.

Hogg, R. V. and Craig, A. T. (1956). Sufficient statistics in elementary distribution theory. Sankhyā 17, 209.

Horvitz, D. G. and Thompson, D. J. (1952). A generalization of sampling without replacement from a finite universe. J. Amer. Statist. Assoc. 47, 663-685.

Jaynes, E. T. (1981). The intuitive inadequacy of classical statistics. Presented at the International Convention on Fundamentals of Probability and Statistics, Luino, Italy.

REFERENCES

Jeffreys, H. (1938). Maximum likelihood, inverse probability and the method of moments. Ann. Eugen. 8, 146-151.

Jeffreys, H. (1973). Scientific Inference (3rd ed.) C. U. P., Cambridge.

Joshi, V. M. (1968). Admissibility of the sample mean as estimate of the mean of a finite population. Ann. Math. Statist. 39, 606-620.

Kalbfleisch, J. D. (1975). Sufficiency and conditionality. Biometrika 62, 251-268.

Kalbfleisch, J. D. (1978). Likelihood methods and nonparametric tests. J. Amer. Statist. Assoc. 73, 167-170.

Kalbfleisch, J. D. and Sprott, D. A. (1970). Application of likelihood methods to models involving large number of parameters (with discussion). J. Roy. Statist. Soc. B 32, 175-208.

Kempthorne, O. (1952). The Design and Analysis of Experiments. John Wiley and Sons, New York.

Kempthorne, O. (1955). The randomization theory of experimental Inference. J. Amer. Statist. Assoc. 50, 946-967.

Kempthorne, O. (1961). The design and analysis of experiments with some reference to educational research. In Research Designs and Analysis, Second Annual Phi Delta Kappa Symposium on Educational Research, 97-126.

Kempthorne, O. (1966). Some aspects of experimental inference. J. Amer. Statist. Assoc. 61, 11-34.

Kempthorne, O. (1974). Sampling inference, experimental inference and observation inference. Paper presented at the Mahalanobis Memorial Symposium on recent trends of research in Statistics, Calcutta, India.

Kempthorne, O. (1975). Inference for experiments and randomization. In A Survey of Statistical Designs and Linear Models, J. N. Srivastava (ed.), 303-331. North Holland Publishing Co., Amsterdam.

Kempthorne, O. (1977). "Why Randomize?" Jl. Statistical Planning and Inference, 1, 1-25.

Kempthorne, O. and Doerfler, T. E. (1969). The behaviour of some significance tests under experimental randomization. Biometrika 56, 231-247.

Kempthorne, O. and Folks, J. L. (1971). Probability, Statistics and Data Analysis. Iowa State University Press, Ames.

Kiefer, J. (1975). Conditional confidence approach in multi-decision problems. In Proceedings of the Fourth Dayton Multivariate Conference, P. R. Krishnaiah (ed.). 143-158. North-Halland, Amsterdam.

Kiefer, J. (1976). Admissibility of conditional confidence procedures. Ann. Math. Statist. 4, 836-865.

Kiefer, J. (1977a). Conditional confidence statements and confidence estimators (with discussion). J. Amer. Statist. Assoc. 72, 789-827.

Kiefer, J. (1977b). The foundations of statistics - are there any? Synthese 36, 161-176.

Kiefer, J. (1980). Conditional inference. In the Encyclopedia of Statistics, S. Kotz and N. Johnson (eds.). Wiley, New York.

Koehn, U. and Thomas, L. D. (1975). On statistics independent of a sufficient statistic: Basu's Lemma. American Statist. 29, 40-43.

Kolmogorov, A. N. (1933). Foundations of the Theory of Probability, translation 1956, Chelsea, New York.

Kolmogorov, A. N. (1942). Determination of the centre of dispersion and degree of accuracy for a limited number of observations, Izv. Akad. Nauk, USSR Ser. Mat 6, 3-32 (In Russian).

Koopman, B. O. (1940). The axioms and algebra of intuitive probability. Ann. Math. 41, 269-292.

Laha, R. G. (1954). On a characterization of the Gamma distribution. Ann. Math. Stat. 25, 784-787.

Lauritzen, S. L. (1973). The probabilistic background of some statistical methods in physical geodesy, Meddelelse nr. 48, Geodastik Institute, Copenhagen.

Lauritzen, S. L. (1974). Sufficiency, prediction and extreme models. Scand. J. Statist. 1, 128-134.

Le Cam, L. (1953). On some asymptotic properties of maximum likelihood estimates and related Bayes estimates. University of California Publ. in Statistics. 1, 227-330.

Le Cam, L. (1964). Sufficiency and approximate sufficiecy. Ann. Math. Statist. 35, 1419-1455.

Lehmann, E. L. (1959). Testing Statistical Hypotheses. John Wiley and Sons, New York.

Lehmann, E. L. and Scheffe, H. (1950). Completeness, similar regions and unbiased estimation. Sankhyā 10, 305-340.

Lindley, D. V. (1958). A survey of the foundations of statistics. Appl. Statist. 7, 186-198.

Lindley, D. V. (1961). Introduction to Probability and Statistics, Part 2, Cambridge University Press.

Lindley, D. V. (1965). Introduction to Probability and Statistics from a Bayesian Viewpoint, Parts I and II, Cambridge University Press, London and New York.

Lindley, D. V. (1971). Bayesian Statistics Review. S.I.A.M., Philadelphia.

Lindley, D. V. (1982). Scoring rules and the inevitability of probability. Int. Statist. Rev. 50, 1-26.

Lindley, D. V. and Novick, M. (1981). The role of exchangeability in inference. Ann. Statist. 9, 45-58.

Lindley, D. V. and Phillips, L. D. (1976). Inference for a Bernoulli process (a Bayesian view). American Statist. 30, 112-119.

Linnik, Yu. V. (1965). On the elimination of nuisance parameters in statistical problems, Proceeding of the Fifth Berkeley Symposium on Mathematical Statics and Probability, 1, 267-280, Berkeley.

Linnik, Yu. V. (1968). Statistical problems with nuisance parameters, translations of Mathematical Monographs, Vol. 20, American Mathematical Society, Providence, Rhode Island.

Midzuno, H. (1952). On the sampling system with probability proportionate to sum of sizes. Ann. Inst. Math. (Japan). 3, 99-107.

Murthy, M. N. (1958). On ordered and unordered estimators. Sankhyā, A. 20, 254-262.

Neyman, J. (1935a). On a theorem concerning the concept of sufficient statistics, Giorn. Inst. Ital. Attuari, 6, 320-334 (in Italian).

Neyman, J. (1935b). Statistical problems in agricultural experimentation. J. Roy. Statist. Soc. II, 2, 107-154.

Neyman, J. (1937). Outline of a theory of statistical estimation based on the classical theory of Probability. Phil. Trans. R. S. of London, Ser A, No. 767, 236, 333-380.

Neyman, J. (1957). 'Inductive behavior' as a basic concept of philosophy of science. Rev. Intl. Statist. Inst. 25, 7-22.

Neyman, J. (1967). A Selection of Early Statistical Papers of J. Neyman. University of California Press, Berkeley.

Neyman, J. (1977). Frequentist probability and frequentist statistics. Synthese 36, 97-131.

Neyman, J. and Pearson, E. S. (1936). Sufficient statistics and uniformly most powerful tests of statistical hypotheses, Stat. Res. Memoirs, 1, 133-137.

Neyman, J. and Scott, E. L. (1948). Consistent estimates based on partially consistent observations, Econometrica 16, 1-32.

Olshevsky, L. (1940). Two properties of sufficient statistics. Ann. Math. Statist. 11, 104-106.

Owen, A. R. G. (1948). Ancillary statistics and fiducial distribution. Sankhyā 9.

Pathak, P. K. (1964). Sufficiency in sampling theory. Ann. Math. Statist. 35, 785-809.

Pitcher, T. S. (1957). Sets of measures not admitting necessary and sufficient statistics or subfields. Ann. Math. Statist. 28, 267-268.

Pitcher, T. S. (1965). A more general property than domination for sets of probability measures. Pac. Jl. Math. 15, 597-611.

Pitman, E. J. G. (1937). Significance tests which can be applied to samples from any population III : the analysis of variance test. Biometrika 29, 322-335.

Planck Max (1949). Scientific Autobiography and Other Papers, Greenwood Press, New York.

Plante, A. (1971). Counter-examples and likelihood. In Foundations of Statistical Inference, V.P. Godambe and D.A. Sprott (eds.).Holt Rinehart, and Winston, Toronto.

REFERENCES

Pratt, J. W. (1961). Review of Lehmann's Testing Statistical Hypotheses. J. Amer. Statist. Assoc. 56, 163-166.

Pratt, J. W. (1965). Bayesian interpretation of standard inference statements (with discussion). J. Roy. Statist. Soc. B 27, 169-203.

Pratt, J. W. (1976). A discussion of the question: for what use are tests of hypotheses and tests of significance. Commun. Statist.-Theor. Meth. A5, 779-787.

Pratt, J. W. (1977). 'Decisions' as statistical evidence and Birnbaum's 'confidence concept'. Synthese 36, 59-69.

Prohorov, Yu. V. (1965). Some characterization problems in statistics. Proc. Fifth Berkeley Symp. on Math. Stat. and Prob. 1, 341-350.

Raiffa, H. and Schlaifer, R. (1961). Applied Statistical Decision Theory. Graduate School of Business Administration, Harvard University.

Rao, C. R. (1952). Minimum variance estimation in distributions admitting ancillary statistics. Sankhya, 12, 53-56.

Rao, C. R. (1971). Some Aspects of Statistical Inference in Problems of Sampling from Finite Population. In Foundations of Statistical Inference, V. P. Godambe and D. A. Sprott (eds.). Holt, New York.

Robinson, G. K. (1975). Some counterexamples to the theory of confidence intervals. Biometrika 62, 155-161.

Robinson, G. K. (1976). Properties of Student's t and of the Behrens-Fisher solution to the two means problem. Ann. Statist. 5, 963-971.

Robinson, G. K. (1979). Conditional properties of statistical procedures. Ann. Statist. 7, 742-755.

Roy, J. and Chakravarti, I. M. (1960). Estimating the mean of a finite population. Ann. Math. Statist. 31, 392-398.

Royall, R. (1971). Linear regression models in finite population sampling theory. In Foundations of Statistical Inference, V. P. Godambe and D.A. Sprott (eds.). Holt, Rinehart, and Winston, Toronto.

Royall, R. (1976). Likelihood functions in finite population sampling survey. Biometrika 63, 605-617.

Rubin, D. B. (1978). Bayesian inference for causal effects: the role of randomization. Ann. Statist. 6, 34-58.

Sandved, E. (1966). A principle for conditioning on an ancillary statistic. Skandinavisk Aktuarietidskrift 50, 39-47.

Sandved, E. (1972). Ancillary statistics and models without and with nuisance parameters. Skandinavisk Aktuarietidskrift 55, 81-91.

Savage, L. J. (1954). The Foundations of Statistics. John Wiley and Sons. New York.

Savage, L. J. (1961). The foundations of statistics reconsidered. Proc. Fourth Berkeley Symp. University of California Press, 1, 575-586.

REFERENCES

Savage, L. J. (1976). On rereading R. A. Fisher (with discussion). Ann. Statist. 4, 441-500.

Savage, L. J., et. al. (1962). The Foundations of Statistical Inference. Methuen, London.

Seheult, A. (1980). Fiducial distribution induced by experimental randomization. Unpublished report. Durham University.

Seidenfeld, T. (1979). Philosophical Problems of Statistical Inference. Reidel, Boston.

Smith, T. M. F. (1976). The foundations of survey sampling: a review. J. Roy. Statist. Soc. A 139, 183-204.

Sprott, D.A. (1973). Normal likelihood and their relation to large sample theory of estimation. Biometrika 60, 457-465.

Sprott, D. A. (1975). Marginal and conditional sufficiency. Biometrika 62, 599-605.

Stein, C. (1945). A two-sample test for a linear hypothesis whose power is independent of the variance. Ann. Math. Statist. 16, 243-258.

Stein, C. (1950). Unbiased estimates with minimum variance. Ann. Math. Statist. 21, 406-415.

Stein, C. (1962). A remark on the likelihood principle. J. Roy. Statist. Soc. A 125, 565-568.

Stone, M. (1976). Strong inconsistency from uniform priors (with discussion). J. Amer. Statist. Assoc. 71, 114-125.

Student (1931). The Lanarkshire milk experiment. Biometrika 23, 398.

Tedin, O. (1931). The influence of systematic plot arrangements upon the estimate of error in field experiments. Journal of Agricultural Science, Cambridge, 21, 191-208.

Thompson, M.E. (1980). Likelihood principle and randomization in survey sampling. Report 78-04, Dept. of Statistics, University of Waterloo.

Wald, A. (1950). Statistical Decision Functions. John Wiley and Sons, New York.

Wallace, D. L. (1959). Conditional confidence level properties. Ann. Math. Statist. 30, 864-876.

Warner, S. L. (1965). Randomized response : A survey technique for eliminating evasive answer bias. J. Amer. Statist. Assoc. 60, 66-69.

Wilkinson, G. N. (1977). On resolving the controversy in statistical inference (with discussions). J. Roy. Statist. Soc. B 39, 119-171.

Yates, F. (1933). The formation of Latin squares for use in field experiments. Empire Journal of Experimental Agriculture, 1, 235-244. (Also reprinted in Yates 1970).

Yates, F. (1934). Contingency tables involving small numbers and the χ^2 test. Suppl. J. Roy. Statist. Soc. 1, 217-235.

REFERENCES

Yates, F. (1964). Fiducial probability, recognizable subsets and Behrens' test. Biometrics 20, 343-360.

Yates, F. (1970). Experimental Design. Selected Papers of Frank Yates, C. B. E., F. R. S., London : Griffin.

Zacks, S. (1969). Bayes sequential designs for sampling finite populations. J. Amer. Statist. Assoc. 64.

Zellner, A. (1971). An Introduction to Bayesian Inference in Econometrics, Wiley; New York.